Smart Materials and Devices for Energy Saving and Harvesting

Smart Materials and Devices for Energy Saving and Harvesting

Editors

Alessandro Cannavale
Ubaldo Ayr

MDPI

Basel • Beijing • Wuhan • Barcelona • Belgrade • Novi Sad • Cluj • Manchester

Editors

Alessandro Cannavale
Polytechnic University
of Bari
Bari
Italy

Ubaldo Ayr
Polytechnic University
of Bari
Bari
Italy

Editorial Office
MDPI AG
Grosspeteranlage 5
4052 Basel, Switzerland

This is a reprint of articles from the Special Issue published online in the open access journal *Energies* (ISSN 1996-1073) (available at: https://www.mdpi.com/journal/energies/special_issues/Material_Device_Harvest).

For citation purposes, cite each article independently as indicated on the article page online and as indicated below:

Lastname, A.A.; Lastname, B.B. Article Title. *Journal Name* **Year**, *Volume Number*, Page Range.

ISBN 978-3-7258-2473-1 (Hbk)
ISBN 978-3-7258-2474-8 (PDF)
doi.org/10.3390/books978-3-7258-2474-8

Contents

About the Editors

Alessandro Cannavale

Alessandro Cannavale is a senior assistant professor at the Polytechnic University of Bari. He has been working in the field of energy and nanotechnologies for energy efficiency throughout his doctoral studies, beginning in 2008. His current research activity covers applied thermodynamics and smart materials/devices for energy saving, heat pumps, and HVAC. He qualified as an associate professor in 2018. He is currently the Associate Editor of Frontiers in Built Environment and an Editorial Board Member for Energies, Discover Energy, Academia Engineering, Buildings, and Clean Technologies.

Ubaldo Ayr

Ubaldo Ayr has a Master's degree, summa cum laude, in Mechanical Engineering from the University of Bari (Italy). After earning his degree, his first job was as an engineer in the research and development division of "Nuovo Pignone S.p.A.", which at that time belonged to the Italian industrial group "ENI" but today forms part of "General Electric". On August 1, 1996, he became an assistant professor at the Polytechnic University of Bari (Italy), and on March 1, 2002, he became an associate professor at the same university. He teaches applied thermodynamics, heat transfer and HVAC systems to mechanical engineering students. Besides teaching, he has taken part in several research projects and scientific collaborations with private companies and public agencies, leading projects in applied energetics and acoustics. In particular, as a researcher, he has studied energy-saving problems, FEM analysis, computational heat transfer and acoustics, solar energy models, wind energy models, thermo-photovoltaic converters, thermo-acoustic heat pumps and engines, and acoustic control systems (active noise control). In energy saving, his research activities concern the application of smart materials and devices. Furthermore, he previously worked as the scientific director of the Acoustics Laboratory of the Polytechnic University of Bari (Italy), which carries out research activities and acoustic testing on behalf of external customers.

energies

MDPI

Editorial

Smart Materials and Devices for Energy Saving and Harvesting

Alessandro Cannavale * and Ubaldo Ayr

Dipartimento di Architettura, Costruzione e Design, Politecnico di Bari, Via Orabona 4, 70125 Bari, Italy; ubaldo.ayr@poliba.it
* Correspondence: alessandro.cannavale@poliba.it; Tel.: +39-080-5963718

Citation: Cannavale, A.; Ayr, U. Smart Materials and Devices for Energy Saving and Harvesting. *Energies* **2024**, *17*, 4684. https://doi.org/10.3390/en17184684

Received: 18 July 2024
Accepted: 18 September 2024
Published: 20 September 2024

1. Introduction

The decarbonisation objectives set for 2050 compel us to envisage any effective strategy in order to reduce the human impact on the planet, drastically reducing greenhouse gas emissions to reduce the effects of global warming [1]. To keep the latter to no more than 1.5 °C (as stated by the Paris Agreement), emissions need to be reduced by 45% by 2030 and reach net zero by 2050 [2]. In this urgent process, on the one hand, we will have to rapidly replace fossil energy sources with renewable ones; on the other hand, it will be essential to improve the efficiency of systems and processes to reduce the energy consumption required. Energy harvesting and energy saving by means of novel materials and technologies represent an intriguing challenge for researchers worldwide. In this roadmap, smart materials may play a pivotal role, providing unprecedented performances and properties, as well as the chance to enhance design opportunities and energy saving either in the fabrication process or in operation. Smart materials are defined as highly engineered materials that respond intelligently to their environment with specialized performance to meet specific users' needs in a reversible fashion. Several classes of materials and devices have been proposed so far, showing highly adaptive properties, such as shape memory materials; piezoelectric materials; chromogenics; new materials for more efficient solar energy conversion; electro-active polymers for energy harvesting, to cite some.

2. An Overview of Published Articles

George Syrrokostas et al. proposed the study (contribution 1) of hybrid electrochromic devices with an iodide/triiodide (I^-/I_3^-) redox couple. The application of 1 V was sufficient to achieve a contrast ratio of 8:1 in a time of about 5 min. The authors also proposed a new method for calculating the loss current and clarified the calculation method for coloration efficiency.

Minzhi Ye proposed a study (contribution 2) to optimize the design of a suspended open-type ceiling radiant cooling panel with a curved and segmented structure to enhance heat transfer. The cooling capacity and heat transfer coefficient of the ceiling radiant cooling panel were investigated using a three-dimensional CFD model. Eventually, they found that the cooling capacity and heat transfer coefficient tend to rise with increasing curvature radius and decreasing curvature width.

Dariusz Obracaj presented a work (contribution 3) dealing with heat stress in deep hot mines acting as a practical limiting factor in the mining of natural resources. This research group studied the dewatering system of the mine; they observed that water flow from the mine dewatering system may be used to effectively reject heat in compressor chillers in underground refrigeration plants.

Paulo Santos and co-authors (contribution 4) studied thermal break strips for steel studs' flanges to mitigate thermal bridges. A validated bidimensional numerical model was adopted, finding that a key parameter is represented by the thickness of thermal break strips.

Andrea Rocchetti et al. investigated the use of metal–organic framework materials to improve the energy efficiency of HVAC systems (contribution 5). In fact, such materials may be exploited to reduce the latent load due to the moisture content in airflows. A psychrometric transformation consisting of cooling and dehumidification was studied for a HVAC system embodying metal–organic framework materials; an energy saving of 30–50% was estimated (in the location of Florence, Italy) compared to a traditional system.

The contribution by Xinhan Qiao and co-authors (contribution 6) deals with the aging of polymeric insulators under various environmental conditions. Silicone rubber indeed tends to lose its initial insulation characteristics, affecting the safety and reliability of the whole power system. In this review paper, several aging and characterization techniques of the polymeric insulators and their aging performance under the action of multiple factors are described.

Chenjue Wang et al. (contribution 7) proposed a new design for a sorption-selective catalytic reduction system to improve ammonia storage density and meet the ammonia demand for high NO_x conversion efficiency at relatively lower temperatures compared to other existing systems.

Zhuyong Li's work (contribution 8) investigated superconducting multi-stage cables, showing higher transmission power and lower energy loss at the same time. In this work, it was observed that the twisted structure of cables reduces the AC loss by up to 80%, compared with the thin strip model.

Minseon Kong co-authored a paper (contribution 9) regarding the performance of dye-sensitized solar cells embodying a solid-state redox mediator based on an ionic liquid and hole-transporting triphenylamine compound. The resulting mediator was used in several devices, reporting a short circuit current density of 4.61 mA/cm^2 and a photovoltaic conversion efficiency of 1.80%. A further increase in the electric parameters of these cells was observed due to the addition of 4-tert butylpyridine.

The paper by Jakub Bernat et al. (contribution 10) deals with the design improvement of dielectric electroactive polymer actuators by assuming modifications in the shape of the actuators. Generally, circular shapes are used; however, in this work, the authors investigated the properties of elliptical shapes in dielectric electroactive polymer actuators. Finite element modeling of the actuators was carried out and an experimental comparison validated the results obtained, showing that the elliptical shape of the actuators allows for a wider range of movement.

Alessandro Cannavale and co-authors reviewed the state of the art of building integrated radiative coolers (contribution 11), with the aim of quantifying the effectiveness of radiative systems applied to the construction sector, with special reference to their operation during daytime.

3. Conclusions

This compilation of articles encompasses a diverse range of research, spanning from innovative photovoltaic cells to electroactive polymer actuators, aging of polymeric insulators, metal–organic framework materials to improve the energy efficiency of HVAC systems and other topics that elucidate the richness of the research field regarding smart materials and devices for energy saving and harvesting.

Conflicts of Interest: The authors declare no conflicts of interest.

List of Contributions:

1. Syrrokostas, G.; Tsamoglou, S.; Leftheriotis, G. Limitations Imposed Using an Iodide/Triiodide Redox Couple in Solar-Powered Electrochromic Devices. *Energies* **2023**, *16*, 7084. https://doi.org/10.3390/en16207084.
2. Ye, M.; Serageldin, A.A.; Nagano, K. Numerical and Parametric Study on Open-Type Ceiling Radiant Cooling Panel with Curved and Segmented Structure. *Energies* **2023**, *16*, 2705. https://doi.org/10.3390/en16062705.
3. Obracaj, D.; Szlązak, N.; Korzec, M. Using a Mine Dewatering System to Increase Cooling Capacity and Energy Recovery of Underground Refrigeration Plant: A Case Study. *Energies* **2022**, *15*, 9481. https://doi.org/10.3390/en15249481.

4. Santos, P.; Lopes, P.; Abrantes, D. Thermal Performance of Load-Bearing, Lightweight, Steel-Framed Partition Walls Using Thermal Break Strips: A Parametric Study. *Energies* **2022**, *15*, 9271. https://doi.org/10.3390/en15249271.
5. Rocchetti, A.; Lippi, M.; Socci, L.; Gullo, P.; Khorshidi, V.; Talluri, L. Metal-Organic Framework Adsorbent Materials in HVAC Systems: General Survey and Theoretical Assessment. *Energies* **2022**, *15*, 8908. https://doi.org/10.3390/en15238908.
6. Qiao, X.; Ming, Y.; Xu, K.; Yi, N.; Sundararajan, R. Aging of Polymeric Insulators under Various Conditions and Environments: Another Look. *Energies* **2022**, *15*, 8809. https://doi.org/10.3390/en15238809.
7. Wang, C.; Chen, S.; Jiang, L.; Wang, Y.; Bao, H. Solid–Gas Sorption System for Ammonia Storage and Delivery Driven by Engine Waste Heat for NOx Reduction of Diesel Engine. *Energies* **2022**, *15*, 8376. https://doi.org/10.3390/en15228376.
8. Li, Z.; Zhang, Z.; Wang, M.; Lv, Y.; Ryu, K. AC Loss Characteristics of HTS Novel Twisted Cables Composed of Soldered-Stacked-Square (3S) Wires. *Energies* **2022**, *15*, 7454. https://doi.org/10.3390/en15207454.
9. Kong, M.; Oh, D.H.; Choi, B.; Han, Y.S. Photovoltaic Performance of Dye-Sensitized Solar Cells with a Solid-State Redox Mediator Based on an Ionic Liquid and Hole-Transporting Triphenylamine Compound. *Energies* **2022**, *15*, 2765. https://doi.org/10.3390/en15082765.
10. Bernat, J.; Kołota, J. Modeling of Dielectric Electroactive Polymer Actuators with Elliptical Shapes. *Energies* **2021**, *14*, 5633. https://doi.org/10.3390/en14185633.
11. Cannavale, A.; Pugliese, M.; Stasi, R.; Liuzzi, S.; Martellotta, F.; Maiorano, V.; Ayr, U. Effectiveness of Daytime Radiative Sky Cooling in Constructions. *Energies* **2024**, *17*, 3210. https://doi.org/10.3390/en17133210.

References

1. United Nations. United Nations—Climate Actions. Available online: https://www.un.org/en/climatechange/net-zero-coalition (accessed on 17 September 2024).
2. United Nations Adoption of the Paris Agreement. Available online: https://unfccc.int/process-and-meetings/the-paris-agreement#:~:text=What%20is%20the%20Paris%20Agreement?&text=The%20Paris%20Agreement%20is%20a,France,%20on%2012%20December%202015 (accessed on 17 September 2024).

energies

MDPI

Article

Limitations Imposed Using an Iodide/Triiodide Redox Couple in Solar-Powered Electrochromic Devices

George Syrrokostas *, Sarantis Tsamoglou and George Leftheriotis

Renewable Energy Laboratory, Department of Physics, University of Patras, GR-26504 Patra, Greece; tsamogloys@gmail.com (S.T.); glefther@upatras.gr (G.L.)
* Correspondence: gesirrokos@upatras.gr

Abstract: In the present study, an iodide/triiodide ($I^-/I_3{}^-$) redox couple is used in hybrid electrochromic devices (ECDs), and the effects of the applied bias potential and bias time on device performance are studied. An applied bias potential of ~1 V is sufficient to achieve an initial contrast ratio of 8:1 in less than 5 min. Increasing both the bias potential and bias time results in an enhancement in loss reactions at the WO_3/electrolyte interface, rather than improving optical performance. Moreover, long-term performance depends on the testing procedure (regularly cycling or after storage), while the formation of iodine (I_2) decreases the initial transparency of the ECDs and affects their overall performance. However, its formation cannot be avoided, even without cycling the ECDs, and the restoration of the optical performance can take place only when the electrolyte is replaced with a fresh one. Finally, a new methodology is applied for calculating the loss current, and a suggestion is made to avoid a common mistake in calculating the coloration efficiency of these hybrid ECDs.

Keywords: electrochromic; hybrid; redox electrolyte; loss current; stability

Citation: Syrrokostas, G.; Tsamoglou, S.; Leftheriotis, G. Limitations Imposed Using an Iodide/Triiodide Redox Couple in Solar-Powered Electrochromic Devices. *Energies* **2023**, *16*, 7084. https://doi.org/10.3390/en16207084

Academic Editors: Alessandro Cannavale and Ubaldo Ayr

Received: 31 August 2023
Revised: 29 September 2023
Accepted: 5 October 2023
Published: 13 October 2023

1. Introduction

Electrochromic (EC) smart windows are energy-saving devices that can change their optical properties with the application of an external bias potential. In this regard, they offer the dynamic control of light and heat passing through them. Their performance depends on a variety of parameters, ranging from material aspects to installation conditions, such as climate zones, control strategies, and the orientation of the building [1–5]. Thus, it has been estimated that their use can lead to a large building energy-saving value of up to ~170 kWh m^{-2} per year, while the energy needed for their operation is only 0.08 kWh m^{-2} per year [6,7]. Further advantages include occupant visual comfort, reduced glare, providing an uninterrupted view of the external environment, and a constant connection with nature, which is not always possible using other competing technologies, like building-integrated photovoltaics (BIPVs). Moreover, the use of EC smart windows can lead to higher productivity in the workplace and better job satisfaction, as stated by SAGE Electrochromics [8].

Of the different types of electrochromic devices (ECDs) that have been proposed [9], the hybrid type consists of two transparent conductive electrodes (e.g., fluorine-doped tin oxide (FTO)-coated glass or indium tin oxide (ITO)-coated glass), on which usually a WO_3 EC layer and a highly transparent catalytic platinum film are deposited, respectively. Between them, a liquid or a solid redox electrolyte is introduced. Such an ECD with an iodide/triiodide ($I^-/I_3{}^-$) redox couple in the electrolyte was first proposed by Georg et al. [10]. During coloration, the following Reactions (1) and (2) take place at the anode and at the cathode, respectively, and their direction is reversed during bleaching:

$$\text{Anode } (\mathrm{WO_3/FTO}):\ x\,Li^+ + x\,e^- + WO_3(bleached) \rightarrow Li_xWO_3(colored) \quad \text{(R1)}$$

$$\text{Cathode (Pt/FTO)}: \ 3I^- \rightarrow I_3^- + 2e^- \quad \text{(R2)}$$

As a result, electrons injected into the WO_3 film from the external circuit at the anode during coloration are balanced by electrons transferred from the cathode to the external circuit and provided via the oxidation of iodide ions to triiodide. Therefore, the role of a redox couple is to counterbalance a redox reaction at the EC layer (anode) during the coloration or bleaching of the ECD (Reaction 1), offering practically an unlimited charge capacity and diminishing the need for the deposition of an ion storage layer at the cathode [11]. Thus, the device fabrication process is simplified, coloration proceeds at a lower bias potential [12,13], and bleaching can proceed even under a short circuit, without the need to apply an opposite bias potential. Moreover, the catalytic layer minimizes the charge transfer resistance at the cathode, facilitating Reaction 2, even though its high transparency is crucial for achieving a highly transparent initial state of the ECD.

However, these "hybrid" or "redox" ECDs have the disadvantage of a permanent internal short circuit due to the direct contact of the EC layer with the redox electrolyte, leading to a so-called "loss current" (I_{loss}), restricting their optical performance. For example, when the $I^-/I_3{}^-$ redox couple is used, the "loss current" is due to Reaction 3, which is the reduction of $I_3{}^-$ ions at the anode:

$$\text{Anode}: \ I_3^- + 2e^- \rightarrow 3I^- \quad \text{(R3)}$$

In other words, electrons that arrive at the anode during the coloration procedure can cause optical modulation (Reaction 1) or reduce the triiodide ions in the electrolyte (Reaction 3) at the EC layer/electrolyte interface, even though it is expected that this a slow process in the case where WO_3 is used as the EC layer [10,14]. A high value of "loss current" increases the amount of energy consumed for the optical modulation of the ECD, restricts the coloration depth, and increases the coloration time, while the open circuit memory is also diminished. According to [14], an acceptable value of I_{loss} should be below 10 $\mu A/cm^2$. Therefore, the total current passing through the ECD during coloration (I_{total}) can be analyzed as the sum of the coloration current (I_{col}, Reaction 1) and the "loss current" (I_{loss}, Reaction 3). Of course, if the electrolyte is in direct contact with the FTO conductive substrate, for example, in the case of a porous EC layer, then the reduction of $I_3{}^-$ ions can also take place at the FTO/electrolyte interface, leading to an extra loss mechanism ($I_{loss,FTO}$), which also contributes to the value of I_{total}. As a result, I_{total} can be expressed using Equation (1):

$$I_{total} = I_{col} + I_{loss} + I_{loss,FTO} \quad (1)$$

Another crucial parameter that affects their performance is the relative position of the electrochemical potential of the as-prepared WO_3 layer and that of the redox electrolyte, which determines the direction of charge transfer at the WO_3/electrolyte interface. If the electrochemical potential (vs NHE) of the as-prepared WO_3 layer is higher (more positive) than that of the redox electrolyte, then electrons will move to the WO_3 layer, inducing coloration without the need for a bias potential. This phenomenon is called "self-coloration". However, during coloration, the electrochemical potential of Li_xWO_3 moves below that of the redox electrolyte (more negative) [15]. Consequently, due to the interfacial loss reactions (Reaction 3), self-bleaching occurs. The higher the difference between the electrochemical potentials, the more intense the above mechanism is. Furthermore, coloration depth depends on the bias potential with respect to the redox potential of the electrolyte, where in general a higher redox potential (more positive) demands a higher value of cathodic potential for the coloration of the device but on the other hand improves charge storage. For that reason, in [16], the bias potential for coloration of the ECDs was chosen with respect to the redox potential of the different redox electrolytes used.

Moreover, the value of the charge transfer resistance at both electrodes is of great importance. A high charge transfer resistance at the EC layer/electrolyte interface and a low one at the Pt/electrolyte interface are desirable. For example, a high value of the charge transfer resistance at the platinized counter electrode may impede both coloration (depth, time) and bleaching [11,13,16], whereas a low one at the EC layer/electrolyte interface will lead to a high "loss current", enhancing self-bleaching, and high power consumption [16]. Finally, absorption in the blue region from the redox couple, especially in the case of an I^-/I_3^- one, leads to a yellowish tint of the devices in the bleached state.

Therefore, efforts have been made to optimize the performance of these hybrid devices, focusing on decreasing the loss reactions at the WO_3/electrolyte interface and on the catalytic properties of the counter electrode. Apart from the I^-/I_3^- redox couple, a variety of other redox couples have been used (Table S1), such as ferrocene/ferrocenium ($Fc^{0/+}$) [9,12], potassium hexacyanoferrate(II)/(III) (KHCF(II)/(III)) [9], tetrathiafulvalene (TTF) [17], tetramethylthiourea/tetramethylformaminium disulfide (TMTU/$TMFDS^{2+}$) [9,13,14,18,19], and recently thiolate/disulfide (T^-/T_2) [11], dissolved in a liquid or a solid electrolyte. For example, a thiolate/disulfide (T^-/T_2) redox couple was used in [11], showing a slow self-bleaching process, while a solid redox (TMTU/$[TMFDS]^{2+}$) electrolyte was developed in [18], exhibiting suppressed loss reactions at the WO_3/electrolyte interface. In [12], a correlation between device performance and Fc concentration in the electrolyte was found; above a certain Fc concentration, the self-bleaching process was enhanced. An optimized concentration in the case of a TTF redox couple, where the ECDs exhibited not only an improved optical performance, but also a reversible behavior, was reported in [17], while the effect of the I^-/I_3^- ratio was examined in [20]. Moreover, a few barrier layers have been deposited at the WO_3/electrolyte interface [14,21], whereas in [13] and in [22], a counter electrode based on a cobalt sulfide film and on an iridium oxide film, respectively, was examined in conjunction with a TMTU/$TMFDS^{2+}$ redox couple instead of a platinized electrode.

However, there is a lack of systematic studies regarding the performance of these hybrid ECDs under prolonged testing under real operating conditions. Furthermore, their stability has been examined so far only by performing continuous coloration–bleaching cycles. For example, in [12], a ~20% contrast attenuation was observed after cycling an ECD having an $Fc^{0/+}$ redox couple for 10 h, while in [13], more than 100,000 cycles were performed with less than 5% contrast attenuation for an ECD with a TMTU/$TMFDS^{2+}$ redox couple and a CoS counter electrode. Moreover, in [23], 11,000 cycles were performed, and transformation of species in the electrolyte was observed, whereas in [11], the change in the optical properties of an I^-/I_3^- redox electrolyte was reported after 50 consecutive cyclic voltammetry scans.

In this work, we have fabricated hybrid ECDs having an I^-/I_3^- redox couple and tested them regularly or after storage for a total period of more than one year. Thus, useful conclusions were found about the stability of the redox electrolyte, regarding its optical and electrochemical properties, while its substitution with a fresh electrolyte led to the restoration of the device's performance. Moreover, emphasis was given to examining how the loss reactions at the EC layer/electrolyte interface evolve with time, where the knowledge so far is limited.

2. Experimental Section

2.1. Fabrication of Hybrid Electrochromic Devices

Amorphous tungsten oxide films (WO_3) with high optical transparency in the visible range were prepared by electron beam gun (e-gun) evaporation. Initially, high-purity WO_3 powder (99.99%) and FTO glass substrates (15 Ohm/sq) were appropriately placed inside a vacuum chamber. Then, evaporation took place at about 10^{-5} mbar and at room temperature. In addition, the deposited film thickness (600–700 nm) was controlled using a quartz thickness controller [24]. The crystal structure, morphology, and stoichiometry of the as-prepared WO_3 films were examined previously in [25] and in [26].

Platinum films were electrodeposited using an aqueous H_2PtCl_6 solution (0.002 M) and an Autolab PGSTAT 204 potentiostat. During a typical three-electrode electrodeposition procedure, a constant potential of -400 mV relative to an Ag/AgCl reference electrode was applied for 60 s. Deposition took place on the conductive side of an FTO glass substrate

(15 Ohm/sq), while a Pt wire served as the counter electrode [27]. As a result, nearly spherical Pt nanoparticles, covering a fraction of the FTO glass substrate, were deposited, as shown in our previous work [27].

For assembling the ECD, the two electrodes (i.e., WO_3/FTO and Pt/FTO) were arranged facing each other, slightly displaced to preserve space for electrical contacts. The devices were sealed peripherally using a low-temperature thermoplastic sealant (Surlyn, 50 μm) at 120 °C for 10 min under pressure, which acted also as a spacer. Then, the space between the two electrodes was filled with a liquid redox electrolyte composed of 0.5 M lithium iodide (*LiI*) and 0.005 M iodine (I_2) in acetonitrile (*ACN*). More specifically, the electrolyte was inserted from one of the two small holes pre-drilled in the counter electrode. The other hole served for air leaking during filling with the electrolyte. Finally, the holes were sealed using Surlyn (supplied by Greatcell Solar Materials, Queanbeyan, Australia) and small pieces of glass.

2.2. Characterization of Redox Electrochromic Devices

The coloration of the ECDs was performed using an electrical circuit comprising mini silicon solar cells (BPW34) as the power source, an LED lamp for illumination, a 0.5 kOhm resistor, and of course the ECD, all connected in series. To achieve the desired applied bias potential (V_{PV}), up to 4 mini silicon solar cells were connected in series, and the current passing through the ECD during coloration was calculated from the potential difference across the 0.5 kOhm resistor.

Similarly, bleaching of the ECDs was performed by connecting their terminals with a 1 kOhm resistor, where again the resulting bleaching current was calculated from the potential difference across the 1 kOhm resistor. All the electrical signs were recorded using a datalogger, connected to a PC.

The typical testing procedure comprised 3 steps, as described in Table 1, except otherwise mentioned. At the end of each step, the transmittance spectrum was recorded.

Table 1. Steps of a typical coloration–bleaching cycle.

Steps	1	2	3
	Bias coloration (5 min)	Open circuit (5 min)	Bleaching through a 1 kOhm resistor (10 min)

In general, the transmittance $T(\lambda)$ spectra of the ECDs were recorded using a Perkin Elmer Lambda 650 UV/VIS Spectrometer, at normal incidence, while calculation of the different optical performance indicators was carried out as described in our previous work [24]. Cyclic voltammetry of complete ECDs was performed using the above potentiostat with a scan rate of 50 mV/s.

The incident photon to current efficiency (IPCE) spectrum of a BPW34 mini silicon solar cell was obtained using a QE-PV-Si quantum efficiency/IPCE measurement kit produced by Oriel. An Oriel 94011A solar simulator, equipped with an AM1.5G filter, in conjunction with a Keithley 2601 source measure unit was used for recording the characteristic I-V curves of the BPW34 mini silicon solar cells.

3. Results

3.1. Effect of the Applied Bias Potential

Figure 1a shows the variation in the luminous transmittance (T_{lum}) during a testing procedure of a hybrid ECD for the different values of the applied bias potential (V_{PV}), while the corresponding full transmittance spectra in the visible and near-IR regions appear in Figure S1. To vary V_{PV}, up to four BPW34 mini silicon solar cells were connected in series. A typical IPCE spectrum and characteristic I–V curves of BPW34 mini silicon solar cells connected in series appear in Figure S2, and their characteristic photovoltaic parameters appear in Table S2.

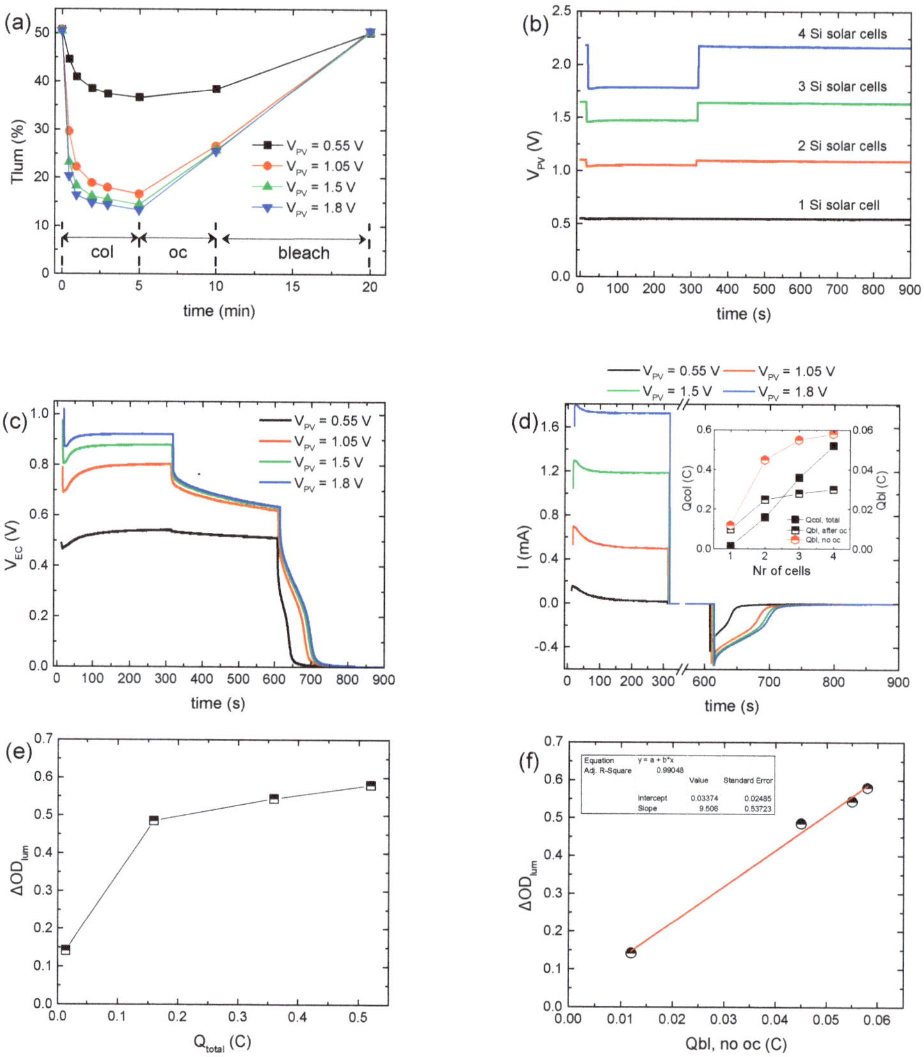

Figure 1. Variation in the luminous transmittance (T_{lum}) (**a**), the potential at the terminals of the BPW34 mini-Si solar cells (V_{PV}) (**b**), the potential at the terminals of the ECD (V_{EC}) (**c**), the total current density (I_{total}) passing through the ECD during a coloration–bleaching cycle (**d**), the luminous optical density modulation with respect to the total charge (**e**), and the luminous optical density modulation with respect to the charge released during bleaching, skipping the open circuit step in the testing procedure for calculating the coloration efficiency (**f**), for the different values of the applied bias potential.

We observe that with V_{PV} increasing from 0.55 to 1.05 V (Figure 1b), the coloration depth increased significantly from 36.6 to 16.6% (Figure 1a), while the coloration time decreased slightly from less than 3 min to 1–2 min. Note that the coloration time was defined as the time needed for the ECD to attain 90% of its maximum optical modulation (ΔT).

Subsequently, a minor improvement was observed in the coloration depth for V_{PV} ~1.5 V, while it remained nearly constant when V_{PV} increased further (~1.8 V) (Figure 1a,b), even though the coloration time for V_{PV} ~1.8 V decreased even further to less than 1 min.

At the same time, the voltage at the ECD terminals (V_{EC}) exhibited a sub-linear increment with V_{PV} (Figure 1c). In particular, its value at the end of the coloration step tends to an upper bound value of nearly 1 V (Figure S3a). Meanwhile, the total current passing through the ECD during coloration (I_{total}) increased linearly with V_{PV} (Figure 1d and Figure S3b).

Even though ΔOD_{lum} increased linearly with V_{EC}, as expected [10,14] (Figure S3c), its variation with the total charge during coloration (Q_{total}) was sub-linear (Figure 1e). Due to the fact that both ΔOD_{lum} and V_{EC} tend to an upper bound value when V_{PV} is increased above ~1 V, we can assume that initially, I_{col} will also show the same behavior with V_{PV}. Finally, for a certain V_{PV}, I_{total} attains a maximum value, after an initial increment, and then a lower steady-state one (Figure 1d), as observed also in [12]. The time needed for the steady state is approximately the same as the coloration time. Therefore, after I_{total} has reached its steady-state value, we can assume that $I_{total} \approx I_{loss}$ [28], and thus, the major part of the increment in I_{total} with V_{PV} can be attributed to the increment in I_{loss}, showing enhanced interfacial losses with an increase in V_{PV} [14]. As a result, I_{loss} increased from 7.65 to 194, 464, and 676 $\mu A/cm^2$ as V_{PV} increased from 0.55 to 1.05, 1.5, and 1.8 V, respectively. In other words, when the ECD reaches its final colored (charged) state, then all the current passing through the ECD is due to the interfacial loss reactions (Reaction 3) and I_{col} tends to zero. If this mechanism is not possible, for example, due to the presence of a barrier layer, then the total current should fall to zero, resembling the behavior of a battery during charging.

Under open circuit conditions, partial bleaching of the ECD (Figure 1a) and a significant voltage drop at the ECD terminals (Figure 1c) were observed, due to the loss reactions at the EC layer/electrolyte interface (Reaction 3). Moreover, the initial voltage drop was more pronounced as V_{PV} increased, due to the higher difference between the electrochemical potential of Li_xWO_3 and the redox potential of the electrolyte. This is due to the lower value (more negative) of the electrochemical potential of Li_xWO_3 vs NHE since its value becomes more negative as the optical density of the EC layer increases [14]. However, after 5 min under open circuit conditions, both T_{lum} and V_{EC} converge when $V_{PV} \geq 1.05$ V. Finally, after the ECD terminals were connected with a 1 kOhm resistor, the ECD was bleached (discharged) (Figure 1c), returning in all cases to its initial optical state (Figure 1a). As a result, an opposite bleaching current (Figure 1d) was measured, from which the charge released during bleaching ($Q_{bl,\ after\ oc}$) was calculated.

As expected, the values of $Q_{bl,\ after\ oc}$ when $V_{PV} \geq 1.05$ V are nearly the same (Figure 1d, inset). Moreover, a large difference was observed between the values of Q_{total} and $Q_{bl,\ after\ oc}$ for the same V_{PV} value (Figure 1d, inset) for two reasons: firstly, as explained above, Q_{total} includes also the charge responsible for triiodide reduction (Reaction 3) at the EC layer/electrolyte interface (Q_{loss}), and secondly, part of the stored charge in the WO_3 film is lost during the open circuit step, as explained above. By skipping the open circuit step in the testing procedure (Figure S4a–c), a more accurate estimation of the overall stored charge in the WO_3 film ($Q_{bl,\ no\ oc}$) was possible (Figure 1d, inset). As a result, ΔOD_{lum} now varies linearly with $Q_{bl,\ no\ oc}$ (Figure 1f), and from the linear regression analysis, the coloration efficiency could be calculated, being 24.3 $cm^2\ C^{-1}$, considering that the area of the ECD was 1.6 × 1.6 cm^2. In this manner, a better estimation of the coloration efficiency is possible, since when using Q_{total}, as is common in the relevant literature [11–13], an underestimation takes place.

Finally, since the ECD returns to its initial optical state after bleaching, we can assume that $Q_{col} \approx Q_{bl,\ no\ oc}$. Therefore, Q_{loss} can be calculated indirectly from Equation (2):

$$Q_{loss} = Q_{total} - Q_{bl,no\ oc} \quad (2)$$

and as appears in Figure S4d, Q_{loss} improved significantly with V_{PV}, as assumed above. It is characteristic that the Q_{loss}/Q_{total} ratio varies from 0.14 to 0.72, 0.85, and 0.89 when V_{PV} increases from 0.55 to 1.05, 1.5, and 1.8, respectively. Alternatively, I_{loss} (or equally Q_{loss}) can be calculated by the difference between I_{total} and the coloration or optical current (I_{opt}), as proposed by Bogati et al. in [14,16]. However, a previously known and constant coloration efficiency is a prerequisite.

3.2. Effect of Bias Time

Subsequently, the applied bias potential (V_{PV}) was kept constant at ~1.5 V, using three series-connected BPW34 mini silicon solar cells, and the bias time was varied from 5 to 20 min. As shown in Figure 2a, when the bias time is prolonged above 5 min, a minor improvement in the coloration depth results. The same also holds for the V_{EC} values (Figure 2b). At the same time, I_{total} exhibited a steady-state value of around 1.2 mA (Figure 2c), irrespective of the bias time. If this current is provided continuously by an external source, a constant V_{EC} and a uniform coloration will be maintained. As a result, Q_{total} increased from 0.36 to 1.4 C, while ΔOD_{lum} was practically constant (Figure 2d). Therefore, since the optical modulation of the ECD is the same in all cases, we can assume that Q_{col} is also the same for the different bias times, and therefore the increment in Q_{total} is attributed solely to the contribution of Q_{loss}. Finally, the T_{lum} and V_{EC} values at the end of the open circuit step (Figure 2a,b) and the values of $Q_{bl,\ after\ oc}$ (Figure S5), as calculated from Figure 2c, were the same, irrespectively of the bias time, whereas in all cases the ECD returned to its initial optical state.

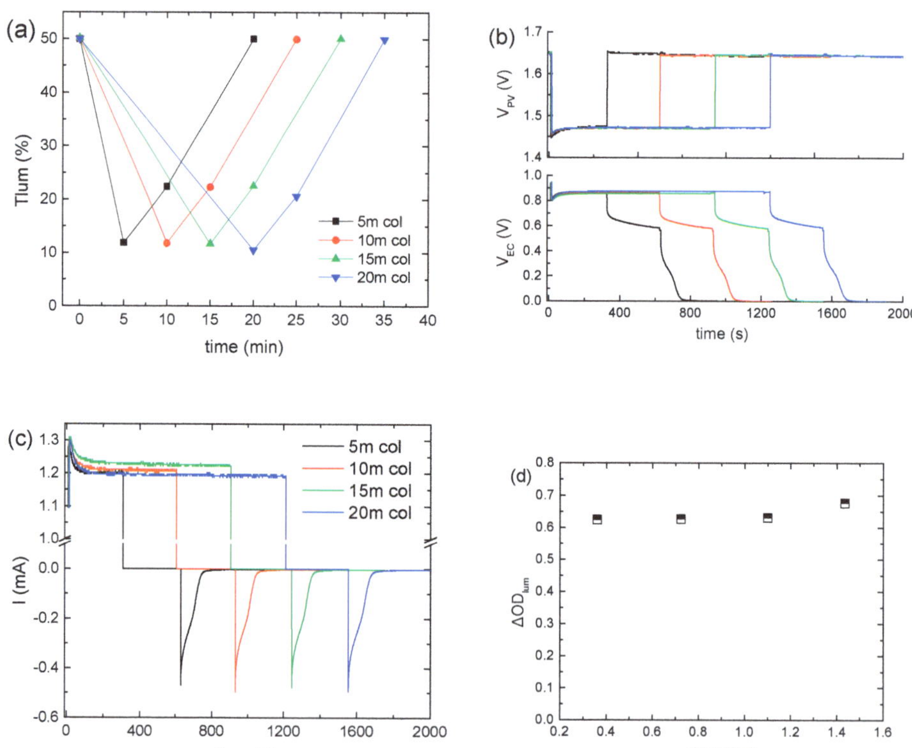

Figure 2. Variation in the luminous transmittance (T_{lum}) (**a**), the potential at the terminals of the BPW34 mini-Si solar cells (V_{PV}) (**b**) and the potential at the terminals of the ECD (V_{EC}) (**b**), the total current density (I_{total}) passing through the ECD during a coloration–bleaching cycle (**c**), and the luminous optical density modulation with respect to the total charge (**d**) for the different values of the bias time.

3.3. Stability of the Hybrid ECD

The stability of the hybrid ECD was also examined by performing a prolonged aging test for at least one year. Initially, the hybrid ECD remained under open circuit conditions in the dark for 3 days to examine if self-coloration takes place. Indeed, an overall reduction in the transmittance was observed (Figure 3a), with the luminous transmittance in the colored state ($T_{lum,\,col}$) decreasing from 55 to 47.2% (Figure 3b). This fact shows that the electrochemical potential (vs NHE) of the as-prepared WO_3 layer is higher (more positive) than that of the redox electrolyte. As a result, the transfer of electrons from the redox electrolyte to the WO_3 film is possible, inducing self-coloration [14].

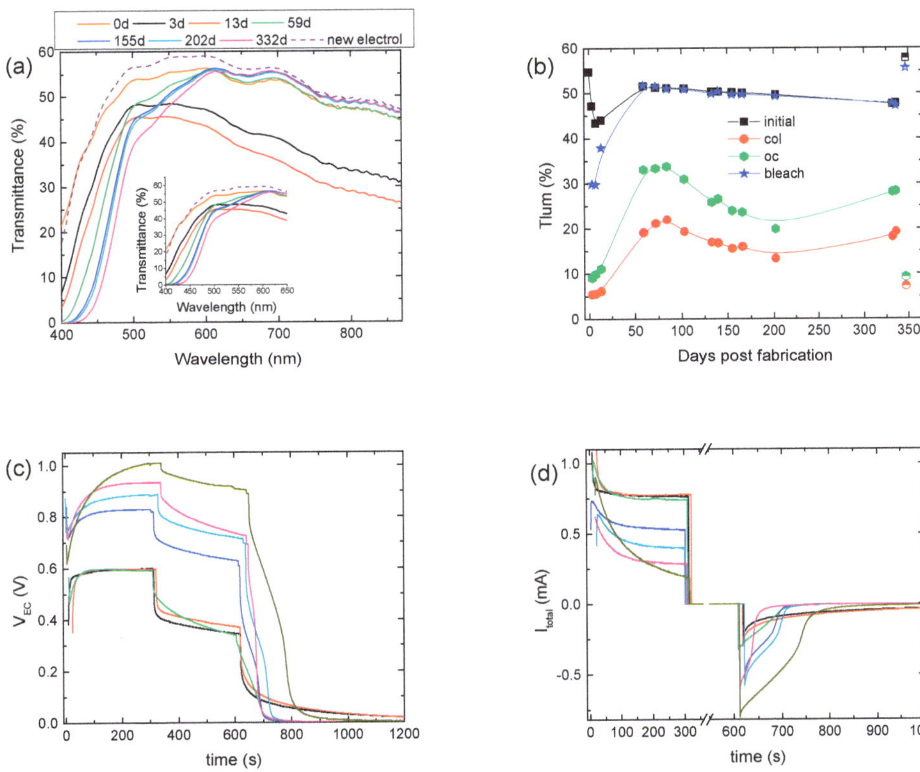

Figure 3. Transmittance spectra of a hybrid ECD for different days post-fabrication (**a**); variation in the luminous transmittance at different stages (**b**), the potential at the terminals of the ECD (V_{EC}) (**c**), and the total current (I_{total}) passing through the ECD (**d**) for different days post-fabrication.

Then, for the following 10 days, we observed an intense coloration, where the device exhibited a $T_{lum,\,col}$ of 5–6% and a contrast ratio above 7:1 (Figure 3b, Table 2). Moreover, the electrical properties of the ECD did not show any significant changes (Figure 3c,d). Optical and electrical losses during the open circuit step were also observed, as expected (Figure 3c,d). Finally, it is worth mentioning that the ECD could not return to its initial optical state and the values of both V_{EC} and I_{total} were not equal to zero at the end of the bleaching step (Figure 3c,d), showing a remaining coloration of the ECD.

Table 2. Characteristic optical performance indicators, total charge, and charge released during bleaching for different days post-fabrication of the ECD.

Days	ΔOD_{lum}	CR	Q_{total} (mC)	ΔV_{EC} (%)	ΔT_{lum} (%)	$Q_{bl,\,after\,oc}$ (mC)
3	0.935	8.6:1	237.7	42.5	66.1	31.87
13	0.850	7.1:1	209.4	38.9	78.9	34.8
59	0.430	2.4:1	231.8	45	72.6	17.8
100	0.422	2.6:1	182.1	35.6	60.2	14.17
200	0.569	3.7:1	133.9	21.5	48.8	28.8
330	0.418	2.6:1	99.87	25.9	55	16.7
Fresh electrolyte	0.900	7.9:1	111.9	12.1	27.2	78.5

Afterwards, the ECD was stored under open circuit conditions for nearly 45 days, and for the following 150 days, the ECD was tested regularly. As appears in Figure 3a, a significant change in the transmittance spectra occurred. More specifically, a gradual reduction in the transmittance of the ECD up to 600 nm was observed, as compared with the transmittance measured immediately after assembling the device (day 0), possibly due to the formation of free iodine (I_2). Similarly, a reduction in the transmittance in the case of an $I^-/I_3{}^-$ redox electrolyte in the 400–700 nm region after 50 consecutive voltammetry scans was observed in [11]. As the authors stated, the absorption in the blue region resulted initially from the production of $I_3{}^-$ ions, having an absorption maximum at 360 nm, and then from their further oxidation to iodine, which shows a strong absorption for $\lambda > 430$ nm [18].

During this period, $T_{lum,\,col}$ varied within the range of 22–13%, showing a trend of improving the coloration depth and the ΔOD_{lum}, as the device was cycled (Figure 3b) (Table 2). Moreover, during coloration, V_{EC} increased considerably. More specifically, its value at the end of the coloration step increased from 0.6 V to nearly 0.9 V (Figure 3c and Figure S6a), while at the same time, only a minor increment in the V_{PV} value occurred, from 0.98 V to 1.1 V (Figure S6b), depicting a small change in the operating point of the mini-Si solar cells. Therefore, the increment in the V_{EC} value was not due to the increment in the V_{PV} value.

A decrement in I_{total} could explain the V_{EC} increment, since the voltage drop at the terminals of the 0.5 kOhm resistor would decrease, leaving a larger part of V_{PV} to be applied at the terminals of the ECD. Indeed, both the maximum value of I_{total} and its steady-state value at the end of the coloration step, which correlates well with I_{loss}, decreased (Figure 3d). For example, the steady-state value of I_{total} was 0.73 mA after 59 days and decreased to 0.4 mA after 200 days. As a result, Q_{total} showed a clear trend for reduction (Figure S6c) (Table 2), and the improvement in the optical performance can be attributed to the reduced losses and the training effect of the WO_3 film due to the continuous operation.

Moreover, the increment percentage of T_{lum} during the open circuit step varied from 73% (day 55) to 50% (day 200), and the voltage drop percentage varied from 45% (59th day) to nearly 20% (200th day) (Figure S6a), both showing reduced optical and electrical losses. As a result, $Q_{bl,\,after\,oc}$ showed an increment from 17.8 mC (59th day) to 28.8 mC (200th day) (Figure S6b) (Table 2). Finally, the device could return to its initial optical state, and both V_{EC} and I_{total} were equal to zero after nearly 100 s of bleaching (Figure 3c,d), showing a complete bleaching (discharge) of the ECD.

A second and longer storage period under open circuit conditions followed for 130 days. After that, the initial transmittance of the ECD decreased further (Figure 3a) for $\lambda < 600$ nm, $T_{lum,\,col}$ increased to 18.2% (Figure 3b), and in general, the optical performance deteriorated (Table 2). A further drop in I_{total} or equally in Q_{total} was also observed (Figures 3d and S6c). During the open circuit step, the increment percentage of T_{lum} and the voltage drop percentage were 55% and 26%, respectively (Figures 3b and S6a) (Table 2), and the value of $Q_{bl,\, after\, oc}$ was 16.7 mC, less than that before the storage period. It seems that, as after the first storage period, the performance of the ECD was affected negatively by storage under open circuit conditions. Nevertheless, the performance drop was not so intense after the second storage period. The exact influence of storage under open circuit conditions is not clear and needs to be further examined.

Nevertheless, by comparing days, before and after the second storage period, where the ECD exhibited the same optical density modulation (ΔOD_{lum}) (i.e., for days 100 and 330), reduced values of ΔT_{lum}, ΔV_{EC}, and Q_{total} and an improved value of $Q_{bl,\, after\, oc}$ were observed (Table 2). That is another indication regarding the reduction in losses at the EC layer/electrolyte interface, after a prolonged testing period.

We believe that one of the reasons behind the reduction in losses, after a prolonged testing period, is the instability of the redox electrolyte. More specifically, the reduction in the initial transmittance of the ECD for $\lambda < 600$ nm (Figure 3a), as discussed above, indicates the formation of free iodine [11,18,20]. Initially, the concentration of free I_2 is very low, since iodine is almost transformed to triiodide according to Reaction 4 [29]:

$$I_2 + \ I^- \ \leftrightarrow \ I_3^- \tag{R4}$$

After prolonged testing, the direction of Reaction 4 is reversed, leading to the consumption of $I_3{}^-$ ions and the formation of free I_2. Another possibility is the oxidation of triiodide ions to iodine according to Reaction 5 [11]:

$$2I_3^- \ \leftrightarrow \ 3I_2 \ + 2e^- \tag{R5}$$

Reaction 5 takes place at more positive potentials than Reaction 2; therefore, it can be avoided by properly selecting the applied bias potential [10]. The presence of I_2 can also negatively affect the conductivity of the electrolyte [20]. Therefore, due to the reduced availability of $I_3{}^-$ ions, the losses at the WO_3/electrolyte interface are reduced. At the same time, a slight variation in the redox potential of the electrolyte to lower values is expected [29].

To verify our above assumptions regarding the role of the electrolyte, the electrolyte was replaced with a fresh one. First, we observed that the transmittance of the ECD returned to its initial state (day 0) (Figure 3a,b) and the device exhibited optical performance (optical density modulation, coloration depth, contrast ratio) similar to that during its first cycle (3rd day). However, a significantly reduced value of I_{total} was measured during coloration, leading to a reduced value of Q_{total} (Figures 3c and S6c). Moreover, reduced optical and electrical losses during the open circuit step (Table 2) were also observed. All the above resulted in a $Q_{bl,\, after\, oc}$ value 2.5 times larger than that in the first cycle (3rd day) (Table 2). Finally, the device could return to its initial optical state after bleaching. Therefore, we can assume that under prolonged testing, changes occur in not only the concentration of the electrolyte but also the properties of the EC layer/electrolyte interface. An increment in the charge transfer resistance at the WO_3/electrolyte interface, due to possible adsorption of electrolyte species, could explain the above results well.

It is interesting to note that the CE is not affected by the replacement of the electrolyte with a fresh one, since the slope in the ΔOD_{lum} vs. $Q_{bl,\, no\, oc}$ graph remains the same (Figure 4a). Moreover, Figure 4b shows cyclic voltammograms of the above ECD for different days post-fabrication. Their shape is typical in the case where an amorphous WO_3 film is used as the EC layer [18]. Moreover, WO_3 is a well-known cathodic electrochromic material, meaning that it is colored during the cathodic scan. During the cathodic scan, a negative shift of the voltage value where coloration begins (the point where current changes sign from positive to negative) was observed, showing that a higher applied bias potential was necessary for coloration. The voltammogram area during the cathodic pulse decreases considerably, revealing an overall decrement in charge density exchanged during coloration (Figure 4c). Replacement of the electrolyte causes a slight positive shift of the voltage value where coloration begins, and an increment in Q_{col} by almost 80% (Figure 4c). Also, the ratio of Q_{col}/Q_{bl} is equal to 1, showing a reversible coloration–bleaching procedure. Finally, the potential value, where the current tends to zero during the anodic pulse, was found to decrease, showing that the deintercalation of Li^+ ions was made easier due to the training of the WO_3 film.

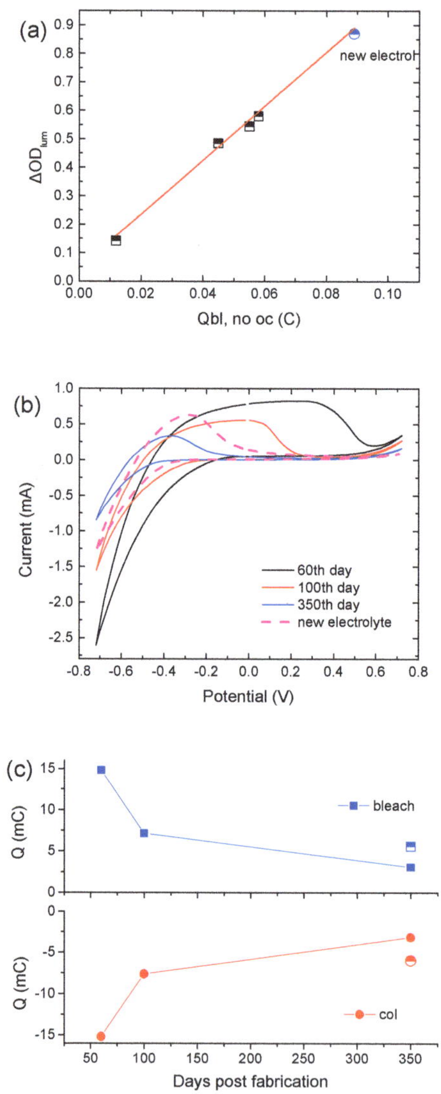

Figure 4. (**a**) Variation in the luminous optical density modulation with respect to the charge released during bleaching, skipping the open circuit step in the testing procedure for calculating the coloration efficiency, (**b**) typical cyclic voltammograms of a hybrid ECD different days post-fabrication, and (**c**) calculated charges from the cyclic voltammograms.

To examine further the effect of the prolonged testing on the overall performance of a hybrid ECD, the first coloration–bleaching cycle was performed 70 days after its fabrication, followed by only four more coloration–bleaching cycles, for a total period of nearly 18 months. In the meantime, only the transmittance was measured, and the

ECD remained under open circuit conditions. Initially, due to self-coloration, an overall decrement in the transmittance was observed, while after 50 days, the transmittance of the ECD for wavelengths above 600 nm returned to its initial value (day 0), showing that self-coloration was no longer possible (Figure 5a). Nevertheless, the transmittance of the ECD at short wavelengths reduced continuously, due to the absorption of the electrolyte, even before the first coloration–bleaching cycle was performed. For example, the transmittance at 500 nm was reduced from its initial value of 53.1% to 47.8% and then to 38.4% after 70 days and 560 days, respectively. Therefore, the formation of free I_2 can take place even without the application of a bias potential, due to Reaction 4.

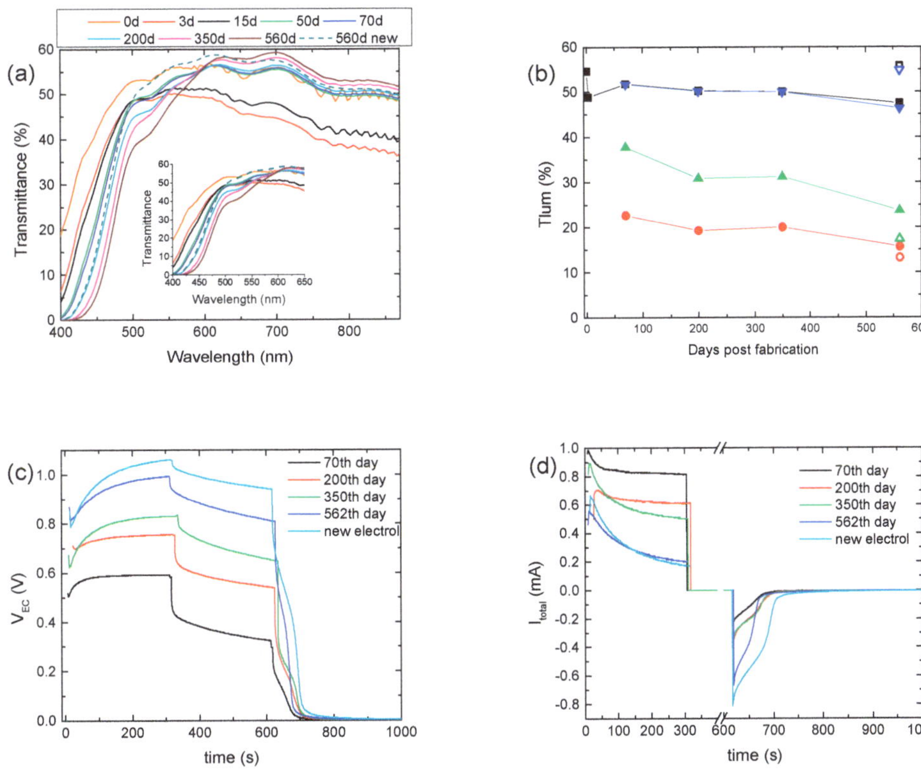

Figure 5. Transmittance spectra of a hybrid ECD, where its first coloration–bleaching cycle was performed 70 days after its fabrication, for different days post-fabrication (**a**); variation in the luminous transmittance at different stages (**b**), the potential at the terminals of the ECD (V_{EC}), (**c**) and the total current (I_{total}) passing through the ECD (**d**) for different days post-fabrication.

During the first operation cycle, an intense coloration and optical irreversibility were not observed, as before. In general, the optical performance of the device was comparable to that of the previously aged ECD, showing a small improvement after 560 days (Figures 5b and S7, Table 3). Again, an increment in V_{EC} values during coloration occurred; i.e., the V_{EC} value at the end of the coloration step increased from 0.59 V to 0.99 V (Figures 5c and S6a), whereas both the maximum value of I_{total} and its value after 5 min of coloration decreased (Figure 5d), with a small exception for $I_{total,\,max}$ of 350 days. Accordingly, Q_{total} showed a clear trend for reduction (Figure S7b and Table 3). Optical and electrical losses during the open circuit step also showed a trend for reduction (Figures 5b,c and S7b). All the above resulted in an improved value of $Q_{bl,\,after\ oc}$, being

3 times larger after 560 days, compared with its value after the first coloration–bleaching cycle (70th day) (Figure S7b and Table 3). Finally, the substitution of the electrolyte with a fresh one remarkably improves the performance of the ECD, where optical and electrical losses were minimized. Note that the fresh electrolyte used here was the EL-HPE high-performance electrolyte, supplied by GreatSellSolar materials.

Table 3. Characteristic optical performance indicators, total charge, and charge released during bleaching for different days post-fabrication of an ECD, where its first coloration–bleaching cycle was performed 70 days after its fabrication.

Days	ΔOD_{lum}	CR	Q_{total} (mC)	ΔV_{EC} (%)	ΔT_{lum} (%)	$Q_{bl,\,after\,oc}$ (mC)
70	0.358	2.3:1	251.4	45.4	66.3	8.1
200	0.416	2.6:1	191	28.7	60.2	14.5
350	0.398	2.5:1	177.8	22.1	56.1	15.3
560	0.482	3.0:1	91.8	18.4	51.6	23.3
Fresh electrolyte	0.6437	4.4:1	93	11.3	23.6	45

Finally, Figure 6 shows images of hybrid ECDs, immediately after assembling and after one year, where the more intense yellowish tint is obvious.

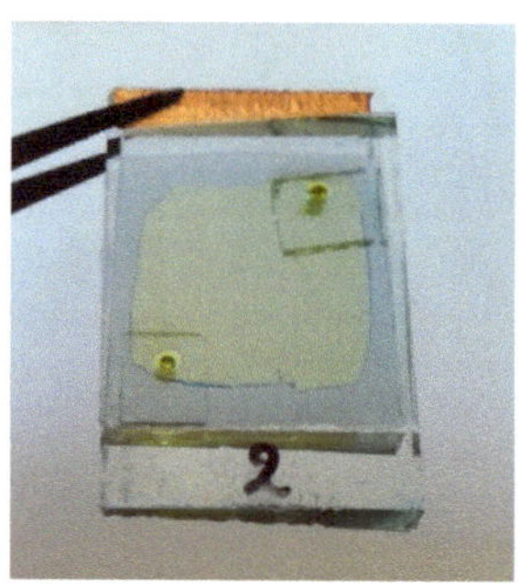

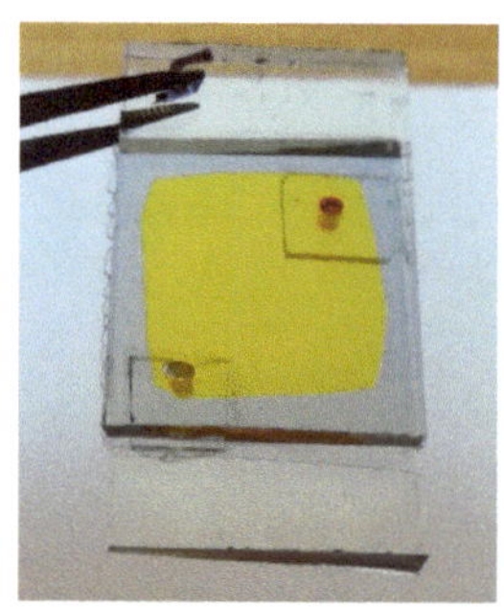

Figure 6. A hybrid ECD with an I^-/I_3^- redox electrolyte immediately after its fabrication (**left**) and after one year (**right**). Only three coloration–bleaching cycles have been performed in this case.

In summary, the above-described mechanisms that are responsible for the performance variation of hybrid ECDs, and especially the instability of the redox electrolyte, take place irrespective of whether the device is cycled or not. Therefore, this is a significant limitation imposed using an iodide/triiodide redox electrolyte.

4. Conclusions

In the present study, we have examined the effect of the applied bias potential and the duration of the bias on the performance of a hybrid EC device having an iodide/triiodide redox couple. We found that an applied bias potential of ~1 V was enough to achieve a contrast ratio of nearly 8:1 and a coloration time of a few minutes. Further increasing the applied bias potential up to 1.8 V resulted mainly in the enhancement of the loss reactions at the WO_3/electrolyte interface, rather than in the improvement in the optical performance (coloration depth and time) of the ECD. These loss reactions take place also during the coloration of the ECD and must be considered to correctly estimate the coloration efficiency. Therefore, for an optically reversible device, it is better to use the charge deintercalated from the WO_3 film during bleaching ($Q_{bl,\,no\,oc}$), rather than using the charge measured during coloration (Q_{total}), due to the significant contribution of Q_{loss} in the value of Q_{total}, which can be up to 90% depending on the applied bias potential value. Optical modulation cannot be improved further by prolonging the bias time up to 20 min, even though a significant value of loss current (~1.2 mA) passes through the ECD, which must be maintained to keep a uniform coloration. Nevertheless, after nearly two months of testing under real operating conditions, the contrast ratio declined from 3:1 to 2:1, even though a reduction in losses at the WO_3/electrolyte interface was observed, and remained nearly constant for the next ten months. The above findings were attributed mainly to the instability of the electrolyte (i.e., formation of free I_2) and the possible increment in the charge transfer resistance at the WO_3/electrolyte interface. The replacement of the electrolyte was the only way to restore the performance of the ECDs, since a similar behavior was observed even without cycling the device. This is a significant limitation when an I^-/I_3^- redox couple is used, affecting the performance not only of hybrid ECDs but also of photoelectrochromic devices. The use of a fresh electrolyte resulted in optically reversible devices even from the first day of their operation.

Supplementary Materials: The following supporting information can be downloaded at: https://www.mdpi.com/article/10.3390/en16207084/s1, Table S1: Comparative table with hybrid ECDs having different redox electrolytes; Figure S1: Transmittance spectra of a hybrid ECD during a coloration–bleaching cycle for the different values of the applied bias potential (V_{PV}), which was increased using series-connected mini silicon solar cells: (a) 1 mini silicon solar cell, (b) 2 mini silicon solar cells, (c) 3 mini silicon solar cells, and (d) 4 mini silicon solar cells; Figure S2: (a) Typical I-V curves of up to 4 series-connected BPW34 mini-Si solar cells and (b) a typical IPCE spectrum; Table S2: Characteristic photovoltaic properties of series-connected BPW34 mini silicon solar cells; Figure S3: (a) Variation in the voltage at the ECD terminals (V_{EC}) with respect to the applied bias potential (V_{PV}), (b) linear increment in I_{total} passing through the ECD with the applied bias potential (V_{PV}), (c) linear increment in the luminous optical density modulation with the voltage at the ECD terminals (V_{EC}); Figure S4: Variation in the applied bias potential V_{PV} (a), the potential at the terminals of the ECD (V_{EC}) (b), and the total current density (I_{total}) passing through the ECD (c) during a coloration–bleaching cycle; (d) variation in Q_{loss} with the applied bias potential (V_{PV}); Figure S5: Variation in the total charge (Q_{total}) and the charge released from the WO_3 layer during bleaching ($Q_{bl,\, after\, oc}$) with the bias time, after an open circuit step; Figure S6: (a) Variation in the potential at the terminals of the ECD (V_{EC}) at the end of each step during a coloration–bleaching cycle with days post-fabrication, (b) variation in the applied bias potential (V_{PV}) for specific days post-fabrication during coloration of the ECD, (c) variation in ΔOD_{lum}, Q_{total}, and $Q_{bl,\, after\, oc}$ with days post-fabrication (open symbols are used in the case of the fresh electrolyte); Figure S7: (a) Variation in the potential at the terminals of the ECD (V_{EC}) at the end of each step during a coloration–bleaching cycle with days post-fabrication; (b) variation in ΔOD_{lum}, Q_{total}, and $Q_{bl,\, after\, oc}$ with days post-fabrication (open symbols are used in the case of the fresh electrolyte).

Author Contributions: Conceptualization, G.S.; formal analysis, G.S.; writing—original draft, G.S.; visualization, G.S.; funding acquisition, G.S.; investigation, G.S. and S.T.; resources, G.L.; writing—review and editing, G.S. and G.L. All authors have read and agreed to the published version of the manuscript.

Funding: This research was funded by the University of Patras, funding program "MEDICUS", grant number 82557.

Data Availability Statement: The data presented in this study are available on request from the corresponding author. The data are not publicly available due to their large volume.

Conflicts of Interest: The authors declare no conflict of interest.

References

1. Papaefthimiou, S. Chromogenic technologies: Towards the realization of smart electrochromic glazing for energy-saving applications in buildings. *Adv. Build. Energy Res.* **2010**, *4*, 77–126. [CrossRef]
2. Granqvist, C.; Arvizu, M.; Pehlivan, I.B.; Qu, H.-Y.; Wen, R.-T.; Niklasson, G. Electrochromic materials and devices for energy efficiency and human comfort in buildings: A critical review. *Electrochim. Acta* **2018**, *259*, 1170–1182. [CrossRef]
3. Cannavale, A.; Ayr, U.; Fiorito, F.; Martellotta, F. Smart Electrochromic Windows to Enhance Building Energy Efficiency and Visual Comfort. *Energies* **2020**, *13*, 1449. [CrossRef]

4. Park, B.R.; Hong, J.; Choi, E.J.; Choi, Y.J.; Lee, C.; Moon, J.W. Improvement in Energy Performance of Building Envelope Incorporating Electrochromic Windows (ECWs). *Energies* **2019**, *12*, 1181. [CrossRef]
5. Lahmar, I.; Cannavale, A.; Martellotta, F.; Zemmouri, N. The Impact of Building Orientation and Window-to-Wall Ratio on the Performance of Electrochromic Glazing in Hot Arid Climates: A Parametric Assessment. *Buildings* **2022**, *12*, 724. [CrossRef]
6. Tong, Z.; Tian, Y.; Zhang, H.; Li, X.; Ji, J.; Qu, H.; Li, N.; Zhao, J.; Li, Y. Recent advances in multifunctional electrochromic energy storage devices and photoelectrochromic devices. *Sci. China Chem.* **2016**, *60*, 13–37. [CrossRef]
7. Granqvist, C.G. Electrochromics for smart windows: Oxide-based thin films and devices. *Thin Solid Films* **2014**, *564*, 1–38. [CrossRef]
8. SageGlass | Electrochromic Glass | Smart Windows, (n.d.). Available online: https://www.sageglass.com/ (accessed on 31 August 2023).
9. Niklaus, L.; Schott, M.; Posset, U.; Giffin, G.A. Redox Electrolytes for Hybrid Type II Electrochromic Devices with Fe−MEPE or $Ni_{1-x}O$ as Electrode Materials. *ChemElectroChem* **2020**, *7*, 3274–3283. [CrossRef]
10. Georg, A.; Georg, A. Electrochromic device with a redox electrolyte. *Sol. Energy Mater. Sol. Cells* **2009**, *93*, 1329–1337. [CrossRef]
11. Giannuzzi, R.; Prontera, C.T.; Primiceri, V.; Capodilupo, A.L.; Pugliese, M.; Mariano, F.; Maggiore, A.; Gigli, G.; Maiorano, V. Hybrid electrochromic device with transparent electrolyte. *Sol. Energy Mater. Sol. Cells* **2023**, *257*, 112346. [CrossRef]
12. Bae, J.; Kim, H.; Moon, H.C.; Kim, S.H. Low-voltage, simple WO_3-based electrochromic devices by directly incorporating an anodic species into the electrolyte. *J. Mater. Chem. C* **2016**, *4*, 10887–10892. [CrossRef]
13. Wang, Z.; Shen, K.; Xie, H.; Xue, B.; Zheng, J.; Xu, C. Robust non-complementary electrochromic device based on WO3 film and CoS catalytic counter electrode with TMTU/$TMFDS^{2+}$ redox couple. *Chem. Eng. J.* **2021**, *426*, 131314. [CrossRef]
14. Bogati, S.; Georg, A.; Graf, W. Sputtered Si_3N_4 and SiO_2 electron barrier layer between a redox electrolyte and the WO_3 film in electrochromic devices. *Sol. Energy Mater. Sol. Cells* **2017**, *159*, 395–404. [CrossRef]
15. Hauch, A.; Georg, A.; Krašovec, U.O.; Orel, B. Comparison of Photoelectrochromic Devices with Different Layer Configurations. *J. Electrochem. Soc.* **2002**, *149*, H159–H163. [CrossRef]
16. Bogati, S.; Georg, A.; Jerg, C.; Graf, W. Tetramethylthiourea (TMTU) as an alternative redox mediator for electrochromic devices. *Sol. Energy Mater. Sol. Cells* **2016**, *157*, 454–461. [CrossRef]
17. Kim, Y.M.; Li, X.; Kim, K.-W.; Kim, S.H.; Moon, H.C. Tetrathiafulvalene: Effective organic anodic materials for WO_3-based electrochromic devices. *RSC Adv.* **2019**, *9*, 19450–19456. [CrossRef] [PubMed]
18. Hočevar, M.; Krašovec, U.O. Solid electrolyte containing a colorless redox couple for electrochromic device. *Sol. Energy Mater. Sol. Cells* **2019**, *196*, 9–15. [CrossRef]
19. Ke, Y.; Wang, Z.; Xie, H.; Khalifa, M.A.; Zheng, J.; Xu, C. Long-Term Stable Complementary Electrochromic Device Based on WO_3 Working Electrode and NiO-Pt Counter Electrode. *Membranes* **2023**, *13*, 601. [CrossRef]
20. Wang, S.; Wang, Y.; Wang, T.; Han, Z.; Cho, C.; Kim, E. Charge-Balancing Redox Mediators for High Color Contrast Electrochromism on Polyoxometalates. *Adv. Mater. Technol.* **2020**, *5*, 326. [CrossRef]
21. Huang, Q.; Dong, G.; Xiao, Y.; Diao, X. Electrochemical Studies of Silicon Nitride Electron Blocking Layer for All-Solid-State Inorganic Electrochromic Device. *Electrochim. Acta* **2017**, *252*, 331–337. [CrossRef]
22. Bogati, S.; Basnet, R.; Georg, A. Iridium oxide catalyst for hybrid electrochromic device based on tetramethylthiourea (TMTU) redox electrolyte. *Sol. Energy Mater. Sol. Cells* **2018**, *189*, 206–213. [CrossRef]
23. Čolović, M.; Hajzeri, M.; Tramšek, M.; Orel, B.; Surca, A.K. In situ Raman and UV–visible study of hybrid electrochromic devices with bis end-capped designed trialkoxysilyl-functionalized ionic liquid based electrolytes. *Sol. Energy Mater. Sol. Cells* **2020**, *220*, 110863. [CrossRef]
24. Syrrokostas, G.; Dokouzis, A.; Yannopoulos, S.N.; Leftheriotis, G. Novel photoelectrochromic devices incorporating carbon-based perovskite solar cells. *Nano Energy* **2020**, *77*, 105243. [CrossRef]
25. Syrrokostas, G.; Leftheriotis, G.; Yianoulis, P. Performance and stability of "partly covered" photoelectrochromic devices for energy saving and power production. *Solid State Ionics* **2015**, *277*, 11–22. [CrossRef]
26. Leftheriotis, G.; Papaefthimiou, S.; Yianoulis, P.; Siokou, A. Effect of the tungsten oxidation states in the thermal coloration and bleaching of amorphous WO_3 films. *Thin Solid Films* **2001**, *384*, 298–306. [CrossRef]
27. Syrrokostas, G.; Bhorkar, K.; Sygellou, L.; Dracopoulos, V.; Leftheriotis, G.; Yannopoulos, S. PtSe2 outperforms Pt as a counter electrode in dye sensitized solar cells. *Mater. Chem. Phys.* **2023**, *305*, 127994. [CrossRef]
28. Bogati, S.; Georg, A.; Graf, W. Photoelectrochromic devices based on sputtered WO_3 and TiO_2 films. *Sol. Energy Mater. Sol. Cells* **2017**, *163*, 170–177. [CrossRef]
29. Boschloo, G.; Hagfeldt, A. Characteristics of the Iodide/Triiodide Redox Mediator in Dye-Sensitized Solar Cells. *Accounts Chem. Res.* **2009**, *42*, 1819–1826. [CrossRef] [PubMed]

energies

MDPI

Article

Numerical and Parametric Study on Open-Type Ceiling Radiant Cooling Panel with Curved and Segmented Structure

Minzhi Ye [1], Ahmed A. Serageldin [2,3,*] and Katsunori Nagano [2,*]

1 Graduate School of Engineering, Hokkaido University, Sapporo 060-8628, Japan
2 Division of Human Environmental System, Faculty of Engineering, Hokkaido University, N13-W8, Kita Ku, Sapporo 060-8628, Japan
3 Department of Mechanical Engineering, Shoubra Faculty of Engineering, Benha University, Cairo 11629, Egypt
* Correspondence: ahmed.serageldin@eng.hokudai.ac.jp (A.A.S.); nagano@eng.hokudai.ac.jp (K.N.); Tel.: +81-080-5747-5714 (A.A.S.)

Abstract: A suspended open-type ceiling radiant cooling panel (CRCP) has been proposed recently. The main challenge is improving its cooling performance to overcome limitations for extensive use. Therefore, this study aims to optimize the design of CRCPs with curved and segmented structure to enhance heat transfer. A three-dimensional CFD model was developed to investigate the cooling capacity and heat transfer coefficient of the CRCPs installed inside a single enclosed room. Panel structure was determined based on four dependent parameters: the panel curvature width (L, m), the panel curvature radius (r, m), the void distance (d, m) between each panel or panel segment, and the panel coverage area (A_c, m^2). The panel surface area (A_s, m^2) and the ratio of panel curvature width to radius (L/r) were also examined. A total of 35 designs were compared under 7 different cooling load conditions, and 245 cases were carried out. The results show that the nominal cooling capacity and heat transfer coefficient rise with increasing curvature radius and decreasing curvature width. The void distance plays the most crucial role in influencing cooling performance. It is possible to simultaneously improve cooling performance, achieve uniform temperature distribution, and reduce the number of panels through structure optimization.

Keywords: ceiling radiant cooling panel; parametric analysis; CFD simulation; cooling capacity

Citation: Ye, M.; Serageldin, A.A.; Nagano, K. Numerical and Parametric Study on Open-Type Ceiling Radiant Cooling Panel with Curved and Segmented Structure. *Energies* **2023**, *16*, 2705. https://doi.org/10.3390/en16062705

Academic Editors: Alessandro Cannavale and Ubaldo Ayr

Received: 14 February 2023
Revised: 28 February 2023
Accepted: 9 March 2023
Published: 14 March 2023

1. Introduction

Radiant ceiling panel systems have been a matter of great concern [1] in recent decades, considering their benefits of high thermal comfort level and energy-saving potential [2]. Such systems can be combined with renewable energy, as they generally use water as the thermal medium for space heating and cooling. Therefore, the use of radiant ceiling panel systems is a popular alternative heating and cooling method to conventional air source systems, and such systems are widely applied as energy efficiency technologies. These systems have been applied in various building types, such as high-volume halls (e.g., vehicle repair shops and markets) [3], office buildings [4,5], schools [6], hospitals [7], and residential buildings [8]. Ceiling radiant cooling panel (CRCP) systems are typically used for cooling, commonly consisting of a metal panel, a water pipe directly or indirectly touching the panel, and insulation on the top of the panel surface [9,10]. The insulation layer above the panel is expected to insulate the heat transfer between the panel and the ceiling to activate heat transfer on the bottom surface towards the conditioned space. However, upward heat loss is inevitable. Thus, some researchers have focused on improving the insulation layer to reduce the heat flux to increase the cooling capacity of the CRCP [11]. On the other hand, an open-type CRCP, also called a suspended radiant ceiling panel (SRCP) [12] or suspended ceiling radiant panel (CRP) [13] has recently come to public attention. Unlike the top-insulated type, an open-type CRCP is installed separately from the ceiling, excluding insulation between the panel top and ceiling. Due to the characteristic of convenient and flexible installation, this type is also widely used in existing or new buildings in practice [14].

Nevertheless, CRCPs are still associated with risk of dew condensation on the cooling surface, as in other radiant cooling systems. Furthermore, high manufacturing and installation costs also limit their practical applications. Therefore, it is urgently required to address these issues by improving system performance through the use of advanced strategies [15]. In recent studies, combining CRCPs with an air circulating strategy has been proposed as an effective method. Shakya et al. [16] proposed a hybrid system coupling natural ventilation and a desiccant dehumidification system with a CRCP. The cooling capacity and thermal comfort can be significantly enhanced by the proposed hybrid system. The proposed hybrid system can save 77% and 61% primary energy consumption compared to a conventional all-air system and radiant cooling system, respectively. Jeong and Mumma [17] investigated the mixed convection effect on the cooling capacity of a CRCP. The results revealed that the total cooling capacity of a CRCP can be increased by 5–35% by the combination of natural and mechanical ventilation. Shin et al. [18] combined an open-type CRCP with air circulators to enhance cooling capacity and energy performance. The results showed that the cooling capacity was enhanced by 26.4%, and energy consumption was reduced by 26.4% compared with a conventional CRCP system. In summary, increasing the convection in the conditioned space of a CRCP is impactful in improving the cooling performance and reducing the energy consumption of CRCP systems. Another possibility is to reform the structural design of the CRCP to achieve the same effect of convection enhancement [19], which has been investigated in many studies.

Moreover, it is crucial to investigate the effect of each parameter through parametric studies to achieve an optimal overall panel structure and arrangement. Statistical analysis is necessary to optimize and prioritize the design considering the conflict between different desires in terms of cooling capacity, panel surface temperature, temperature distribution, or other factors. For instance, with respect to the design of a thermoelectric radiant panel (TCRP) system, Lim et al. [20] carried out a parametric study consider spacing, panel and insulation thickness, outdoor air temperature, and heat sink. The authors evaluated the effects of design factors and operation conditions on the cooling performance. The results proved that the spacing and outdoor air temperature are the main factors that influence of the cooling performance of the TCRP. Luo et al. [21] conducted a parametric study on the thickness of aluminum panels and insulation. The optimum thickness of aluminum panels and insulation was found to be around 1–2 mm and 40–50 mm, respectively.

In this study, an open-type CRCP with a segmented and curved shape was proposed, which is expected to achieve better cooling performance through the structural optimization of the design. The objectives of this study are to evaluate the effects of the panel design parameters of the novel panel structure and explore the ideal design to maximize the cooling capacity of CRCP in terms of energy and cost reduction and achieving the required thermal comfort level. A parametric analysis was conducted of the proposed novel open-type ceiling radiant cooling panel (CRCP), which has curved shapes and voids between the adjacent panels. A three-dimensional CFD simulation model was developed to study the cooling capacity and heat transfer coefficient of the panel in an enclosed space. Thirty-five panel designs were examined and compared with the reference panels. The influence of four independent and two dependent panel design parameters were investigated based on the calculation of nominal cooling capacity, heat transfer coefficient, airflow, and indoor temperature distribution. Finally, the optimal panel design was discussed and recommended according to both cooling performance and thermal comfort.

2. Design of Ceiling Radiant Ceiling Panels (CRCPs)

2.1. Literature Review of CRCP Design

As stated above, it is important to increase the efficiency and cooling performance of CRCPs system through the use of advanced strategies. Table 1 lists previous studies that aimed to enhance the cooling capacity of panels by via panel design. The designs of panels included the water tube/water channel configuration design, panel configuration design, panel distribution arrangement design, and panel material design. The results showed that the cooling capacity or even the uniformity of the indoor air temperature field can be substantially improved by applying novel design strategies. Serageldin et al. [22] compared nine different open-type CRCPs including two curved CRCPs with voids by numerical simulation, considering the heat flux, heat transfer coefficient, and indoor operative temperature. The results illustrated that changing the panel shape to a curved design and adding voids can increase the radiation heat transfer coefficient by 31% and convection heat transfer by 174% and decrease the indoor air temperature by 1 K.

Table 1. Studies on CRCP structure design.

Literature	Panel Type	Methodology	Design Strategy	Improvement Results
Mosa et al. [23,24]	SRCP	Numerical simulation	Serpentine and dendritic flow channel design	The dendritic architecture allows for a significant improvement in the cooling panel performance.
Hassan and Kaood [25]	SRCP	CFD simulation	Application of internal longitudinal fins	The presented balanced design enhances the cooling capacity and cooling rate by 1.54 and 17.7%, respectively.
Radzai et al. [26]	RCP	CFD simulations	New RCP serpentine-based flow configuration	The proposed designs have the potential to improve the overall efficiency of RCP in terms of temperature distribution, cooling capacity, and pressure.
Labat et al. [12]	SRCP	Genetic algorithm	Arrangement of multiple panels on the ceiling	The uniformity of the temperature field can be significantly improved by using 10 panels or more compared with using a single panel.
Shin et al. [27]	Open type	Experiments and CFD simulation	Open-type CRCP installed with void areas between adjacent ceiling panels	The open-type CRCP can provide 54–80% higher nominal cooling capacity than a conventional closed-type CRCP.
Radwan et al. [28]	Open type	CFD simulation	New multisegmented mini-channel-based CRCP	The design can accomplish the same cooling capacity and identical indoor air temperature by using a higher panel surface temperature.
Zhang et al. [13]	Open type	Experiments	A new type of CRCP with inclined aluminum fins	The cooling capacity of the CRCP with inclined fins is about 19% higher than that of a suspended panel.
Lv et al. [29]	RCP	Experiments	A novel grooved radiant ceiling panel filled with heat transfer liquid	The cooling capacity of this radiant panel was 18–25% higher than that of traditional metal radiant panels.
Ning et al. [30]	CRCP	CFD simulation	CRCP with a thin air layer	The cooling capacities are increased by 43–46% compared to the original CRCP.
Xing and Li [11]	CRCP	Experiments	Replacement of the radiation shield with a convection shield	The improved inbuilt air gap has a better synergy in improving cooling capacity and anticondensation ability.

2.2. Parameter Design of Curved Open-Type CRCPs

The aim of this study is to explore the inter-relationship between panel design (including curved structure and panel distribution designs) and cooling performance. It was urgently required to determine the optimal design, which can balance contradictory goals in terms of enhancing cooling capacity and maintaining thermal comfort. Therefore, four independent parameters were used to determine the panel shape, as shown in Figure 1: the panel curvature width (L, m), the curvature radius (r, m), the void distance between each panel or panel segment (d, m), and the panel coverage area (A_c, m^2). Two dependent parameters were also investigated: the ratio of panel curvature width to radius (L/r) and panel surface area (A_s, m^2). The L/r ratio is a parameter affecting the curvature shape of the panel. The curvature of the panel increases with an increase in the value of L/r. The panel surface area (A_s) is a parameter depending on all four independent parameters and calculated according to Equation (1). It closely relates to the manufacturing cost. Therefore, it is preferred to minimize the panel surface area in the design stage.

$$A_s = \frac{2\sin^{-1}\left(\frac{L}{2r}\right)}{180} \cdot \pi r \cdot l \cdot n \quad (1)$$

where l is the panel length (m), and n is the panel number related to the panel curvature width (L), void distance (d), and panel coverage area (A_c).

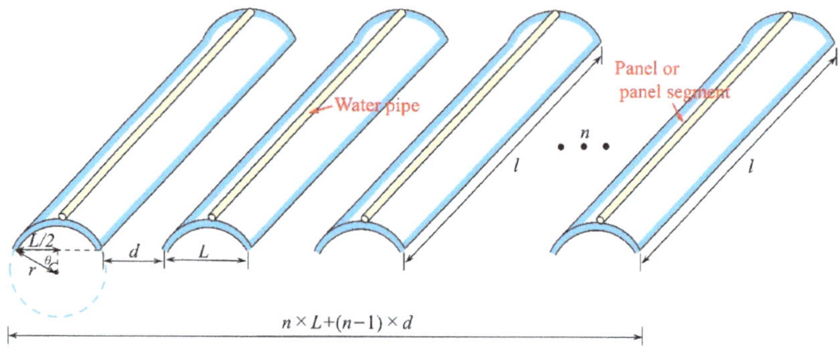

Figure 1. Schematic diagram of the panel and design parameters.

Thirty-five designs were created and compared to show the effect of each parameter. The values of each parameter in each case are summarized in Table 2. The panel curvature width (L) was altered from 0.03 to 0.12 m, the panel curvature radius (r) was altered from 0.03 m to 0.3 m, the void distance (d) was altered from 0 m to 0.33 m, and the panel cover area (A_c) was altered from 7.58 m^2 to 12.96 m^2. Accordingly, the L/r ratio varied from 0 to 2, and the panel surface area (A_s) varied from 2.51 m^2 to 16.90 m^2. When the effect of one independent parameter was examined individually, the other three parameters were held constant. Two dependent parameters—L/r ratio and A_s—were investigated with different panel width and curvature radius values, while the coverage area and void distance were the same. Additionally, four replenished designs were proposed for optimization and verification.

Table 2. Design of parameters.

Design	L (m)	r (m)	d (m)	L/r	A_c (m^2)	A_s (m^2)
1	0.03	0.06	0.03	0.5	12.96	8.18
2	0.06	0.06	0.03	1	12.96	10.28
3	0.09	0.06	0.03	1.5	12.96	11.72
4	0.12	0.06	0.03	2	12.96	16.90
5	0.06	0.03	0.03	2	12.96	14.90
6	0.06	0.09	0.03	0.7	12.96	10.06
7	0.06	0.15	0.03	0.4	12.96	9.96
8	0.06	0.2	0.03	0.3	12.96	9.93
9	0.06	0.3	0.03	0.2	12.96	9.91
10	0.06	0.06	0	1	12.96	13.58
11	0.06	0.06	0.01	1	12.96	13.05
12	0.06	0.06	0.03	1	12.96	10.03
13	0.06	0.06	0.06	1	12.96	7.53
14	0.06	0.06	0.1	1	12.96	5.77
15	0.06	0.06	0.14	1	12.96	4.51
16	0.06	0.06	0.21	1	12.96	3.51
17	0.06	0.06	0.33	1	12.96	2.51

Table 2. *Cont.*

Design	*L* (m)	*r* (m)	*d* (m)	*L*/*r*	A_c (m^2)	A_s (m^2)
18	0.06	0.06	0.03	1	11.43	9.05
19	0.06	0.06	0.03	1	10.80	8.53
20	0.06	0.06	0.03	1	10.15	8.02
21	0.06	0.06	0.03	1	8.86	7.02
22	0.06	0.06	0.03	1	7.58	5.77
23	0.1	0.09	0.03	1.1	12.96	11.18
24	0.15	0.2	0.03	0.75	12.96	11.17
25	0.2	0.4	0.03	0.5	12.96	10.85
26	0.3	1	0.03	0.3	12.96	10.80
27	0.16	-	0.03	0	12.96	10.84
28	0.045	0.03	0.03	1.5	12.96	10.09
29	0.135	0.09	0.03	1.5	12.96	12.65
30	0.225	0.15	0.03	1.5	12.96	13.19
31	0.3	0.2	0.03	1.5	12.96	13.73
32	0.03	0.06	0.05	0.5	12.96	6.04
33	0.03	0.06	0.1	0.5	12.96	3.62
34	0.03	0.06	0.05	0.5	11.43	5.38
35	0.03	0.06	0.1	0.5	11.43	3.35

3. Model Development

ANSYS 2020 R2 Fluent commercial software was used to perform computational fluid dynamics (CFD) simulations. A three-dimensional finite-volume model was developed to determine the heat transfer and temperature field. The assumptions applied in this model are listed as follows:

1. The heat transfer is calculated under a steady-state condition;
2. The air density difference is ignored, and only the gravitational force effect is considered;
3. The heat transfer between the water pipe and the panel surface is ignored, and the panel surface temperature is considered uniform;
4. The emissivity is constant and a property of the surface, which is independent of wavelength;
5. The surface is opaque and diffuse, and only the transferred radiation between two surfaces is considered.

3.1. Geometry

The enclosed room is 4 m (Length) × 4 m (Width) × 2.9 m (Height), which has are same dimensions and arrangement as the room model for a suspended flat panel validated by Shin et al. [27]. Twelve cylindrical occupant dummies with dimensions of 0.3 m (D) × 1.1 m (H) are deployed symmetrically in the room to mimic human bodies, generating energy dissipation in the space and representing a cooling load. The panels are suspended 0.3 m beneath the ceiling and arranged along the central line. In the validated case, the panel is one flat, solid panel with dimensions of 3.6 m (Length) × 3.6 m (Width) × 0.03 m (Height), while in other cases with curved and segmented type panels, four independent panel design parameters (L, r, d, and A_c) are set using the design values listed in Table 2. Instead of directly inputting the coverage area (A_c), the number of panels (n) is employed in the geometric drawing, which can be determined according to Equation (2):

$$n = \left\lceil \frac{1}{(L+d)} \cdot \left(\frac{A_c}{l} + d\right) \right\rceil \tag{2}$$

Moreover, the symmetry boundary condition was applied to the middle plane to simplify the modeling and accelerate the simulation speed due to the completely symmetrical characteristic of the room. In this study, all the simulations were conducted in half of the space, as shown in Figure 2.

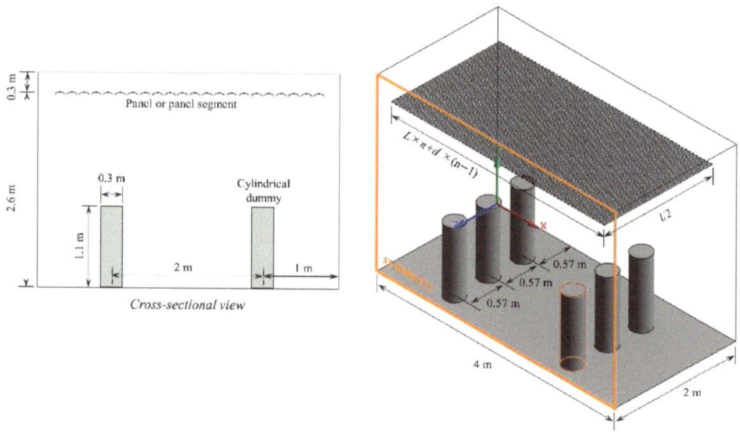

Figure 2. The geometry of the model and a cross-sectional view at the middle plane.

3.2. Mesh

The mesh was generated by the ANSYS Meshing tool using tetrahedron mesh and inflation layers near the cylinder surface (Figure 3a). The grid around the wall, panel, and dummy cylinder surface was refined to address the expected high gradient of temperature by adding an inflation layer with 0.001 m first-layer thickness and a growth rate of 1.2. Mesh-independent analysis was carried out for one curved panel design to minimize the impact of element size on the simulation accuracy. Four different element numbers were selected: 2.2×10^5, 5.9×10^5, 9.3×10^5, 1.4×10^6, 2.1×10^6, 2.6×10^6, and 3.6×10^6. The total heat flux and indoor air temperature varied with an increasing number of elements, as shown in Figure 3b. Therefore, the preferred number of elements was 1,432,275 in the present study, as the results demonstrated a change in heat flux of only 0.17% and a change in average air temperature of only 0.81% compared with the finer mesh.

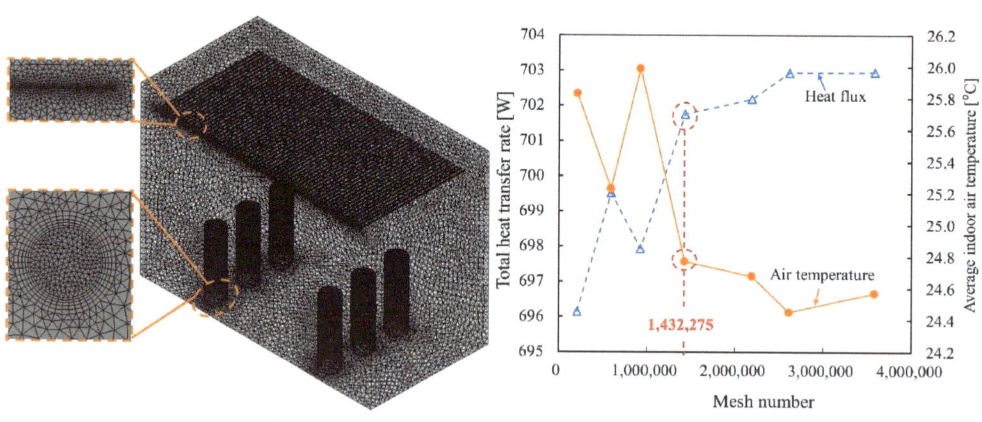

Figure 3. (**a**) Computational mesh used in the present study. (**b**) Mesh-independent analysis results.

3.3. Description of Numerical Equations

3.3.1. Governing Equation

- Mass conservation [31]:

$$\nabla \cdot (\rho \vec{v}) = 0 \tag{3}$$

- Momentum conservation [31]:

$$\nabla \cdot (\rho \vec{v} \vec{v}) = -\nabla p + \nabla \cdot (\overline{\overline{\tau}}) + \rho \vec{g} \tag{4}$$

where the surface stress tensor ($\overline{\overline{\tau}}$, N/m^2) is given by the following equation:

$$\overline{\overline{\tau}} = \mu \left[\left(\nabla \vec{v} + \nabla \vec{v}^{T} \right) - \frac{2}{3} \nabla \cdot \vec{v} I \right] \tag{5}$$

- Energy conservation [31]:

$$\nabla \cdot (\rho C_p \vec{v} T_f) = \nabla \cdot (k_f \nabla T_f) \tag{6}$$

where ρ is the fluid density (kg/m^3), $\vec{v}$ is the velocity vector (m/s), p is the static pressure (N/m^2), $\rho \vec{g}$ is the gravitational body force (N/m^3), μ is the molecular viscosity (kg/m·s), I is the unit tensor, C_p is the specific heat capacity (J/kg·K), T_f is the fluid temperature (K), and k_f is the fluid thermal conductivity (W/m·k).

3.3.2. Turbulent Model

The standard k-ε model proposed by Launder and Spalding (1972) is used to describe the effect of turbulence. The equations for turbulent kinetic energy k (m^2/s^2) and the turbulent dissipation rate ε (m^2/s^3) are expressed as Equations (7) and (8) [31], respectively:

- Turbulent kinetic energy:

$$\frac{\partial}{\partial t}(\rho k) + \frac{\partial}{\partial x_i}(\rho k u_I) = \frac{\partial}{\partial x_j}\left[\left(\mu + \frac{\mu_t}{\sigma_k} \right) \frac{\partial k}{\partial x_j} \right] + G_k + G_b - \rho\varepsilon - Y_M + S_k \tag{7}$$

- Turbulent dissipation rate:

$$\frac{\partial}{\partial t}(\rho \varepsilon) + \frac{\partial}{\partial x_i}(\rho \varepsilon I) = \frac{\partial}{\partial x_j}\left[\left(\mu + \frac{\mu_t}{\sigma_\varepsilon} \right) \frac{\partial \varepsilon}{\partial x_j} \right] + C_{1\varepsilon} \frac{\varepsilon}{k}(G_k + C_{3\varepsilon} G_b) - C_{2\varepsilon} \rho \frac{\varepsilon^2}{k} + S_\varepsilon \tag{8}$$

where μ_t is the turbulent viscosity (kg/m·s), as follows.

$$\mu_t = \rho C_\mu \frac{k^2}{\varepsilon} \tag{9}$$

where u_i is the velocity, G_k is the turbulence kinetic energy generated by the mean velocity gradients, G_b is the turbulence kinetic energy generated by buoyancy, and Y_M is the dilatation dissipation term. C_μ, σ_ε, σ_k, $C_{1\varepsilon}$, and $C_{2\varepsilon}$ are empirical constants with the following default values: C_μ = 0.09, σ_ε = 1.2, σ_k = 1, $C_{1\varepsilon}$ = 1.44, and $C_{2\varepsilon}$ = 1.92. S_k and S_ε are the user-defined source terms.

3.3.3. S2S Model

In radiant systems, radiative heat transfer accounts for a significant portion of total heat transfer. Therefore, the surface-to-surface (S2S) radiation model was used in the simulations. In this model, the radiation of a surface (k) is composed of both emission and reflection [31].

$$q_{k,out} = \varepsilon_k \sigma T_k^4 + \rho_k q_{k,in} \tag{10}$$

where ε_k is the emissivity, σ is Boltzmann's constant, and $q_{k,in}$ is the energy incident on the surface (k) from the surroundings, which is represented as a summation of radiation from the surrounding surface (j), as shown in Equation(11).

$$A_k q_{k,in} = \sum_{j=1}^{N} A_j F_{jk} q_{j,out} \tag{11}$$

where A_k and A_j are the area of surface k and surface j (m^2), respectively; F_{jk} is the view factor between surface j and surface k; and $q_{j,out}$ is the radiative heat flux of the surface j (W/m^2).

3.3.4. Boussinesq Model

The Boussinesq model is used to model the natural convection in the closed space driven by buoyancy force. The model performs with the fluid density as a function of the temperature gradient as follows [31].

$$\left(\rho - \rho_0\right)g \approx -\rho_0\beta(T - T_0)g \tag{12}$$

where ρ_0 is the specified constant density of the flow, T_0 is the operating temperature, and β is the thermal expansion coefficient.

3.3.5. Other Equations

In addition, the total heat flux (q, W/m^2) is defined as the total heat transfer rate through all the panel surfaces divided by the panel surface area (A_s, m^2).

$$Q_{tot} = \frac{Q_{tot}}{A_s} \tag{13}$$

The radiation and convection heat transfer coefficient are then calculated based on the following equations.

$$H_r = \frac{Q_r}{A_s(AUST - T_p)} \tag{14}$$

$$h_c = \frac{Q_c}{A_s(T_a - T_p)} \tag{15}$$

where $AUST$ is the area-weighted uncooled temperature of the surfaces excluding the panel surface (°C), T_a is the air temperature (°C), and T_p is the panel surface temperature (°C).

Then, the operative temperature (T_{op}, °C) can be roughly determined by Equation (16).

$$T_{op} = \frac{h_c T_a + h_r \cdot AUST}{h_c + h_r} \tag{16}$$

3.4. *Numerical Schemes*

The governing equations are iteratively solved at each control volume in the computational domain until convergence is achieved. The Semi-Implicit Method for Pressure Linked Equations (SIMPLE) algorithm was applied for coupling pressure and momentum. The first-order upwind discretization scheme was chosen for turbulent kinetic energy and turbulent dissipation rate. The second-order upwind discretization scheme was used for the pressure, momentum, and energy. Enhanced wall treatment was selected as a wall treatment. The convergence criteria are 10^{-5} for all equations, with the exception of 10^{-6} for the energy equation.

3.5. *Boundary Condition*

Table 3 lists the boundary conditions and emissivity. The room is assumed to be well-insulated without heat transfer so that the envelope is assigned to the adiabatic condition. The non-uniform temperature distribution on the panel surface always occurs from the rise of chilled water or pipe arrangement. However, the temperature difference on the panel surface has less effect than the large difference between the air and the panel. Our study mainly focuses on optimizing the panel shape design based on heat transfer performance to improve the cooling capacity and indoor thermal condition. Therefore, the panel surface temperature is set constant at 15.83 °C, which is the experimentally measured value given by Shin et al. [27]. Each panel design was investigated under seven different cooling load conditions (621.69 W, 746.03 W, 870.37 W, 994.71 W, 1119.04 W, 1243.38 W, and 1405.02 W), owing to differences in heat flux emitted from cylindrical dummies. In summary, 245 cases were simulated for analysis.

Table 3. Boundary condition and emissivity of each surface.

	Condition	Temperature (°C)	Heat Flux (W/m^2)	Emissivity
Wall/ceiling	adiabatic	-	0	0.82
Floor	adiabatic	-	0	0.95
Middle_plane	symmetry	-	-	-
Panel_surface	T = constant	15.83	-	0.92
Cylinder_outer	q = constant	-	50/60/70/80/90/100/113	0.92
Cylinder_upper	adiabatic	-	0	0.92

4. Results and Discussion

4.1. CFD Validation

The CFD model was validated with the experimental results for a flat panel presented by Shin et al. [27]. Figure 4 shows the air temperature distribution of the vertical measured line in three validation cases in which the cooling load was adjusted from 469.92 W to 1409.76 W. The simulated temperature showed marginal differences, with an average error of 1.01% in Case 1, 0.89% in Case 2, and 0.86% in Case 3, indicating that the CFD model agrees with the experimental measurements conducted for the freely suspended panel under different cooling load conditions.

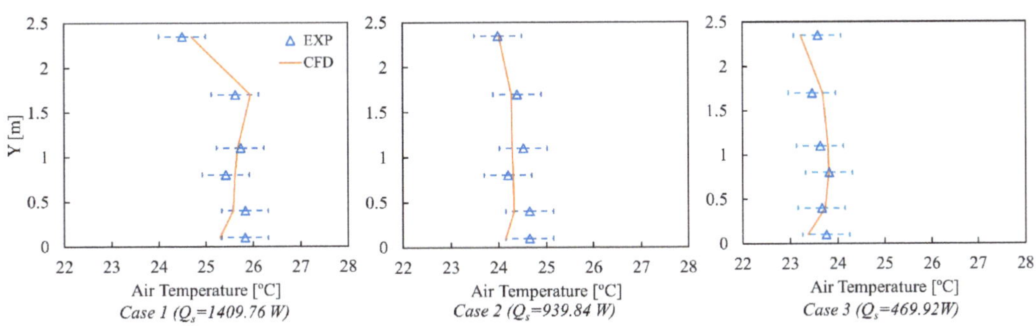

Figure 4. Comparison between CFD and experimental results reported by Shin et al. [27].

4.2. Panel Surface Temperature

Different ceiling radiant cooling panels can be compared and evaluated using the cooling capacity curve present in the standard [32], which is represented by the cooling capacity and the difference between the operative and panel surface temperature (Equation (17)). Previous studies [22,27] reported the curves under different cooling load conditions, adjusting the panel surface temperature to ensure that the indoor air temperature was within a comfort range. Nevertheless, in this study, the panel surface temperature was maintained at 15.83 °C in each case, with the cooling load increasing from 621.69 W to 1405.02 W.

Figure 5 compares the results of heat flux, heat transfer coefficient, average indoor air temperature, and the difference between operative and panel surface temperature, which were obtained under different panel surface temperature conditions. The panel surface temperature was set from 14.83 °C to 19.83 °C with an interval of 1 °C according to the design guidelines presented in [33]. Except for indoor air temperature increasing with the panel surface temperature increase, the heat flux and temperature difference were almost the same, with the difference maintained within 5%, indicating that different panel surface temperature settings only affect indoor thermal conditions but not the panel cooling performance. The amount of heat transferred from the panel is related to the load and the panel itself rather than the surface temperature. Therefore, the boundary setting of this study was simplified.

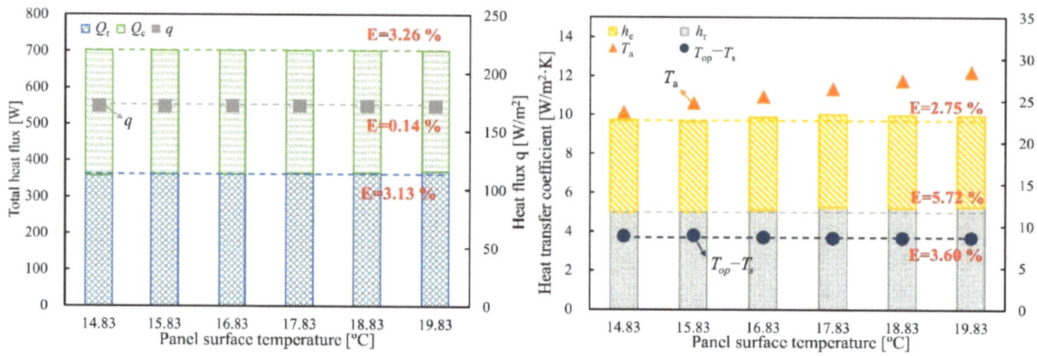

Figure 5. Calculation results under different panel surface temperatures.

4.3. Parametric Study

The cooling capacity of CRCPs is influenced by multiple factors—not only the panel design but also the cooling load, indoor condition, and panel surface temperature—making it difficult to evaluate and compare directly. Therefore, the cooling performance should be compared between different panel designs under a generalized operation condition. In this study, the cooling capacity was analyzed using power regression, which is in a functional relationship with the temperature difference between the operative and panel surface temperature as follows:

$$q = k(T_{\mathrm{op}} - T_{\mathrm{s}})^{\mathrm{n}} \tag{17}$$

The coefficient (k) and exponent (n) of each design are summarized in Appendix A. The nominal cooling capacity was obtained when the temperature difference was 8 K ($T_{\mathrm{op}} - T_{\mathrm{s}}$ = 8 K). All the designs were compared with a closed-type CRCP proposed in the European Standard [32], named 'Standard', and an open-type flat CRCP with a distributed layout proposed by Shin et al. [27], named 'A-d' in the following figures.

4.3.1. Panel Curvature Width

The effect of panel curvature width is shown by comparing the results of Design No.1–No.4. The curvature width (L) of Design 1, Design 2, Design 3, and Design 4 is 0.03 m, 0.06 m, 0.09 m, and 0.12 m, respectively. While the curvature radius is maintained at 0.06 m, and the void distance is maintained at 0.03 m. As shown in Figure 6, the cooling capacity and heat transfer coefficient decrease dramatically with increasing L. With a decrease in L from 0.12 m to 0.03 m, the nominal cooling capacity increases by 35.8% from 113.89 W/m^2 to 154.64 W/m^2. Additionally, both the h_{r} and h_{c} are improved significantly by 49.8% and 35%, respectively, under the same cooling load condition. Compared with Design 2 and Design 3, the convective heat transfer accounts for more in Design 1 and Design 4. Figure 7 compares the velocity contours in Design 1 and Design 4 under the highest cooling load. It is clear that the design with a shorter width contributes to accelerating the air moving through the openings around the panel to promote heat exchange.

4.3.2. Panel Curvature Radius

Figure 8 compares different curvature radii using the results from Design 5, Design 2, and Design No.6–No.9, with curvature radii 0.03 m, 0.06 m, 0.09 m, 0.15 m, 0.2 m, and 0.3 m, respectively. On the other hand, the panel curvature width is constant at 0.06 m, and the void distance is constant at 0.03 m. The results show that the cooling capacity and heat transfer coefficient increase with increasing curvature radius. In particular, the nominal cooling capacity and radiation heat transfer coefficient are obviously improved by 9.1% and 36%, respectively, when r is increased from 0.03 m to 0.06 m. However, when r is larger than 0.06 m, the impact of increasing r is significantly reduced. The nominal cooling capacity and radiation heat transfer coefficient only increase by 4.2% and 3.9%, respectively, when r is increased from 0.06 m to 0.3 m.

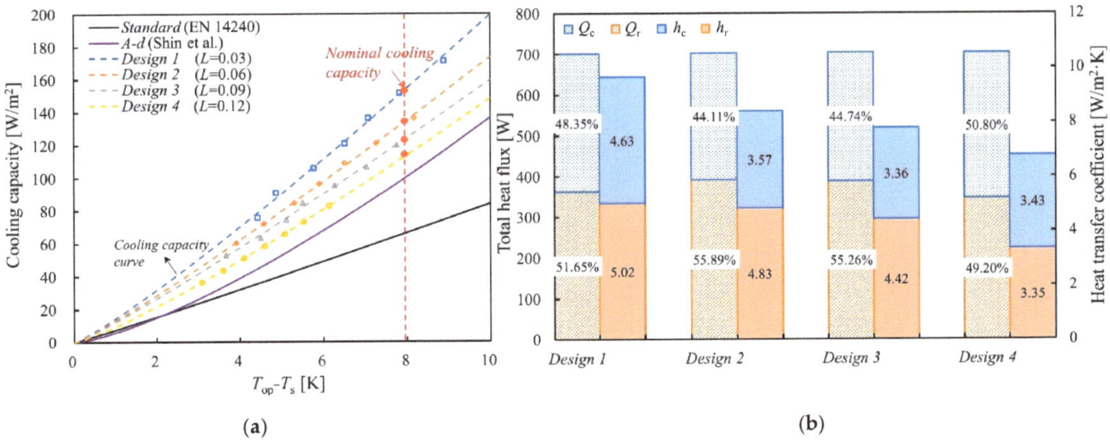

Figure 6. Comparison of different panel curvature lengths (**a**), cooling capacity curves [27,32] (**b**), and thermal performances under a cooling load of 1405.02 W.

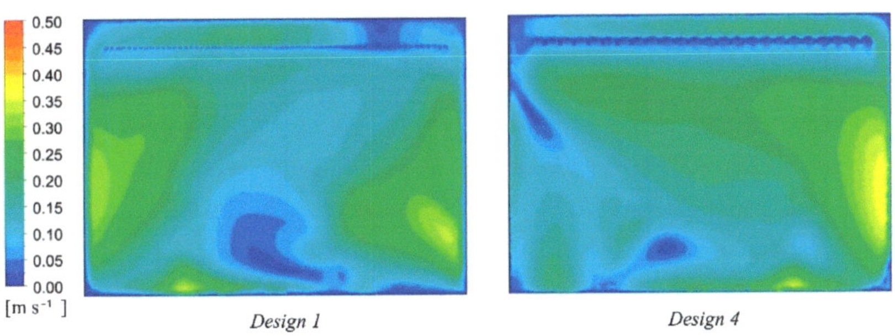

Figure 7. Velocity contours in Design 1 and Design 4 under a cooling load of 1405.02 W.

4.3.3. Void Distance

Figures 9 and 10 illustrate the effect of voids between adjacent panels or panel segments. Eight different distances were compared, varying from 0 m to 0.3 m, with $L = 0.06$ m and $r = 0.06$ m. A solid panel without an opening (Design 10) results in the same nominal cooling capacity as the CRCP proposed by Shin et al. When d is expanded from 0 m to 0.03 m, the nominal cooling capacity increases significantly by 33% from 101.27 W/m^2 to 134.69 W/m^2, and the h_r and h_c are improved by 6.4% and 92.9%, respectively. Including an opening between panels or panel segments can effectively increase convection heat transfer and enhance indoor air movement, as shown in the comparison of airflow distribution between Design 10 and Design 12 in Figure 10. The cooling capacity, h_r, and h_c continue to increase as the distance increases, but the growth slows when d is larger than 0.06 m. In particular, the nominal cooling capacity is highest in Design 16 when d is 0.21 m, which is 4.4% higher than that in Design 13, 13.3% higher than that in Design 12, and 50.1% higher than that in Design 10.

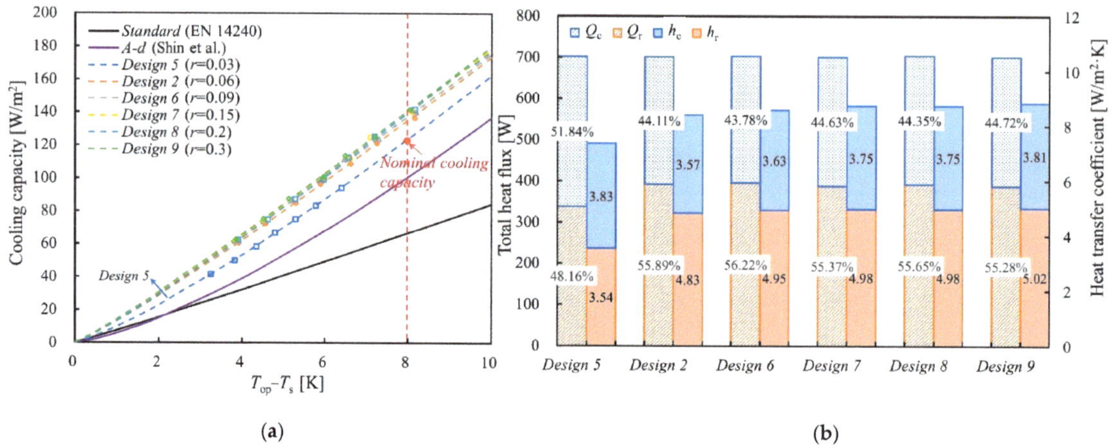

Figure 8. Comparison of different curvature radii (**a**), cooling capacity curves [27,32] (**b**), and thermal performances under a cooling load of 1405.02 W.

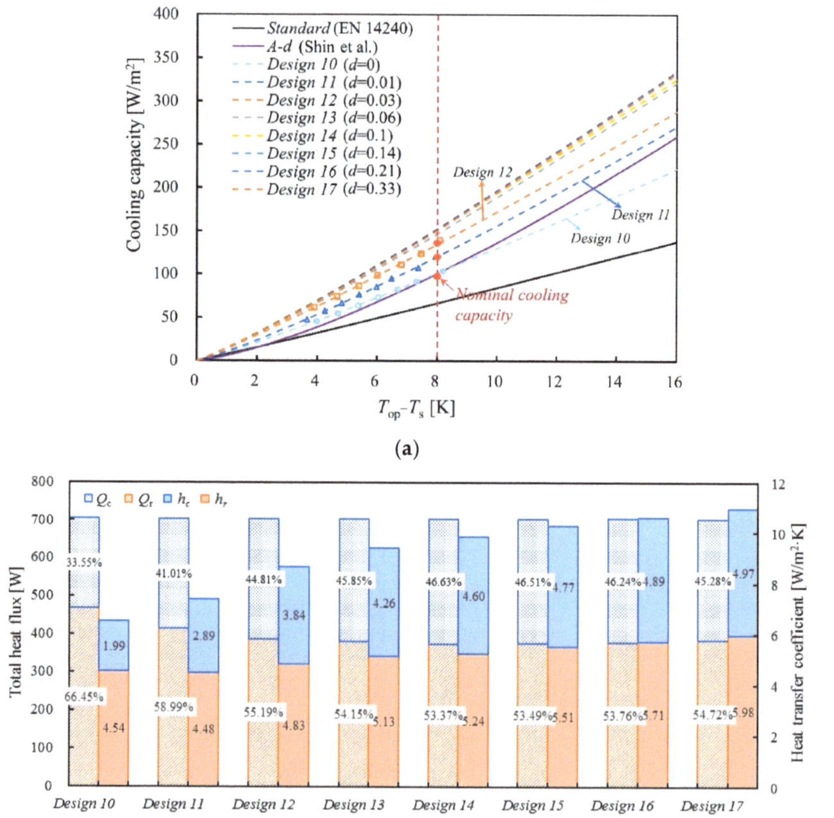

(**b**)

Figure 9. Comparison of different void distances (**a**), cooling capacity curves [27,32] (**b**), and thermal performances under a cooling load of 1405.02 W.

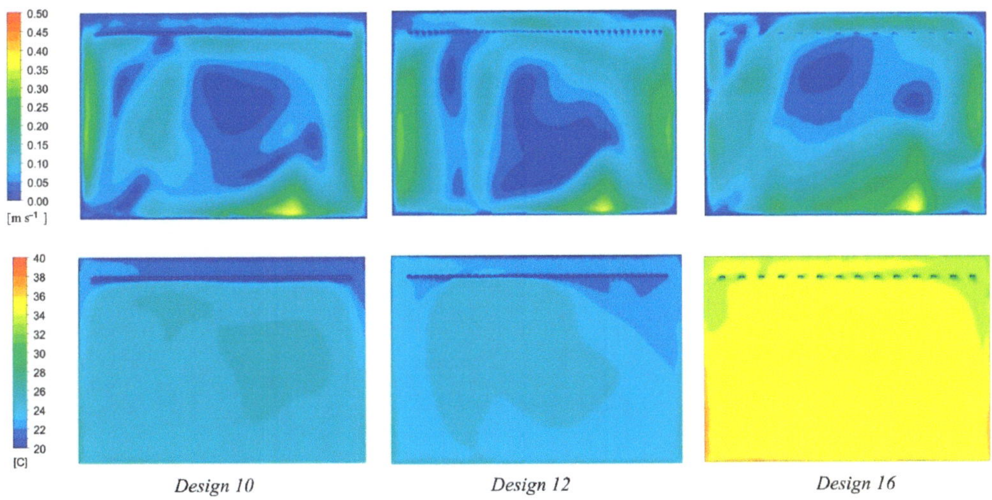

Figure 10. Velocity and temperature contours in Design 10, Design 12, and Design 16 under a cooling load of 1405.02 W.

The indoor air temperature in Design 16 is 10 °C higher than in Design 10 and Design 12, as shown in Figure 10, because the panel number in this design is too small to match the required total heat transfer amount. However, the air temperature uniformity in Design 16 is better than in the other two cases due to active air activity and sufficient heat exchange. The cooled air trapped on the top surface is allowed to move down, resulting in better cooling performance.

4.3.4. Coverage Area

The effect of coverage area was investigated by comparing Design 2 and Design No.18–No.22, as shown in Figure 11. When the coverage area is 11.43 m^2, the nominal cooling capacity is 12.8% higher than 7.58 m^2 and 5.8% higher than 12.96 m^2. The total heat transfer coefficient increases by 4.3% from 8.4 $W/m^2 \cdot K$ to 8.76 $W/m^2 \cdot K$ when the coverage area is reduced from 12.96 m^2 to 7.58 m^2. In other words, expanding the distance between the side of the panel and the wall within an appropriate range can enhance the cooling performance to the same extent as increasing the void distance between adjacent panels.

4.3.5. L/r Ratio and Surface Area

Two dependent parameters—L/r ratio and the panel surface area (A_s)—were examined with the other parameters unchanged. The L/r ratio is discussed by comparing Design No.23–No.27, as shown in Figure 12, which have the same void of 0.03 m and surface area of 11 $m^2 \pm 0.4$ m^2. As a result, when the L/r ratio is 0.5, the nominal cooling capacity is the highest, at 129.65 W/m^2, which is about 5% higher than that of the flat design. On the other hand, the h_c decreases by 12.2% and the h_r increases by 3.6% when the L/r is decreased from 1.1 to 0, illustrating that the curved shape can effectively promote air movement over the top surface and enhance convective heat transfer because the curved structure has a streamlined shape. Notably, when comparing the optimum design of $L/r = 0.5$ with the flat design, the h_c is increased by 5.3%, and h_r is maintained at the same value.

The panel surface area (A_s) was considered with an L/r ratio of 1.5 for the same void distance and panel coverage area, as shown in Figure 13. It should be noted that the L and r values are different in each design, as A_s is a dependent parameter related to all four independent design parameters. When A_s increases from 10.09 m^2 to 13.73 m^2, the nominal cooling capacity decreases by 15.2% from 136.11 W/m^2 to 118.20 W/m^2. The h_r and h_c in Design 28 are also 9.4% and 18.6% higher than those in Design 31, respectively. It is concluded that the short and distributed panel design with less panel surface area achieves better cooling performance than the large solid panel design.

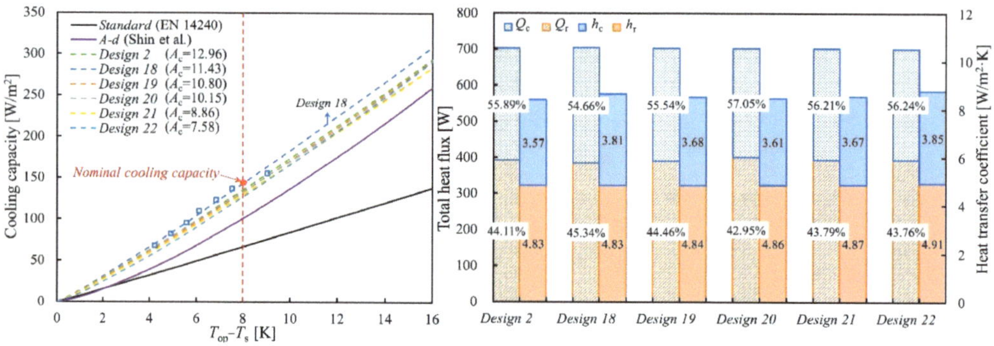

(a) (b)

Figure 11. Comparison of different coverage areas (**a**), cooling capacity curves [27,32] (**b**), and thermal performances under a cooling load of 1405.02 W.

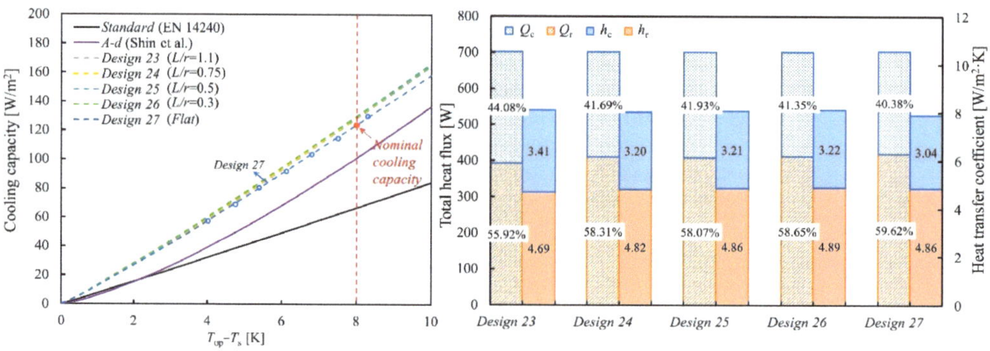

(a) (b)

Figure 12. Comparison of different L/r ratios (**a**), cooling capacity curves [27,32] (**b**), and thermal performances under a cooling load of 1405.02 W.

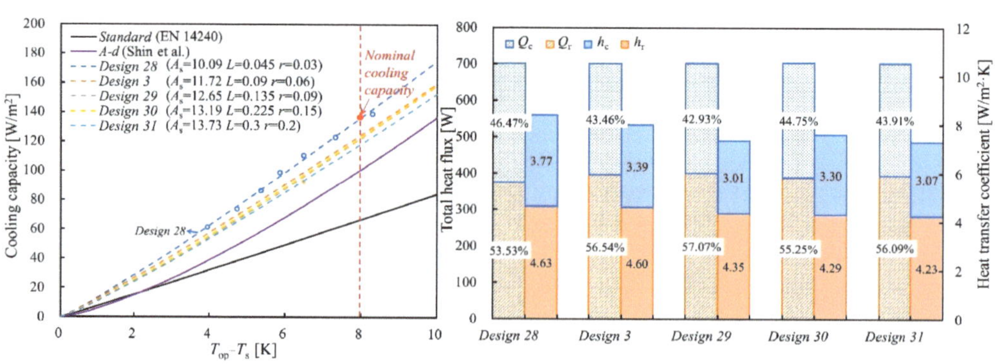

(a) (b)

Figure 13. Comparison of different surface areas (**a**), cooling capacity curves [27,32] (**b**), and thermal performances under a cooling load of 1405.02 W.

4.4. Sensitivity Measures

Based on the parametric analysis, the local sensitivity analysis of the effects of four independent design parameters on cooling capacity was carried out using the manual one-at-a-time (OAT) approach [34]. The sensitivities were measured by monitoring the changes in cooling capacity following the variation of one parameter while all other parameters were held constant. A linear regression equation was derived as a function of cooling capacity and each parameter. The results of sensitivity measures are summarized and compared in Figure 14. Consistent with the above conclusion, the void distance (d) plays the most crucial role in influencing cooling capacity, followed by panel curvature width (L) and radius (r). It is possible to achieve the same or even better indoor thermal conditions by applying fewer panels, demonstrating the potential for cost reduction by optimizing the panel arrangement and construction.

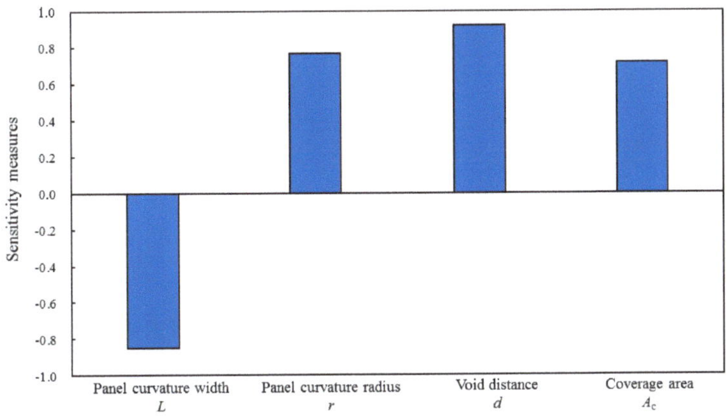

Figure 14. Sensitivity analysis of four dependent design parameters.

5. Discussion

Figure 15 shows the coefficient (k) and exponent (n) in Equation (17) of each panel design. It is clear that different panel designs achieve totally different cooling performances and that structural design is an effective way to improve the efficiency of the CRCP system. All the CRCP designs proposed in this study with curved and segmented shapes can achieve better cooling capacity than those proposed in previous studies under the same condition. Moreover, four designs (Design No.32–No.35) were replenished, combining the concluded optimum panel design of $L = 0.03$ m and $r = 0.06$ m with the void distance and coverage area. Figure 16 illustrates the flow and temperature fields on the symmetry plane in Design 1 and Design No.34–No.35. Design 34, with large openings between panels and between the panels and the wall, is the optimum among all the designs, if it prioritizes increasing the nominal cooling capacity. However, this design is unable to meet thermal comfort requirements under a cooling load of 1405.02 W in practice because of an insufficient number of panels. The temperature fields become more uniform in proposed Design No.34–No.35, but the average indoor temperature in Design 34 and Design 35 is about 4 K and 8 K higher than in Design 1, respectively. In regard to achieving both the thermal comfort conditions included in the ASHRAE standard [35] and maximum cooling capacity, Design 1 is the alternative optimum solution, which can maintain the indoor air temperature at 24.77 °C. Compared with Design 34, Design 1 has the same L and r values but with smaller opening areas. Therefore, (1) $L = 0.03$ m and $r = 0.06$ m represent the ideal panel shape, as concluded by comparing 35 designs in this study; (2) larger void distance and openings between panels and the wall are able to promote cooling capacity. However, preference should be given to thermal comfort and the number of required panels, which can be decided according to cooling capacity and panel surface area.

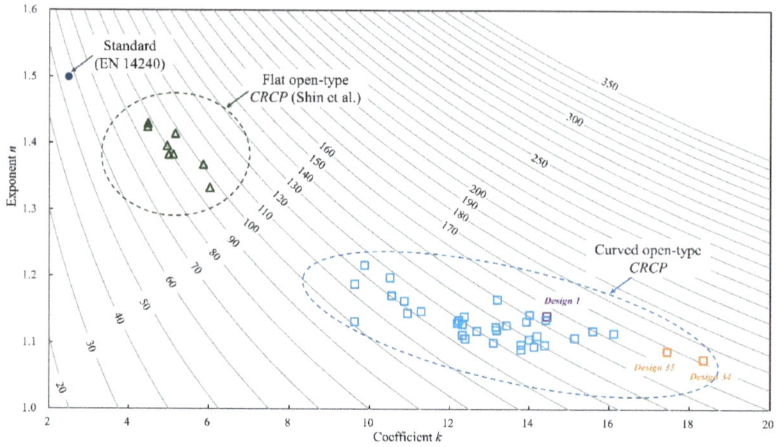

Figure 15. k and n of different CRCP designs ($T_{op} - T_s$ = 8 K) [27,32].

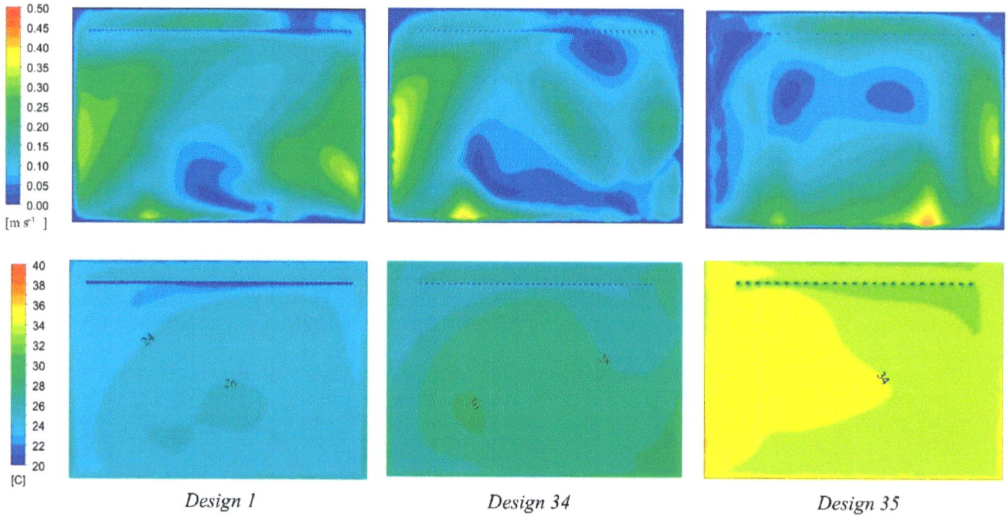

Figure 16. Velocity and temperature contours in Design 1, Design 34, and Design 35 under a cooling load of 1405.02 W.

Finally, we also recommend verifying the generalization of this optimum design by applying other models or experiments in future studies. Further improvements in CRCPs are also expected to be achieved through the application of advanced optimization methodology or the combination of different design strategies.

6. Conclusions

In this study, we carried out a parametric analysis of an open-type CRCP with a curved and segmented structure using CFD simulation to study the effect of panel structure on cooling capacity and heat transfer in comparison with conventional CRCPs presented in the literature. Four independent and two dependent design parameters were investigated by comparing thirty-five panel designs and operating sensitivity analysis. The optimal panel design was then proposed in terms of cooling capacity, heat transfer coefficient, and airflow distribution. The results are as follows:

- The freely suspended CRCP with curved shape and void in proposed this study achieves better cooling performance than the previous reference results. The nominal cooling capacity can be improved by 157.90% compared with the transitional panel design represented in the standard;
- The nominal cooling capacity and heat transfer coefficient increase with increasing panel curvature radius and decreasing curvature width. The nominal cooling capacity is highest when $L = 0.03$ m and $r = 0.06$ m, which is the optimal panel design among the designs proposed in this study;
- Compared with the large solid panel design, the short and distributed panel design with less panel surface area achieves better cooling performance because it can promote air movement around the panel and assist in sufficient heat exchange;
- The distances between adjacent panels and between the panel and the wall play the most significant role in improving the cooling performance of the panel, demonstrating the potential to simultaneously reduce costs and achieve better indoor thermal conditions by optimizing the distribution of CRCPs;
- There should be a balance between improving cooling performance and ensuring the comfort of the indoor environment in practical operations, and total heat transfer should be accounted for based on the panel surface area and cooling capacity.

Author Contributions: Conceptualization, A.A.S. and K.N.; methodology, M.Y.; validation, M.Y. and A.A.S.; formal analysis, M.Y.; investigation, M.Y.; writing—original draft preparation, A.A.S.; writing—review and editing, K.N.; supervision, K.N. All authors have read and agreed to the published version of the manuscript.

Funding: This work was supported by JSPS KAKENHI (grant number JP22J11117).

Data Availability Statement: The data presented in this study are available on request from the corresponding author.

Conflicts of Interest: The authors declare no conflict of interest.

Nomenclature

List of Symbols

L	Panel curvature width (m)
r	Panel curvature radius (m)
d	Void distance (m)
A_c	Panel coverage area (m^2)
A_s	Panel surface area (m^2)
l	Panel length (m)
n	Panel number (-)
ρ	Fluid density (kg/m^3)
$\vec{v}$	Velocity vector (m/s)
p	Static pressure (N/m^2)
$\vec{g}$	Gravitational body force (N/m^3)
$\vec{F}$	External body force (N/m^3)
μ	Molecular viscosity (kg/m·s)
I	Unit tensor
C_p	Specific heat capacity (J/kg·K)
T_f	Fluid temperature (K)
k_f	Fluid thermal conductivity (W/m·K)
k	Turbulent kinetic energy (m^2/s^2)
ε	Turbulent dissipation rate (m^2/s^3)
ε_k	Emissivity
σ	Boltzmann's constant
A_j, A_k	The area of surface j and k, respectively (m^2)
q_j, q_k	Radiative heat flux of surfaces j and k, respectively (W/m^2)
F_{jk}	View factor between surfaces j and k

ρ_0	Specific constant density of the flow (kg/m^3)
T_0	Operating temperature (°C)
β	Thermal expansion coefficient
q_{tot}	Total heat flux (W/m^2)
h_r	Radiation heat transfer coefficient (W/m^2·K)
h_c	Convection heat transfer coefficient (W/m^2·K)
$AUST$	Area-weighted uncooled temperature (°C)
T_p, T_s	Panel surface temperature (°C)
T_a	Air temperature (°C)
T_{op}	Operative temperature (°C)
Q_s	Cooling load (W)
CRCP	Ceiling radiant cooling panel
SRCP	Suspended radiant ceiling panel
CRP	Ceiling radiant panel
TCRP	Thermoelectric radiant panel
RCP	Radiant ceiling panel
CFD	Computational fluid dynamics

Appendix A

Table A1. Coefficient for nominal cooling capacity equation.

	k	n	R^2	$q_{\Delta T = 8K}$ (W/m^2)
Standard [33]	7.489	1.051	-	66.6
A-d [26]	5.8604	1.3674	-	100.6
Design 1	14.448	1.14	0.9949	154.64
Design 2	13.178	1.1193	0.9983	135.11
Design 3	12.321	1.1118	0.9957	124.37
Design 4	9.626	1.1882	1	113.89
Design 5	9.8661	1.2165	0.9997	123.81
Design 6	14.118	1.0942	0.9976	137.38
Design 7	13.994	1.1053	0.9974	139.36
Design 8	13.427	1.1261	0.9991	139.62
Design 9	14.392	1.0968	0.9986	140.81
Design 10	9.6204	1.132	0.9999	101.27
Design 11	10.549	1.1707	0.9984	120.35
Design 12	13.795	1.0974	0.9965	134.69
Design 13	13.927	1.1322	0.9994	146.67
Design 14	15.133	1.1075	0.9981	151.39
Design 15	13.196	1.165	0.9997	148.78
Design 16	14.418	1.1346	0.9994	152.60
Design 17	14.01	1.1421	0.9992	150.61
Design 18	14.198	1.1104	0.9843	142.90
Design 19	12.365	1.1398	0.9987	132.29
Design 20	13.795	1.0901	0.9987	133.10
Design 21	12.23	1.1338	0.9996	129.23
Design 22	10.507	1.1978	0.9978	126.82
Design 23	12.19	1.13	0.9989	127.79
Design 24	12.321	1.128	0.9979	128.63
Design 25	12.68	1.118	0.9951	129.65
Design 26	13.098	1.1004	0.9983	129.11
Design 27	12.39	1.1065	0.9989	123.69
Design 28	13.149	1.1239	0.9943	136.11
Design 29	10.864	1.1631	0.9901	122.00
Design 30	11.279	1.1475	0.9941	122.62
Design 31	10.94	1.1445	0.9998	118.20
Design 32	15.585	1.1182	0.9985	159.42
Design 33	16.117	1.1143	0.9996	163.53
Design 34	18.354	1.0754	0.9932	171.76
Design 35	17.442	1.0877	0.9997	167.45

References

1. Rhee, K.N.; Kim, K.W. A 50 Year Review of Basic and Applied Research in Radiant Heating and Cooling Systems for the Built Environment. *Build. Environ.* **2015**, *91*, 166–190. [CrossRef]
2. Imanari, T.; Omori, T.; Bogaki, K. Thermal Comfort and Energy Consumption of the Radiant Ceiling Panel System. *Energy Build.* **1999**, *30*, 167–175. [CrossRef]
3. Dudkiewicz, E.; Jadwiszczak, P.; Jeżowiecki, J. Examination of Operational Dynamics of Radiant Ceiling Panel. *Cent. Eur. J. Eng.* **2011**, *1*, 159–167. [CrossRef]
4. Miriel, J.; Serres, L.; Trombe, A. Radiant Ceiling Panel Heating-Cooling Systems: Experimental and Simulated Study of the Performances, Thermal Comfort and Energy Consumptions. *Appl. Therm. Eng.* **2002**, *22*, 1861–1873. [CrossRef]
5. Ye, M.; NAGANO, K.; Serageldin, A.A.; Sato, H. Field Study on Energy Consumption and Thermal Comfort of a Nzeb Using Radiant Ceiling Panel and Open-Loop Groundwater Heat Pump System in a Cold Region. *J. Build. Eng.* **2023**, *67*, 105999. [CrossRef]
6. Memon, R.A.; Chirarattananon, S.; Vangtook, P. Thermal Comfort Assessment and Application of Radiant Cooling: A Case Study. *Build. Environ.* **2008**, *43*, 1185–1196. [CrossRef]
7. Valdiserri, P.; Cesari, S.; Coccagna, M.; Romio, P.; Mazzacane, S. Experimental Data and Simulations of Performance and Thermal Comfort in a Patient Room Equipped with Radiant Ceiling Panels. *Buildings* **2020**, *10*, 235. [CrossRef]
8. Tye-Gingras, M.; Gosselin, L. Comfort and Energy Consumption of Hydronic Heating Radiant Ceilings and Walls Based on CFD Analysis. *Build. Environ.* **2012**, *54*, 1–13. [CrossRef]
9. Feng, J.; Bauman, F.; Schiavon, S. Experimental Comparison of Zone Cooling Load between Radiant and Air Systems. *Energy Build.* **2014**, *84*, 152–159. [CrossRef]
10. Park, S.H.; Kim, D.W.; Joe, G.S.; Ryu, S.R.; Yeo, M.S.; Kim, K.W. Establishing Boundary Conditions Considering Influence Factors of the Room Equipped with a Ceiling Radiant Cooling Panel. *Energies* **2020**, *13*, 1684. [CrossRef]
11. Xing, D.; Li, N. Thermal Performance Improvement for the Ceiling Radiant Cooling Panel with an Inbuilt Air Gap by the Convection Shield. *Sustain. Energy Technol. Assessments* **2021**, *44*, 101012. [CrossRef]
12. Labat, M.; Lorente, S.; Mosa, M. Influence of the Arrangement of Multiple Radiant Ceiling Panels on the Radiant Temperature Field. *Int. J. Therm. Sci.* **2020**, *149*, 106184. [CrossRef]
13. Zhang, L.; Liu, X.H.; Jiang, Y. Experimental Evaluation of a Suspended Metal Ceiling Radiant Panel with Inclined Fins. *Energy Build.* **2013**, *62*, 522–529. [CrossRef]
14. Jordan, S.; Hafner, J.; Kuhn, T.E.; Legat, A.; Zbašnik-Senegačnik, M. Indoor Environment in Retrofitted Offices Equipped with Radiant Ceiling Panels. *Gradjevinar* **2016**, *68*, 125–134. [CrossRef]
15. Xing, D.; Li, N.; Zhang, C.; Heiselberg, P. A Critical Review of Passive Condensation Prevention for Radiant Cooling. *Build. Environ.* **2021**, *205*, 108230. [CrossRef]
16. Shakya, P.; Ng, G.; Zhou, X.; Wong, Y.W.; Dubey, S.; Qian, S. Thermal Comfort and Energy Analysis of a Hybrid Cooling System by Coupling Natural Ventilation with Radiant and Indirect Evaporative Cooling. *Energies* **2021**, *14*, 7825. [CrossRef]
17. Jeong, J.W.; Mumma, S.A. Ceiling Radiant Cooling Panel Capacity Enhanced by Mixed Convection in Mechanically Ventilated Spaces. *Appl. Therm. Eng.* **2003**, *23*, 2293–2306. [CrossRef]
18. Shin, M.S.; Choi, J.S.; Rhee, K.N. Cooling Capacity and Energy Performance of Open-Type Ceiling Radiant Cooling Panel System with Air Circulators. *Energies* **2021**, *14*, 5. [CrossRef]
19. Ye, M.; Serageldin, A.A.; Radwan, A.; Sato, H.; Nagano, K. Thermal Performance of Ceiling Radiant Cooling Panel with a Segmented and Concave Surface: Laboratory Analysis. *Appl. Therm. Eng.* **2021**, *196*, 117280. [CrossRef]
20. Lim, H.; Kang, Y.K.; Jeong, J.W. Thermoelectric Radiant Cooling Panel Design: Numerical Simulation and Experimental Validation. *Appl. Therm. Eng.* **2018**, *144*, 248–261. [CrossRef]
21. Luo, Y.; Zhang, L.; Liu, Z.; Wang, Y.; Wu, J.; Wang, X. Dynamic Heat Transfer Modeling and Parametric Study of Thermoelectric Radiant Cooling and Heating Panel System. *Energy Convers. Manag.* **2016**, *124*, 504–516. [CrossRef]
22. Serageldin, A.A.; Ye, M.; Radwan, A.; Sato, H.; Nagano, K. Numerical Investigation of the Thermal Performance of a Radiant Ceiling Cooling Panel with Segmented Concave Surfaces. *J. Build. Eng.* **2021**, *42*, 102450. [CrossRef]
23. Mosa, M.; Labat, M.; Lorente, S. Constructal Design of Flow Channels for Radiant Cooling Panels. *Int. J. Therm. Sci.* **2019**, *145*, 106052. [CrossRef]
24. Mosa, M.; Labat, M.; Lorente, S. Role of Flow Architectures on the Design of Radiant Cooling Panels, a Constructal Approach. *Appl. Therm. Eng.* **2019**, *150*, 1345–1352. [CrossRef]
25. Hassan, M.A.; Kaood, A. Multi-Criteria Assessment of Enhanced Radiant Ceiling Panels Using Internal Longitudinal Fins. *Build. Environ.* **2022**, *224*, 109554. [CrossRef]
26. Radzai, M.H.M.; Yaw, C.T.; Lim, C.W.; Koh, S.P.; Ahmad, N.A. Numerical Analysis on the Performance of a Radiant Cooling Panel with Serpentine-Based Design. *Energies* **2021**, *14*, 4744. [CrossRef]
27. Shin, M.S.; Rhee, K.N.; Park, S.H.; Yeo, M.S.; Kim, K.W. Enhancement of Cooling Capacity through Open-Type Installation of Cooling Radiant Ceiling Panel Systems. *Build. Environ.* **2019**, *148*, 417–432. [CrossRef]
28. Radwan, A.; Katsura, T.; Ding, L.; Serageldin, A.A.; EL-Seesy, A.I.; Nagano, K. Design and Thermal Analysis of a New Multi-Segmented Mini Channel Based Radiant Ceiling Cooling Panel. *J. Build. Eng.* **2021**, *40*, 102330. [CrossRef]

29. Lv, G.; Shen, C.; Han, Z.; Liao, W.; Chen, D. Experimental Investigation on the Cooling Performance of a Novel Grooved Radiant Panel Filled with Heat Transfer Liquid. *Sustain. Cities Soc.* **2019**, *50*, 101638. [CrossRef]
30. Ning, B.; Chen, Y.; Liu, H.; Zhang, S. Cooling Capacity Improvement for a Radiant Ceiling Panel with Uniform Surface Temperature Distribution. *Build. Environ.* **2016**, *102*, 64–72. [CrossRef]
31. ANSYS FLUENT 12.0/12.1 Documentation. Available online: https://www.afs.enea.it/project/neptunius/docs/fluent/index.htm (accessed on 31 January 2023).
32. *EN 14240*; Ventilation for Buildings-Chilled Ceilings-Testing and Rating. CEN: Brussels, Belgium, 2004.
33. *ISO 11855-2:2012(E)*; Building Environment Design-Design, Dimensioning, Installation and Control of Embedded Radiant Heating and Cooling Systems. Part 2: Determination of the Design Heating and Cooling Capacity. International Organization for Standard: Geneva, Switzerland, 2012.
34. Borgonovo, E.; Plischke, E. Sensitivity Analysis: A Review of Recent Advances. *Eur. J. Oper. Res.* **2016**, *248*, 869–887. [CrossRef]
35. *ANSI/ASHRAE 55-2013*; ASHRAE Standard: Thermal Environmental Conditions for Human Occupancy. American Society of Heating, Refrigerating and Air-conditioning Engineers Inc.: Atlanta, GA, USA, 2013.

energies

MDPI

Article

Using a Mine Dewatering System to Increase Cooling Capacity and Energy Recovery of Underground Refrigeration Plant: A Case Study

Dariusz Obracaj *, Nikodem Szlązak and Marek Korzec

Faculty of Civil Engineering and Resource Management, AGH University of Science and Technology, Mickiewicza 30 Av., 30-059 Kraków, Poland

* Correspondence: obracaj@agh.edu.pl; Tel.: +48-12-6172180

Abstract: Heat stress in deep hot mines is a factor that often determines the possibility of technical mining of natural resources. One of the solutions enabling miners to work in such mines is air cooling. Cooling systems vary, and their selection depends on the type of mine and the mining methods used. Limited air cooling capabilities exist in electric-powered coal mines. The main solution for air cooling is based on movable spot air coolers. Such systems commonly use surface or underground refrigeration plants. An underground refrigeration plant (URP) equipped with compressor chillers does not achieve more than 2.5–3.0 MW of cooling capacity due to the limited heat rejection capacity of return air streams in a typical coal mine. The method discussed in this paper, using mine water to discharge waste heat from the underground refrigeration plant, provides a measurable benefit for optimizing the mine air cooling system. The main purpose of this research is to study the feasibility and effect of water diversion from the actual mine drainage system to the underground refrigeration plant. The water drainage system in an underground mine is called the dewatering system of the mine. The heated water in the condensers of the chillers is directed back to the mine's central dewatering system. The recovery from water discharged to the surface contributes to optimising energy consumption for a mine air cooling and the sustainable discharge of wastewater. In addition, using the total water flow from the mine dewatering system to reject heat in compressor chillers, compared with the traditional solution, can improve the cooling capacity of URP. These findings may provide beneficial guidance for practical applications in deep hot mines with small natural water inflow.

Keywords: mine air cooling systems; underground refrigeration plant; mine dewatering system; waste heat recovery; heat pumps

Citation: Obracaj, D.; Szlązak, N.; Korzec, M. Using a Mine Dewatering System to Increase Cooling Capacity and Energy Recovery of Underground Refrigeration Plant: A Case Study. *Energies* **2022**, *15*, 9481. https://doi.org/10.3390/en15249481

Academic Editors: Alessandro Cannavale and Ubaldo Ayr

Received: 22 November 2022
Accepted: 10 December 2022
Published: 14 December 2022

1. Introduction

The exploitation of raw materials in underground mines is associated with many natural hazards. In deep mines, one of the dominant threats is the climate. Its scale depends on many factors that have been presented recently [1–4]. The basic threat is the primary temperature of the rocks, also called virgin rock temperature (VRT). VRT depends on a geothermal temperature at a given mine location. For example, at a depth of 1000 m, during the exploitation of the gold deposit in the West Wits Basin, VTR reaches 20 °C [5] and 44 °C in the Jastrzębie coal mines, Poland [6]. The VTR is related to the different geothermal temperatures.

The state of climatic hazards at workstations in deep mines is also affected by proper planning of underground workings, rational ventilation and organization of the technological process, primarily the transport of mined output, materials and the location of heat-emitting equipment.

The above-mentioned factors influence the fact that various air conditioning systems are used in mines. Their specificity depends on the deposit exploitation systems used but are also adapted to legal and climatic conditions in a given country. In European coal mines, closed refrigeration circuits are used to cool mine air due to the widespread use of electrical machinery and equipment.

The choice of a cooling system is primarily determined by the demand for cooling power in the mining area of a mine. The specificity of air cooling in deep hot mines is the constant demand for cooling power regardless of the season. An important factor is also the type of refrigeration units used and the possibility of heat rejection. A particular limitation in the application of the cooling system may be the distance between the areas of mining work and the locations of the refrigeration units.

Primary, secondary and tertiary cooling systems are used in subsurface mines worldwide [7]. In a typical coal mine, the length of all underground roadways exceeds 100 km. The distance of workplaces from downcast shafts is usually greater than 5 km. Primary and secondary cooling systems are not effective in such vast mining areas. Therefore, a tertiary cooling system is used, consisting of air cooling by movable coil exchangers called spot air-coolers, which are placed close to workplaces. Such cooling devices are also called mobile spot air-coolers [7–10]. Water cooled in a refrigerant plant is distributed to closed-circuit cooling-coil heat exchangers (indirect coolers) in mining districts.

Cooled water circulates in a closed pipe circuit distributed along the roadways. Several dozen spot air-coolers can be connected to the pipeline network several kilometres from the refrigeration plant.

Underground or surface refrigeration plants are used for chilling water. In surface cooling stations, several chillers are connected in a cascade system. Both compressor and absorption units are used. The condensation heat of the refrigerant in compressor chillers is rejected to the atmosphere. In the event of an increase in the demand for cooling power, such surface cooling installations can be expanded to certain limited extents. The limit is the capacity of the chilled water through the vertical pipelines. For technical reasons, pipe diameters in air shafts are not larger than 300 mm. This allows for achieving cooling capacity in the surface refrigeration plant (SRP) up to about 15 MW. Such cooling systems are described in [11–13]. However, in the case of large distances between SRP and workplaces, the temperature of chilled water flowing through pipelines increases [14,15]. This is due to water supply through vertical pipelines and the Joule-Thompson effect. The solutions in the hydrostatic pressure reduction of the water column between vertical and horizontal pipelines are also required.

Therefore, increasingly, the construction of underground refrigeration plants (URP) with a greater cooling capacity than in the systems built so far is being considered, and practical, economically justified removal of the heat of condensation is sought.

In the case of underground refrigeration plants consisting of stationary chilling machines, the heat of condensation is usually rejected into return air flowing in ventilation roadways toward upcast shafts using evaporative water coolers. Until now, URPs equipped with compressor cooling machines have been used in Poland if the cooling power demand does not usually exceed 2.0–3.0 MW [7,16]. URPs with higher cooling power are usually not built because it is impossible to dissipate more condensation heat from the refrigerant. This is due to the thermodynamic parameters of the air and its flow in mine excavation.

The previous experience related to the use of underground plants and the transfer of heat to the air shows that:

1. Rejecting condensation heat into mine air streams is crucial for the correct operation of the underground cooling plant.
2. Heat rejection to the mine air is often associated with the need to install evaporative water coolers at a considerable distance from cooling units (this necessitates building long pipelines connecting evaporative water coolers with cooling units and installing circulation pumps to overcome resistance in these pipelines).
3. The forcing fan must be placed in the evaporative water cooler, which increases the cooling system's energy consumption.

In the case of heat transfer to the mine atmosphere, there are problems related to ensuring appropriate air parameters at the location of evaporative water coolers (too high air temperature, too high humidity, ensuring adequate water flow).

Another way of transferring the heat of condensation from the URP is to transport water through vertical pipelines in the intake air shaft to the surface, cool it, and bring it back down the mine. However, the cost of pumping water has kept such systems from becoming widespread. Xu et al. [3] presented a method of using mine water in the surface-subsurface circulation cycle. From this circulation, part of the cool water from the surface can be transferred to URP, and part is passed through the post-mining goaf.

It is confirmed that the heat condensation from the compressor units can be transferred to the service water or dewatering system of a mine. That integrated system is called the water reticulation system [9]. In the case of electric-powered coal mines, the consumption of service water is minimised, and the inflow of natural water from the strata depends on the geological conditions of a particular mine. In such mines, efforts are made to minimise the mine water circulation. Increasing service water use is out of the question in mines if the dewatering system does not have enough capacity in reserve. Therefore, attention is drawn to using the inflow of natural water from the strata to the workings of the real-life mine. Natural water is also called fissure water.

There are two different sources of water inflow to the dewatering system in mines: ground water (known as fissure water) and mine water (known as service water). Service water consumption in Polish mines is small compared to world mines and constitutes only a small percentage of the total water pumped out to the surface.

The main objective of the dewatering system is to capture fissure water at the outlet points and supply it to the surface of the mine. In mines, most of the water inflows by gravity from the mining areas to the central chamber of the dewatering pumps. The main drainage chambers play a crucial role in the entire system, in which pumps pumping water continuously to the surface are installed. The total water inflow to the pump chamber is related to the inflow of fissure water and service water.

The total water inflow limits the possibility of rejecting condensation heat in mine dewatering systems to the system. Polish mines belong to mines with a small inflow of fissure water. Currently, the largest amounts of water inflow to mines vary widely, from about 20 m^3/h to over 3600 m^3/h [17]. The average inflow is about 700 m^3/h. However, the water inflow's structure is various, so water is often pumped out from different areas of the mine and different levels. Such a drainage system makes it impossible to receive heat rejection from the refrigeration plant fully.

Water from underground settlers can be pumped to the surface indirectly (in two or more stages) or directly. The method of pumping out water in the mine depends on its inflow and the number of active mining levels. In Polish conditions, direct water drainage to the surface is primarily used in large water inflows. In the indirect dewatering system, water is transferred between depth levels and then to the surface. In such a situation, the primary drainage chambers are located at the level with the greatest water inflow. Water can also be transferred by gravity or by pumps between levels.

Water from mine drainage is partly used on the surface. On average, about 1/3 of the water pumped out on the surface is used in mines. The rest of the water is discharged as wastewater. Depending on mine water's salinity degree, some mines are forced to direct water to a desalination plant.

The concept of using water from the dewatering system to reject the refrigerant condensation heat was considered for one of the coal mines. The idea is to transfer mine water through the condensers of underground chillers and pump it to the surface. To develop a technical solution, it was necessary to analyse the possibility of rejecting heat for the maximum cooling capacity of the designed URP, technical possibilities of water purification and pumping, analysis of the temperature of water flowing into the main pumping chamber, analysis of the temperature of water pumped out to the surface, and the possibility of using low-temperature heat from water on the surface.

Cooling is an energy-intensive process. Therefore, such a system must be as efficient as possible [9,18]. Increasing air cooling efficiency can be implemented in many ways, depending on the various solutions used. One way is to recover condensation heat from chillers for heating purposes. For example, in energy infrastructures of mines, there are heating needs related to, among others, heating of rooms, utility water and air inlet to shafts [19–21].

This paper presents the implemented solution for transferring condensation heat from chillers in URP to the mine dewatering system. The objective is a detailed examination of the condenser circuit. The conditions the dewatering system must meet will be analysed to use this method in similar mines. Determining the possibility of utilising recovery heat from cooling chillers is a crucial aspect that should always be considered. The presented example is a kind of energy optimisation in cooling underground excavations with URP.

2. Background behind Using Mine Water to Heat Rejection of Refrigeration Plant

2.1. Cooling Demand in a Real-Life Mine—Case Study

Halemba coal mine is located in the Upper Silesian Coal Basin in Poland, with an area of almost 33 km^2. The mine has four working levels: 380 m, 525 m, 830 m, and 1030 m. Currently, mining is carried out mainly from the 1030 m level. VRT in mining areas ranges from 40 °C to 48 °C. Difficult microclimate conditions at the workplaces make it impossible for the mining crew to work. Therefore, it is necessary to cool the air in the working locations.

A characteristic feature of deep coal mines, including the Halemba mine, is that the demand for cooling power is practically unchanged throughout the year. The surface temperature varies from −20 °C to 30 °C throughout the year, depending on the season. According to the regulations [22], the air in the intake shafts must be heated to a minimum of 1 °C. That means that the air temperature in the shaft varies from 1 °C to 30 °C throughout the year. The impact of seasonal changes in air temperature in the shaft is negligible for consideration of air parameters at working locations [23,24]. This is due to the high VRT, large distances between the workings and shafts, and the presence of many heat-generating machines in the mine. Therefore, the mine has an identical demand for cooling power all year round.

The relatively low efficiency of cooling systems with SRP applied in mines with many widespread working locations means that the URP with compressor units is considered in the Halemba mine. Regarding the prediction of cooling power demand in workstations ranging from 4.5 to 4.8 MW and cooling losses in the closed circuit system, the need for cooling power of centrally located URP is 6.0 MW [25].

High air temperature and humidity of return airways excluded the possibility of transferring condensation heat to the ventilation air using evaporative water coolers. Therefore, it was decided to transfer condensation heat from URP to the mine water.

2.2. Parameters of Dewatering System of the Mine

The dewatering system in the analysed mine drains water from the mining areas to the sump at level 1055 m. First, water flows to the sump, partly by gravity through the sewage system and partly pumped through pipelines from the intermediate pumping stations in mining areas. Next, water is pumped to settlers at level 1030 m near the shaft by pumps with a total capacity of 800 m^3/h. Finally, the main drained pumps in the dewatering chamber pump water from settlers to the surface by shaft piping. Halemba mine does not desalt the mine water before it is discharged into the wastewater due to the low salinity. Figure 1 shows the schematic diagram of the dewatering system.

Depending on the season, the temperature of the pumped water changes in a small range from 26 to 28 °C. Therefore, a higher water temperature value, i.e., 28 °C, was assumed to design the rejection of condensation heat.

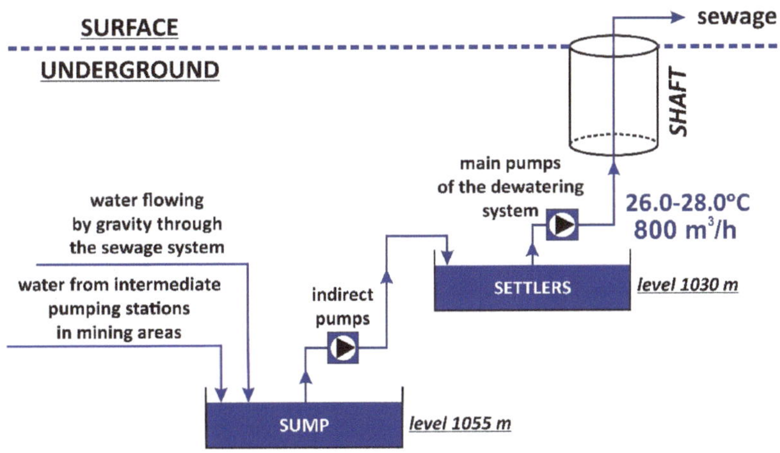

Figure 1. Schematic diagram of the dewatering system in the analysed mine.

For steady waterflow conditions and neglecting changes in kinetic and potential energies in a cooling device, the cooling power is the difference between the outlet and inlet enthalpy flows. Therefore, considering the water enthalpy, the cooling power can be calculated as follows:

$$\dot{Q} = \dot{m}_w c_{pw}(t_1 - t_2) \quad (1)$$

where $\dot{m}_w$ is the mass flow rate of water in kg/s, c_{pw} is the specific heat at constant pressure in J/(kg·°C) and t_1 and t_2 are inlet and outlet temperatures of water from the heat exchange device of the air-cooling system in °C.

Figure 2 shows the change in heat transfers to water depending on its volumetric flow rate and the temperature difference when flowing through the condensers. The presented graph shows that it is necessary to ensure a high flow of cooling water to reject heat from URP, which is often not ensured in the cooling systems of underground mines. Another critical factor is the temperature of the water. If it is too high, the heat rejection potential decreases.

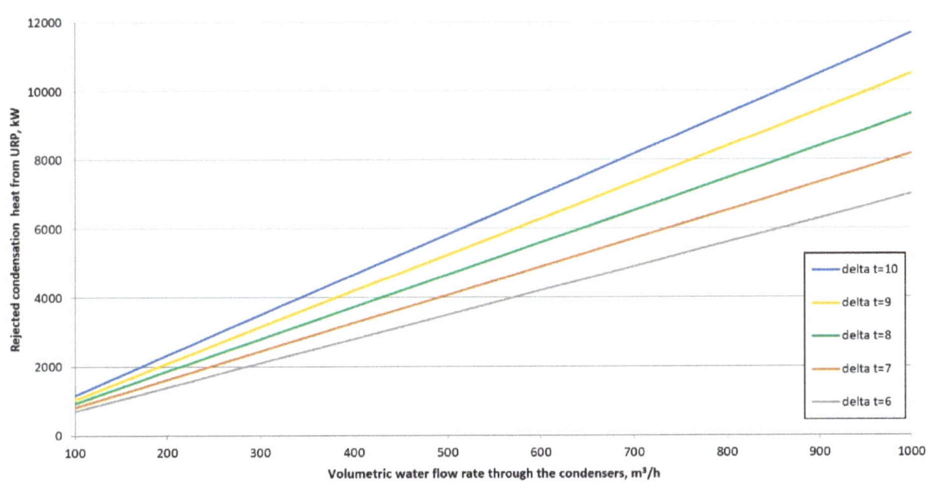

Figure 2. Removal of condensation heat depending on the flow rate and temperature difference of cooling water for condenser chillers in URP.

Such water often contains impurities and cannot be used directly to cool the condensers, as it would cause rapid contamination. However, such water can be used with underground heat exchangers mediating heat exchange. Another factor that may limit the water potential of heat rejection is its temperature. The main facets affecting the water temperature in settlers are parameters of air flow out from the inlet shaft. In addition, the ventilation of settlers causes the water temperature to fluctuate throughout the year.

2.3. Technical Solution of Cooling and Heat Rejection Systems

URP is located in the vicinity of settlers of the dewatering system at level 1030 m.

Cooling-coil heat exchangers have been proposed to cool the air at workstations. Chilled water in the URP will be supplied to spot air-coolers in two operating areas at 1500 m to 6000 m from the shaft [25]. The chilled water will flow through insulated pipelines. After receiving the heat in the air coolers, the water returns through uninsulated pipelines to the URP, where it is cooled again. A diagram of the chilled water pipeline network of the cooling system is shown in Figure 3.

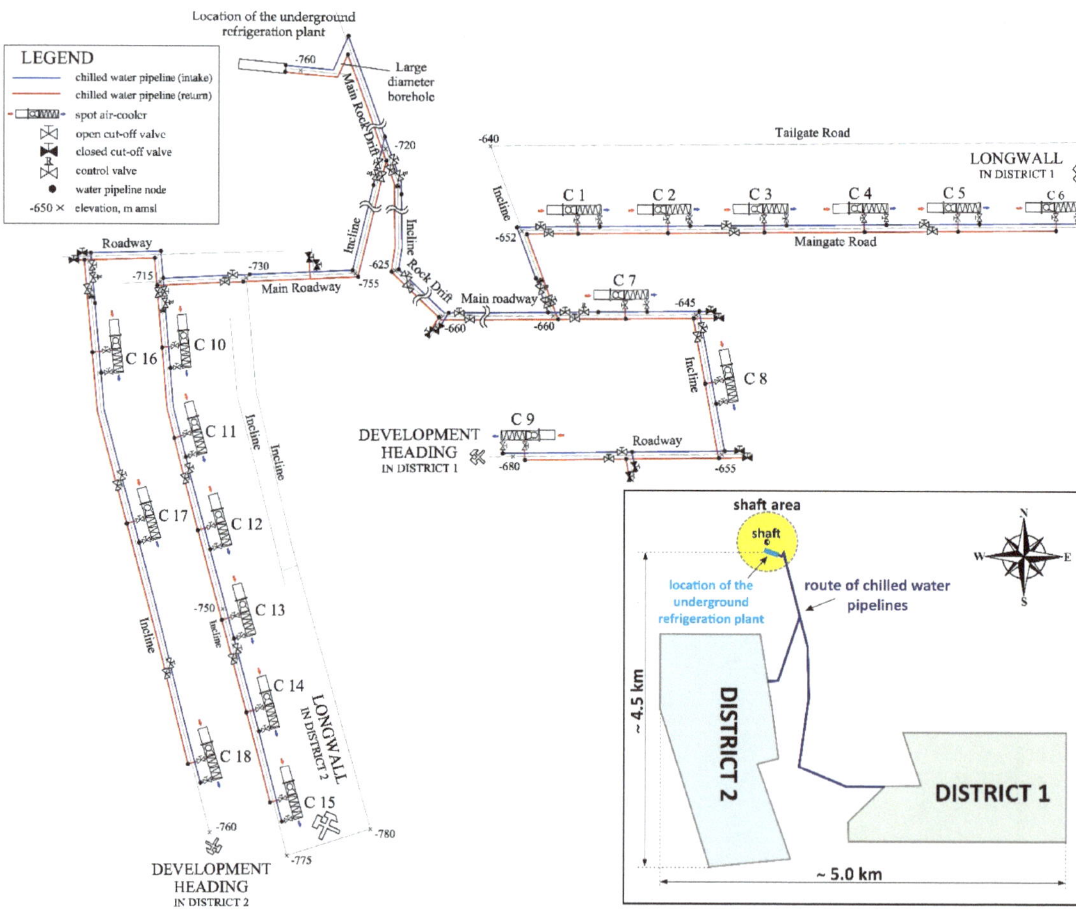

Figure 3. Chilled water circuit in the analysed mine.

The entire cooling system consists of four circuits:

- circuit of chillers' evaporators—circulation of chilled water,
- refrigerant circuit—the internal circuit of chillers,

- primary circuit of condensers of refrigeration units—circulation of cooling water for condensers,
- secondary circuit of chillers' condensers—circulation of mine water receiving heat.

To design the refrigeration system to work properly, it had to be ensured that the heat of condensation of the refrigerant in the chillers of URP was discharged. Figure 4 shows a schematic diagram of chilling and cooling circuits in the URP, together with the inclusion in the dewatering system.

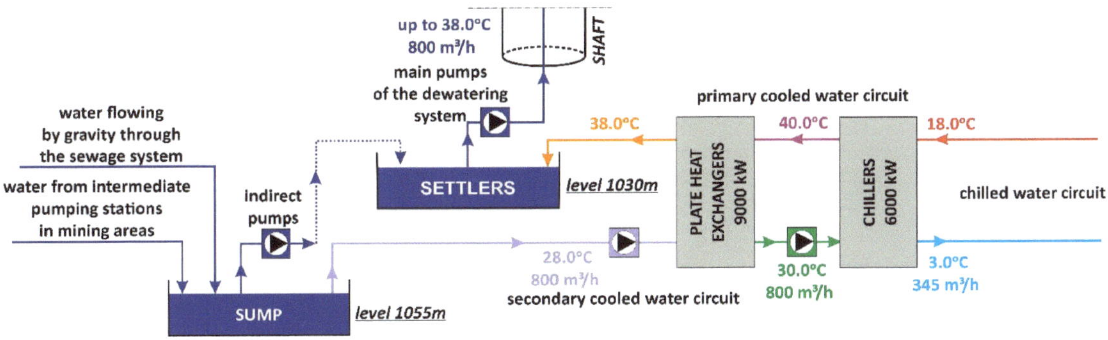

Figure 4. Simplified sketch of the cooling mine water in the analysed mine.

Ensuring the rejection of condensation heat determines the effectiveness of chilling water in the evaporator cycle to the required temperature of 3 °C. The condensers of the refrigeration units will be cooled with water from the mine dewatering system through plate heat exchangers. Water will be transported from the sump to the pipeline and heat exchangers. After discharging the heat in the chillers' condensers, the water will flow through the return pipeline to the settlers, from where it will be pumped to the surface using the mine's dewatering pumps. The distances between the URP and the sump and settlers do not exceed 100 m. Therefore, it allows assuming no heat loss during water transport. For other circuits, heat loss must always be considered.

The energy balance of the analysed installation, assuming the efficiency of the primary heat exchange devices, allows the determination of water temperatures in the individual compartments of the condenser cooling circuit. The results are shown in Figure 2. Considering the temperature and mass of water flowing into the dewatering system, the heat stream from cooling the condensers, the water temperature in the settlers will be 38 °C. The specific enthalpy of water at the inlet to the shaft pipeline will be 160 kJ/kg.

The solution allows the accumulation of condensation heat from URP in mine water. It does not increase the costs because the dewatering system drains water to the surface anyway, and the costs of its pumping must be incurred regardless of the cooling system.

3. Results and Discussion

3.1. Energy Balance of Refrigeration System

Air cooling systems are characterized by high energy consumption. The main energy-intensive devices are motors of chiller compressors and water circulation pumps.

Analysed URP has 5 chiller units, including 1 unit with a capacity of 2000 kW and four units with a capacity of 1000 kW. The units are connected in parallel to the chilled water pipeline network and the condenser cooling pipeline network, respectively. The total condenser power of these chillers is 7800 kW. The electric power of compressor motors in chillers is 2050 kW. The energy efficiency ratio (EER) is 4.18 and 4.22, respectively.

The second significant source of power demand is circulating pump drives. Electric power demand for pump drives is given by Equation (2):

$$P_p = \frac{\dot{V}_w \Delta h \rho_w g}{\eta_p} \tag{2}$$

where $\dot{V}_w$ is the volumetric flow rate of the pumped water in m^3/s, ρ_w is the water density in kg/m^3, Δh is the pump head in m, g is the gravitational acceleration in m/s^2, and η_p is the efficiency of the pump system.

The calculations were carried out in accordance with the water parameters shown in Figure 2. The results of calculations of pump drive power are presented in Table 1.

Table 1. Parameters of circular pumps in URP.

Pump	Flow Rate, m^3/h	Delivery Head, m	Efficiency, %	Electric Power Demand, kW	Rated Power of the Engine, kW
Pumps of chilled water circuit (2 pcs)	348	230	75	298	380
Pump of primary cooling water circuit	760	90	68	274	315
Pump of secondary cooling water circuit	800	60	71	184	200

The additional demand for electric power in the URP is 100 kW, related to the drives of automatic damper actuators, drivers and controllers, and lighting. The total electric power demand in the URP is 2800 kW.

The chilled water installation includes fan drives in 17 spot air-coolers and automatic dampers. The electric power demand in a chilled water circuit is 450 kW.

The designed cooling system is assumed to operate continuously throughout the year, with the same cooling capacity. The cooling power demand is constant regardless of the season and outdoor air parameters.

The total demand for electric power in the cooling system is 3250 kW, resulting in electricity consumption of 78 MWh per day.

If the cooling water from the URP had to be transported to the surface through a 1030 m long vertical pipeline, the chiller condensers would have to be increased by 3350 kW. Such an increase in power demand in the cooling system is unacceptable due to operating costs. Therefore, the solution is to pump the water into the air through the existing mine dewatering system.

3.2. Energy Balance of Dewatering System

A mine dewatering system also generates high costs related to the operation of pumps due to the lifting height and high capacity of pumped water in the shaft. In the analysed case, the power of the pump sets in the dewatering system is about 3500 kW. Every day, pumps consume around 84 MWh of electricity.

Due to the costs incurred, it is reasonable to use water to receive the heat of condensation of the refrigerant in chillers. The temperature of the water pumped out from the bottom shaft at level 1030 m is 38 °C. Therefore, it is necessary to consider the water temperature change in the shaft pipeline. The expected water temperature on the surface is required to determine the possibility of its use and removal.

3.3. Predicted Temperature of Mine Water on the Surface

As air flows from the inlet of the shaft to the bottom of the shaft, the air's temperature and moisture content vary along its length. The reason is the heat and mass exchange between the strata and the airflow in the shaft. The shaft wall temperature varies with depth as a result of the geothermal rock gradient. Masses of moisture or infiltration water can flow from the surrounding rocks into the shaft. In addition, there is heat exchange between the air and devices or pipelines transporting media in the shaft. Air self-compression due to height differences is also an essential factor. In the Halemba mine, a significant internal heat source is the pipeline transporting hot water from level 1030 m to the surface. The nature of airflow in the shaft and water in the pipeline is like a pipe-in-pipe with counter-current flows. However, mass and heat are exchanged between the shaft (outer tube) and the strata. Methods of predicting air temperature and humidity in air shafts are known [26]. The analytical solutions to determine heat and mass transfers in mine shafts and heat exchange between flowing counter-current air in the shaft and water in the pipe were adopted during design. The solution is given by [27–29]. The authors of these works presented a simplified solution to the differential equations of heat and moisture transfer to vertical workings. The assumptions are that the surrounding rocks are isotropic and homogeneous, and the virgin temperature of the surrounding rock varies linearly with depth. The heat and mass transfer from the rocks to the airflow is steady, and their directions are perpendicular to the axis of the shaft. The steady state of the heat flow is related to the shaft's ventilation time by considering the period from the moment of its construction to the present. Seasonal change in outdoor air is taken into consideration. In the analysed shaft, the water inflow from the rock mass is about 3.5 m^3/h. Other air and water flow parameters are presented in Table 2.

Table 2. Parameters of shaft and pipeline.

Parameter	Value
Shaft	
Diameter, m	7.5
Rock parameters at 30 m below the pit-bank level:	-
- Virgin rock temperature, °C	7.5
- Thermal diffusivity, 10^{-6} m^2/s	1.2
- Coefficient of thermal conductivity, W/(m K)	2.8
Rock parameters at the shaft bottom:	-
- Virgin rock temperature, °C	37.8
- Thermal diffusivity, 10^{-6} m^2/s	0.7
- Coefficient of thermal conductivity, W/(m K)	1.9
Average mass flowrate, kg/s	262
Convective heat transfer coefficient, W/(m^2 K)	18.7
Average specific heat of water vapour, J/(kg K)	1860
Average specific heat of dry air, J/(kg K)	1005
Pipeline	
Diameter, m	0.4
Coefficient of thermal conductivity, W/(m K)	48
Mass flowrate, kg/s	222.2
Average density of water, kg/m^3	995
Average specific heat of water, J/(kg K)	4179
Linear heat transfer coefficient, W/(m K)	54.5

The temperature of pumped-out water on the surface depends on the inlet air temperature of the shaft and the temperature of the water at the level of 1030 m. For the prediction of water temperature on the surface, calculations of heat exchange between the water pipeline and the air in the shaft were made based on the models given in [27–29]. Calculations were made for the summer, transitional and winter periods. Assumptions of air parameters for these periods are given in Table 3. The results of the calculations are shown in Figure 5.

Table 3. Medium-term parameters of air inlet to the shaft.

Parameter	Winter	Spring/Autumn	Summer
Air temperature at the inlet of the shaft, °C	1	13	28
Relative humidity of air at the inlet of the shaft, %	34	58	67

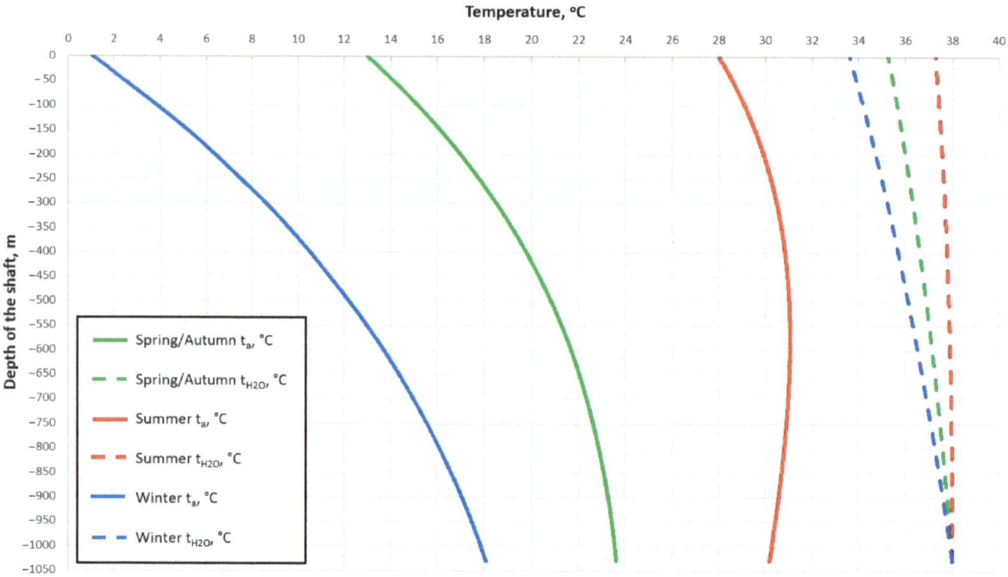

Figure 5. Calculated air temperature t_a along the shaft and water temperature t_{H2O} in the shaft pipeline.

The calculation results show that water temperature on the surface will vary from 33.6 °C in winter to 37.3 °C in summer, depending on the season. The legislation of many countries complains discharging of used water from industrial plants to surface watercourses not only on the composition but also the temperature of the water. For the Halemba mine, the maximum water temperature discharged as wastewater cannot exceed 35 °C [30].

It follows from the above that periodically the temperature of the water pumped to the surface will exceed 35 °C, resulting in the water having to be cooled before discharge to the sewage.

3.4. Heat Demand of the Surface Infrastructure of the Mine

European underground mines are characterized by a high demand for heat in the winter. The heat is mainly demanded in heating the surface infrastructure buildings, producing domestic hot water, and heating the inlet air to the shafts. There are many ways to use waste heat to cover these needs. Very often in mines, waste heat from various processes is considered. However, the difficulties lie in correlating waste heat parameters' quantity and continuity with the plant's energy needs.

In the analysed example, the possibility of using the heat contained in the mine water to cool the chiller condenser in URP was considered. The transfer of condensation heat from the underground chillers to the water pumped to the surface increases its temperature. Periodically, the water is too hot to be discharged into the sewer and needs to be cooled down. Thus, water has a certain energy potential that allows it to be used for heating purposes. This heat can be used directly or through heat pump systems. The article's next section presents the concepts of such water use.

Energy needs on the surface of the mine were analysed to assess the possibility of assessing the use of waste heat. Analyses have shown that in winter, it is possible to use heat to power three installations:

- heating of domestic hot water up to the temperature of 65 °C—constant heat demand throughout the year;
- heating the air in the inlet shafts—the required temperature of the heating water is 90 °C—variable heat demand;
- heating of buildings at 65 °C—variable demand when outdoor temperature drops below 15 °C.

The hot-utility water installation ensures surface facility users' living and technological needs. The largest tap water consumption is caused by employee baths, departmental workshops and the living needs of office buildings. The average daily hot water consumption per employee is 60 dm^3. The mine employs 3125 people, which gives a total water consumption of 187.5 m^3/day. The heat load for domestic hot water is 500 KW. About 12.0 MWh of thermal energy is used to prepare domestic hot water during the day. It is assumed that the demand for heat for domestic hot water preparation is constant throughout the year.

Following the legal requirements in many countries, it is necessary to heat the inlet air to the mine in winter to avoid freezing the shaft hoist elements. The heat demand for air heating in the intake shafts depends on the outdoor temperature and the flow rate of the inlet air into the mine. About 12,000 m^3/min of air (240 kg/s) flows through the inlet shaft, where a pipeline discharges water to the surface. The peak power of the air heaters in the inlet shaft at a design temperature of −20 °C is 5.0 MW. The calculations assume that the air is always heated to the temperature of 1 °C required by law, and the heat demand depends on the outdoor air temperature.

The peak heat load of buildings at an outdoor air temperature of −20 °C is 80 W/m^2. The thermal power demand for heating buildings with an area of 5000 m^2 is 4.0 MW. In calculating the heat demand for the heating system, the heat load depends on the outside air temperature.

The assumptions presented above indicate that, apart from the domestic hot water installation, the demand for heat energy varies throughout the year and depends on the outside air temperature. The peak heat demand for the outdoor air temperature of −20 °C was determined separately for each installation. However, this demand is infrequent. As the outside air temperature increases, it decreases, as shown in Figure 6.

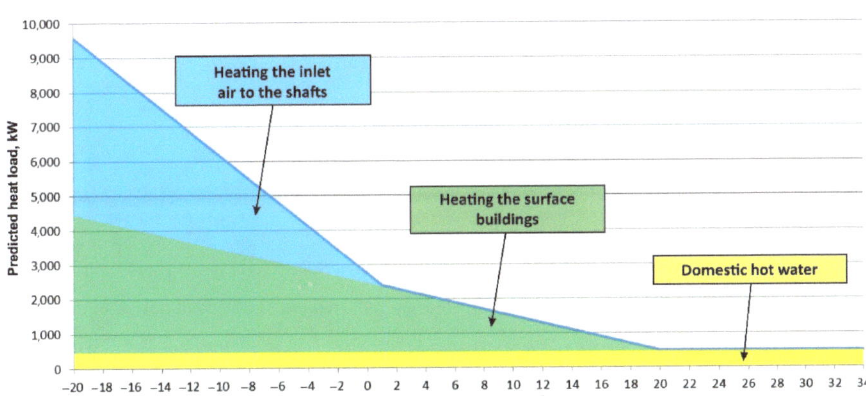

Figure 6. Cumulative demand for heating facilities in the analysed mine.

However, the graph presented in Figure 6 does not show the heating demand; rather, it shows only the heat load variability. Therefore, using statistical climatic data for Katowice, changes in heat demand and outside air temperature were analysed. Figure 7 presents the frequency of occurrence of temperatures in the area of the Halemba mine. The presented graph shows that very low air temperatures are relatively rare, and then the thermal loads of the systems are the greatest.

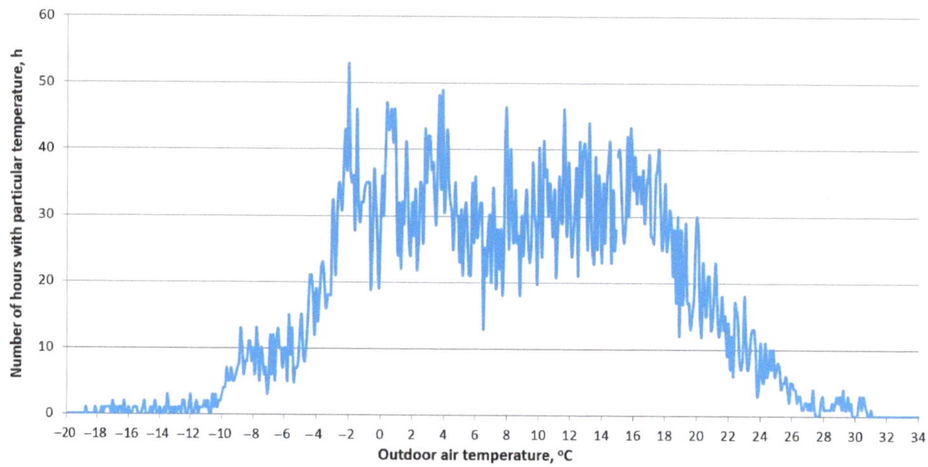

Figure 7. Air temperature frequency for the geographic location of the mine.

Based on the heat demand diagram for the analysed mine and the temperature frequency diagram, an ordered diagram of the hourly energy demand in kWh was developed, shown in Figure 8. The total annual demand for heat in the mine is 17,153 MWh/year (61,750 GJ/year).

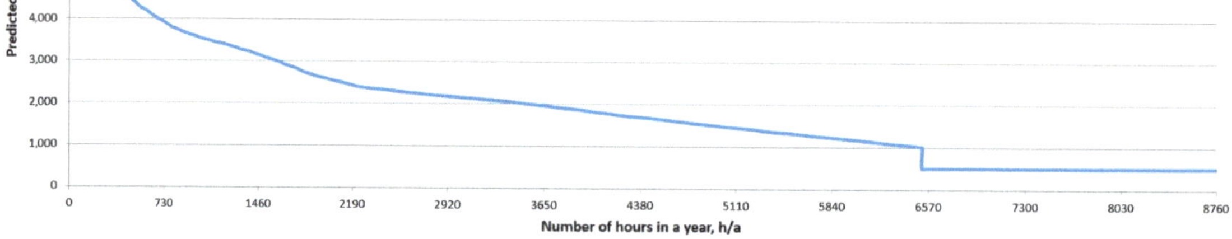

Figure 8. Ordered graph of the annual requirement for thermal power in the mine facilities.

3.5. The Concept of Heat Recovery Utilization from the Refrigeration System

The possibility of using waste heat from the mine dewatering system was analysed based on the assumptions. The temperature of the water pumped by the dewatering system to the surface is too low and does not allow for its direct use. Therefore, to use the waste heat contained in it, the use of water-to-water heat pumps was considered. Due to possible water contamination, and thus the protection of heat pump evaporators, an intermediate heat exchanger was used. In the event of a heat demand lack, a cooling tower is also provided to cool the water before it is discharged into the sewage system. Figure 9 shows a schematic diagram of the solution for obtaining waste heat from mine water pumped on the surface.

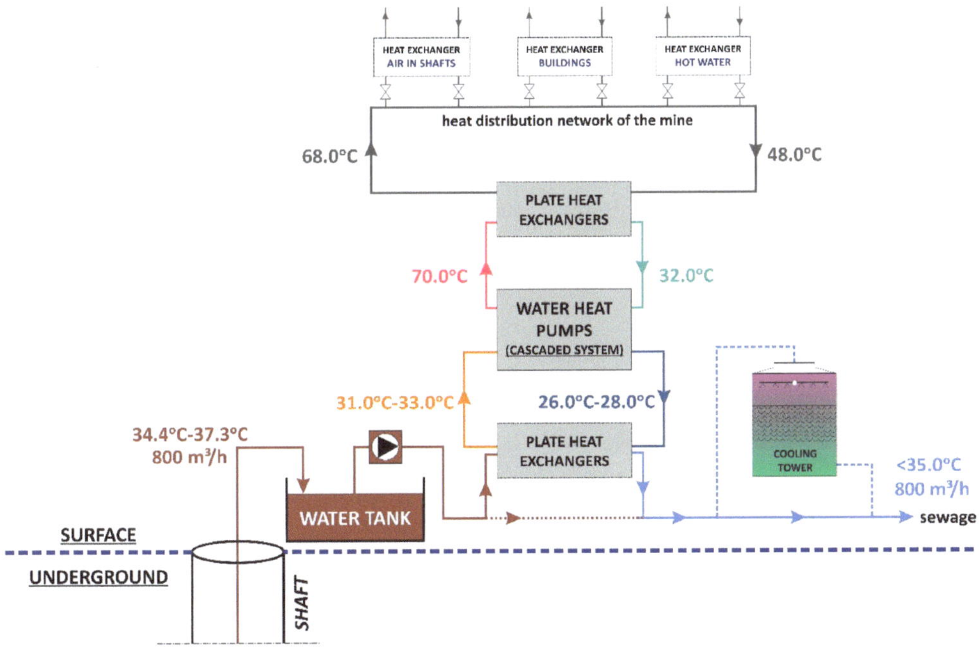

Figure 9. Simplified sketch of the heat recovery solution from the mine water pumped by the dewatering system.

It is assumed that the heat from the mine water will be transferred to the district heating system in the summer to heat the domestic hot water installation. The water will be cooled by 10 °C by a set of heat pumps (cascade system). The temperature difference of the water on the upper heat source will be 38 °C. Therefore, a set of heat pumps with the coefficient of performance COP = 3.8 was selected. Figure 9 shows the heat exchangers included in the mine's heating system.

When it is impossible to transfer heat to the district heating network, the water can be cooled in a cooling tower. A cooling tower with a cooling capacity of 3000 kW was selected. This cooling tower is a reserve installation in case of a heat demand lack during the summer. The water is discharged into the sewage treatment system, which will operate the same way as before. The cooling system in the mine is under construction.

4. Conclusions

The solution is the first underground refrigeration plant of this type with such high cooling power in the Polish mining industry, in which virtually all water from the mine dewatering system is used to receive the condensation heat of refrigerant in an underground refrigeration plant (URP).

The use of mine water, which is constantly pumped to the surface, allows the design of URP with a capacity of more than 3000 MW or the extension of existing plants.

The transfer of condensation heat from the chillers in the URP to the water significantly increases its temperature. Therefore, if the heat demand in the surface infrastructure of the mine is sufficiently high in the summer, heat recovery from the water heated in URP brings definite benefits.

Using such a solution was possible thanks to the specific dewatering system of the analysed mine. The dewatering system has a constant supply of service and fissure water with a constant temperature throughout the year.

In the analysed case, the specific enthalpy of water discharged to the surface is 160 kJ/kg. The heat demand for domestic hot water preparation in summer is 500 kW. In winter, the heat demand in the surface infrastructure is 9700 kW. Depending on the needs, the heat can be used for various installations in the mine.

However, heat demand in the summer is small, and the mine water can exceed the temperature of 35.0 °C. Therefore, reserve water cooling towers are needed. However, the cooling towers are predicted to have a short usage period.

The results show that the energy potential of the UCP condensation heat transferred to the water discharged from the mine allows partial coverage of the heat demand on the surface of the mine in winter by using water-to-water heat pumps. The results also indicated that using mine water for heat rejection from URP in electric-powered mines might be suitable for designing and building underground air cooling systems.

Author Contributions: Conceptualization, D.O., N.S. and M.K.; methodology, D.O., N.S. and M.K.; software, M.K.; validation, D.O.; formal analysis, M.K.; investigation, D.O. and M.K.; resources, N.S., D.O. and M.K.; data curation, D.O.; writing—original draft preparation, D.O.; writing—review and editing, D.O. and M.K.; visualization, M.K.; supervision, D.O. and M.K.; project administration, D.O. All authors have read and agreed to the published version of the manuscript.

Funding: The article was prepared as part of the Subsidy for the Maintenance and Development of Research Potential at Faculty of Civil Engineering and Resource Management AGH University of Science and Technology No. 16.16.100.215.

Conflicts of Interest: The authors declare no conflict of interest.

References

1. Maurya, T.; Kailash, K.; Vardhan, H.; Aruna, M.; Raj, G.M. Potential Sources of Heat in Underground Mines—A Review. *Procedia Earth Planet. Sci.* **2015**, *11*, 463–468. [CrossRef]
2. Li, Z.; Xu, Y.U.; Li, R.; Jia, M.; Wang, Q.; Chen, Y.; Cai, R.; Han, Z. Impact of the water evaporation on the heat and moisture transfer in a high-temperature underground roadway. *Case Stud. Therm. Eng.* **2021**, *28*, 101551. [CrossRef]
3. Xu, Y.U.; Li, Z.; Wang, J.; Chen, Y.; Li, R.; Wang, Q.; Jia, M. Ventilation and heat exchange characteristics in high geotemperature tunnels considering buoyancy-driven flow and groundwater flow. *Int. J. Therm. Sci.* **2022**, *173*, 107400. [CrossRef]
4. Wei, D.; Du, C.; Lin, Y.; Chang, B.; Wang, Y. Thermal environment assessment of deep mine based on analytic hierarchy process and fuzzy comprehensive evaluation. *Case Stud. Therm. Eng.* **2020**, *19*, 100618. [CrossRef]
5. Biffi, M.; Stanton, D.; Rose, H.; Pienaar, D. Ventilation Strategies to Meet Future Needs of the South African Platinum Industry. *J. S. Afr. Inst. Min. Metall.* **2007**, *107*, 59–66.
6. Doležal, L.; Knechtel, J.; Taufer, A.; Trávníček, L. Primary Rock Temperature Fields in Czech and Polish Part of the Upper of the Upper Silesian Coal Basin. *Arch. Min. Sci.* **2013**, *58*, 55–72. [CrossRef]
7. Kamyar, A.; Aminossadati, S.M.; Leonardi, C.; Sasmito, A. Current Developments and Challenges of Underground Mine Ventilation and Cooling Methods. In Proceedings of the 2016 Coal Operators' Conference, Mining Engineering, University of Wollongong, Wollongong, Australia, 18–20 February 2019. Available online: https://ro.uow.edu.au/coal/615 (accessed on 22 November 2022).
8. Belle, B.; Biffi, M. Cooling pathways for deep Australian longwall coal mines of the future. *Int. J. Min. Sci. Technol.* **2018**, *28*, 865–875. [CrossRef]
9. Vosloo, J.; Liebenberg, L.; Velleman, D. Case study: Energy savings for a deep-mine water reticulation system. *Appl. Energy* **2012**, *92*, 328–335. [CrossRef]
10. Dosa, I.; Sorina, A.; Codrut, D.P.; Urdea, G.U. Influence of ambient temperature on the efficiency of underground spot cooling system. In *MATEC Web of Conferences*; EDP Sciences: Les Ulis, France, 2022; Volume 354. [CrossRef]
11. Wojciechowski, J. Application of the GMC-1000 and GMC-2000 mine cooling units for central air-conditioning in underground mines. *Arch. Min. Sci.* **2013**, *58*, 199–216.
12. Chu, Z.; Ji, J.; Zhang, X.; Yang, H.; Dong, H.; Liu, J. Development of ZL400 mine cooling unit using semi-hermetic screw compressor and its application on local air conditioning in underground long-wall face. *Arch. Min. Sci.* **2016**, *61*, 949–966. [CrossRef]
13. Guo, P.; Wang, Y.; Duan, M.; Pang, D.; Li, N. Research and application of methods for effectiveness evaluation of mine cooling system. *Int. J. Min. Sci. Technol.* **2015**, *25*, 649–654. [CrossRef]

14. Szlązak, N.; Obracaj, D.; Swolkień, J.; Piergies, K. Controlling the distribution of cold water in air cooling systems of underground mines. *Arch. Min. Sci.* **2016**, *61*, 793–807. [CrossRef]
15. Qi, Y.D.; Cheng, W.M.; Xin, S. Theoretical analysis of temperature rising for chilled water in the long distance transport pipelines in coal mine. *Arch. Min. Sci.* **2019**, *64*, 785–796.
16. Szlązak, N.; Obracaj, D.; Swolkień, J. An evaluation of the functioning of cooling systems in the Polish coal mine industry. *Energies* **2018**, *11*, 2267. [CrossRef]
17. Sas, S. Assessment of the Possibility of Reducing Primary Energy Consumption for Heating the Inlet Air to a Mine. Doctoral Thesis, AGH University of Science Technology, Cracow, Poland, 2022.
18. duPlessis, G.E.D.; Arndt, D.C.; Mathews, E.H. The development and integrated simulation of a variable water flow energy saving strategy for deep-mine cooling systems. *Sustain. Energ. Tech. Assess.* **2015**, *10*, 71–78.
19. Obracaj, D.; Sas, S. *Possibilities of Using Energy Recovery in Underground Mines*; EDP Sciences: Les Ulis, France, 2018; Volume 29. [CrossRef]
20. Feng, X.-P.; Jia, Z.; Liang, H.; Wang, Z.; Wang, B.; Jiang, X.; Cao, H.; Sun, X. A full air cooling and heating system based on mine water source. *Appl. Therm. Eng.* **2018**, *145*, 610–617. [CrossRef]
21. Guo, P.; He, M.; Zheng, L.; Zhang, N. A geothermal recycling system for cooling and heating in deep mines. *Appl. Therm. Eng.* **2017**, *116*, 833–839. [CrossRef]
22. Water Law Act of 20 July 2017 (Journal of Laws of 2017, Item 1566, as Amended), Polish Act of Law. Available online: https://isap.sejm.gov.pl/isap.nsf/DocDetails.xsp?id=WDU20170001566 (accessed on 22 November 2022).
23. Yi, X.; Ren, L.; Ma, L.I.; Wei, G.; Yu, W.; Deng, J.; Shu, C. Effects of seasonal air temperature variation on airflow and surrounding rock temperature of mines. *Int. J. Coal Sci. Technol.* **2019**, *6*, 388–398. [CrossRef]
24. Szlązak, N.; Obracaj, D.; Korzec, M.; Swolkień, J. Efficiency of cooling methods in Polish underground mines. In Proceedings of the 11th International Mine Ventilation Congress, Xi'an, China, 14–20 September 2018; Science Press: Beijing, China; Springer Nature Singapore Pte Ltd., Crop.: Beijing, China, 2019; pp. 713–733.
25. Szlązak, N.; Obracaj, D.; Korzec, M. Technical design of the air-cooling system in Halemba Coal Mine. 2019; *non-publised studies*.
26. Li, W.J.; Zou, S.H.; Yang, W.X.; Hu, Q. Model of Heat and Mass Exchange between a Downcast Shaft and the Air Flow to the Mine. *Geofluids* **2020**, *2020*, 8853839. [CrossRef]
27. Wacławik, J. Forecasting of mine climatic conditions with the effect of humidity taken into consideration. *Min. Bull. Sci. Bull. Stanislaw Staszic Acad. Min. Metall.* **1971**, *295*, 5–95.
28. Wacławik, J.; Mikołajczyk, W.; Branny, M. 1983: Temperatura i wilgotność powietrza kopalnianego. *Arch. Min. Sci.* **1983**, *18*, nr 4.
29. Filek, K. Water transport by shaft lines in air-cooling installations in underground mine excavations. *Górnictwo Geoinżynieria* **2011**, *35*, 57–68.
30. Regulation of the Minister of Maritime Economy and Inland NAVIGATION of July 12, 2019, on Substances That Are PARTICULARLY Harmful to the Aquatic Environment and the Conditions to be Met When Introducing Sewage into Waters or into the Ground, as Well as When Discharging Rainwater or Meltwater into Waters or for Water Devices, Polish Act of Law. Available online: https://isap.sejm.gov.pl/isap.nsf/DocDetails.xsp?id=WDU20190001311 (accessed on 22 November 2022).

energies

Article

Thermal Performance of Load-Bearing, Lightweight, Steel-Framed Partition Walls Using Thermal Break Strips: A Parametric Study

Paulo Santos *, Paulo Lopes and David Abrantes

ISISE, Department of Civil Engineering, University of Coimbra, 3030-788 Coimbra, Portugal
* Correspondence: pfsantos@dec.uc.pt

Abstract: Thermal bridges are a very relevant issue for lightweight steel-framed (LSF) construction systems given the high thermal conductivity of steel, which can negatively compromise their thermal behaviour, reduce their durability, and decrease the building energy efficiency. Several thermal bridge mitigation strategies exist, including the attachment of thermal break strips (TBS) to the steel studs' flanges as one of the most widely employed techniques. In this research, the relevance of TBS to the thermal performance improvement of load-bearing LSF partition walls was assessed by performing a parametric study, making use of a validated 2D numerical model. A sensitivity analysis was performed for five different key parameters, and their importance was evaluated. The assessed parameters included the number of TBS and their thickness, width, and thermal conductivity, as well as the vertical steel stud spacing. We found that these parameters were all relevant. Moreover, regardless of the TBS thermal conductivity, it is always worth increasing their thickness. However, the increase in the TBS width does not always lead to increased thermal resistance; a thermal performance reduction was noted when increasing the width of the TBS at higher thermal conductivities. Therefore, it was concluded that it is more efficient to increase TBS thickness than their width.

Keywords: thermal performance; lightweight steel framed; partition walls; thermal break strips; parametric study; cross-section dimensions; thermal conductivity; stud spacing; number

Citation: Santos, P.; Lopes, P.; Abrantes, D. Thermal Performance of Load-Bearing, Lightweight, Steel-Framed Partition Walls Using Thermal Break Strips: A Parametric Study. *Energies* **2022**, *15*, 9271. https://doi.org/10.3390/en15249271

Academic Editors: Alessandro Cannavale and Ubaldo Ayr

Received: 2 November 2022
Accepted: 4 December 2022
Published: 7 December 2022

1. Introduction

One of the activity sectors with higher energy consumption is the building sector, which is responsible for 40% of consumption and 36% of the emissions of greenhouse gases in the European Union (EU) [1]. Globally, the values for this industry are 36% and 37% [2], respectively. In the last 2 years, there was a decline in both of these values due to the COVID-19 pandemic [2]. The energy consumed in the EU for space heating and cooling corresponds to 50% of the total, of which 80% is used in buildings [3]. Therefore, the EU promotes the use of renewable energy for heating and cooling [4], as well as renovation of building stock, prioritizing energy efficiency to achieve the EU's energy and climate goals [3]. In the EU, 75% of the building stock is not energy efficient; thus, the renovation of existing buildings could reduce total energy consumption by 5–6% [1], corresponding to a reduction in gas imports of 13–15.6% [3].

The construction sector has been adapting and changing to further industrialized and lightweight substitutes. The light steel-framed (LSF) system is one of these alternatives, which has been proliferating given its remarkable advantages relative to traditional heavyweight brick masonry and reinforced concrete systems. These advantages include [5–7]: easier handling and transportation, given the very reduced weight; high mechanical strength; suitability for prefabrication, ensuring improved quality-control of building elements; adequacy for modular construction; time savings resulting from a faster construction process onsite; high potential for reuse and recyclability (greater than 95%), representing more sustainable construction; excellent stability of dimensions and shape in contact with moisture; decrease in waste; water consumption reduction (dry construction); and reduced need for an intensive workforce and heavy machinery.

However, the LSF system, when not well designed, given the high thermal conductivity of the steel structure, is disposed to forming thermal bridges, which may lead to a thermally inefficient building envelope with condensation problems and thermal discomfort of the occupants [8].

Thermal bridges in buildings can be caused by materials with high thermal conductivity (e.g., steel studs) trespassing layers with high thermal resistance (e.g., batt insulation) within the assemblies [9], or due to junctions, such as connections between the wall and wall, wall and roof, or wall and floor [10]. Thermal bridges decrease the energy efficiency of buildings, impacting the heating and cooling energy needs by up to 30% [11] and 20% [10], respectively. Several studies have been undertaken on reinforced concrete and brick masonry walls. Al-Sanea and Zedan [12] verified that mortar joints, corresponding to 9.1% of the total wall area, reduce the *R*-value by 51% and increase the transmission load by 103%. The energy consumption increase due to existing thermal bridges of representative wall configurations used in Greece was shown to be up to 35% more than estimated during the design stage, as shown by Theodosiou and Papadopoulos [13]. Hua Ge and Baba [10] verified that thermal bridges would increase the annual heating load by 30% and 18% for hot climates and cold climates, respectively. The effect of thermal bridges on buildings was also studied by Jedidi and Benjeddou [14], who concluded that to minimize the problems from thermal bridges, insulation placement on the outside was the preferable option, decreasing the risk of condensation, mould growth, and heat loss.

Due to steel's high thermal conductivity, the importance of thermal bridges in LSF structures can be even greater [15], making this issue an important topic for many researchers in diverse fields, such as parametric studies [8,16,17]; development of measurement methods [18,19]; analytical assessment methodology [20]; experimental methods [21–23]; and numerical simulations [23,24].

Some thermal bridge mitigation strategies have been studied for LSF buildings, such as the use of external thermal insulation composite systems (ETICS) [8,16,25,26]; the use of TBS [16,17,21–23]; the use of steel stud flanges' indentation [17]; and the use of slotted steel studs [16,27]. However, regarding the use of TBS, a systematic study of relevant related parameters (e.g., TBS cross-section dimensions and material thermal conductivities) to enhance the thermal performance of load-bearing partition LSF walls was absent within the existing literature.

Each thermal bridge mitigation strategy has its intrinsic advantages and/or drawbacks. The use of ETICS is very common, with their main advantage being reduced LSF wall thickness, as the extra ETICS thickness can be located outside the floor area and does not compromise the net floor area [26]. The use of TBS has the main advantage of very cost-effective material use, as these strips are placed in a very localized way (along the studs' flanges) and the small amount of material is located where it is most needed, i.e., along the steel frame [21]. One possible drawback is the reduced mechanical resistance to shear actions due to lateral loading [28], because the sheathing panels are now further way from the steel frame. The use of steel stud flanges with an indentation to decrease the contact area between the steel and the sheeting panels has the main advantage of not increasing the LSF thickness, and it is also possible to fill the original flange gap with insulation material [17]. The main drawback of the slotted steel studs is their consequent load-bearing capacity reduction [16].

In this paper, the thermal performance enhancement achieved by the use of TBS on load-bearing partition LSF walls was evaluated. A parametric study was performed where two steel stud spacings were evaluated (400 and 600 mm), two TBS positions were considered (on the exterior steel stud flange and on both flanges), and five thermal conductivities of the TBS material (ranging from 7.5 up to 120 mW/m/K) were assessed. Moreover, regarding the TBS cross-section geometry, five thicknesses (ranging from 5 up to 15 mm) and five widths (ranging from 30 up to 70 mm) were studied. These LSF wall configurations corresponded to 202 numerical models that were evaluated in this parametric study. These numerical models were successfully validated experimentally, and their accuracy was also verified by comparing the results with the ISO 10211 test cases, the ISO 6946 analytical approach, and a three-dimensional finite element model.

Following this short introduction, in the next section (Materials and Methods), the assessed load-bearing partition LSF walls are presented, starting with a description of the reference LSF wall, followed by the parameters evaluated within this parametric study, and characterization of the materials and the numerical simulations carried out. After this, the results are shown and discussed for the evaluated parameters. Finally, the main conclusions of the work are presented.

2. Materials and Methods

In this section, the LSF walls are detailed, starting with the reference. Moreover, the parameters of this study are defined, and the geometry of the assessed TBS—namely their thickness and width—is also presented. This is followed by the characterization of the materials used in this research in relation to their thermal conductivities. Next, the numerical simulations carried out in this study are explained, including the discretization of the evaluated domain, boundary conditions, and verification of the model accuracy and its validation.

2.1. Reference Partition LSF Wall

In this parametric study, a load-bearing LSF partition wall was assumed as a reference. Its cross-section is displayed in Figure 1. All evaluated LSF walls were made with commercial C90 × 43 × 15 × 1.5 cold-formed steel studs with the following dimensions: web (90 mm), flange (43 mm), lip return (15 mm), and steel sheet thickness (1.5 mm). Figure 1 shows the cross-section of a typical partition, used here as a reference LSF wall. There was a 90 mm thick batt insulation made with mineral wool (MW) filling the wall cavity. A 12 mm thick oriented strand board (OSB) structural sheathing panel was placed on both sides of the vertical steel studs. Moreover, there was an additional inner sheathing layer made with gypsum plasterboard (GPB), with a thickness of 12.5 mm.

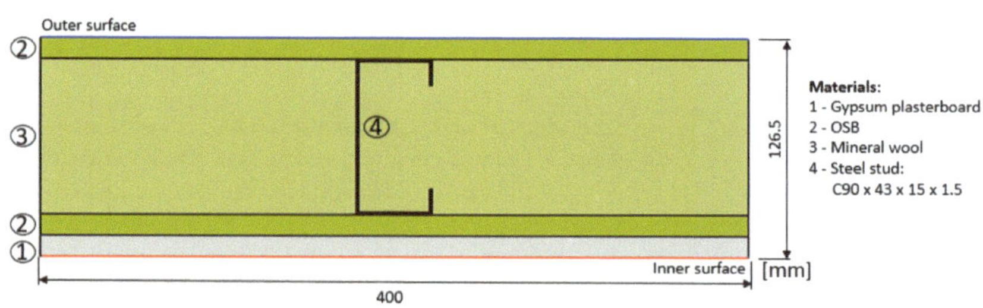

Figure 1. Reference load-bearing partition LSF wall: horizontal cross-section.

2.2. Evaluated Parameters and Values

Table 1 displays the five evaluated parameters in this sensitivity analysis, as well as their respective values. These five parameters, which are also illustrated in Figure 2, were the TBS thickness, width, and thermal conductivity and number, as well as the steel studs' spacing.

Table 1. Parameters to be evaluated and assigned values.

Parameter	"Values"
Thermal break strips:	
- Thickness [mm]	5, 7.5, **10** *, 12.5, 15
- Width [mm]	30, 40, **50** *, 60, 70
- Conductivity [mW/(m·K)]	7.5, 15, 30, 60, 120
- Number	**Zero** *, one [1], two [2]
Steel stud spacing [mm]	**400** *, 600

* Reference values in bold; [1] one: outer flange; [2] two: both inner and outer flanges.

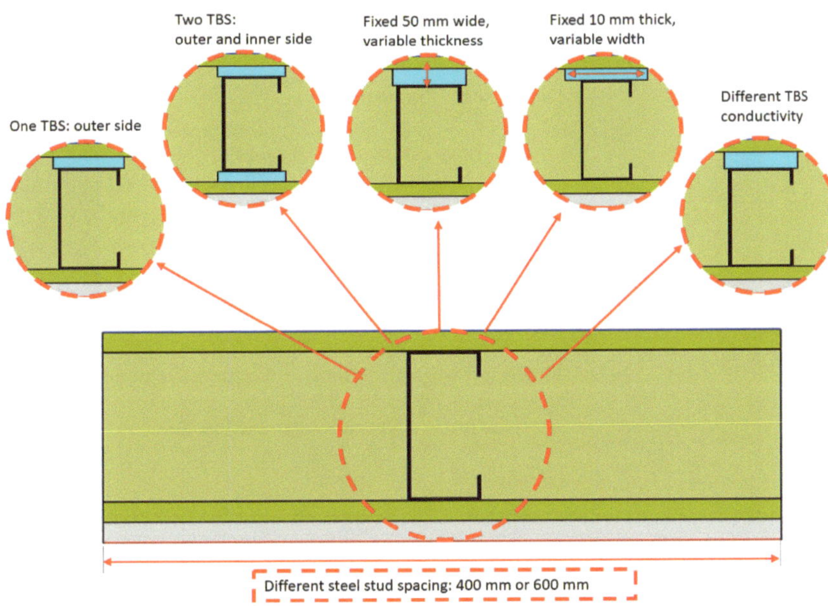

Figure 2. Evaluated parameters in the sensitivity analysis.

Regarding the TBS evaluated values:

(1) their cross-section width was fixed (50 mm) and their thickness was changed within the interval of 5–15 mm, with an increment of 2.5 mm;
(2) their thickness was fixed (10 mm) and their width ranged within the interval of 30–70 mm, with an increment of 10 mm;
(3) for the two previous evaluated parameters, five different TBS thermal conductivities were studied (7.5, 15, 30, 60, and 120 mW/m/K);
(4) the number of TBS was zero (without TBS), one TBS (outer flange side), and two TBS (both inner and outer flange side).

The fifth parameter was the vertical steel stud spacing on the LSF partition walls, which was 400 and 600 mm.

In this parametric study it was assumed that the mineral wool, given its high expansion capacity, was able to fill the increased air cavity layer created by introducing the TBS along the steel stud flange, as shown in Figure 2.

2.3. Characterization of Materials

The utilized materials in the LSF walls are displayed in Table 2, with their respective thicknesses, t, as well as their thermal conductivities, λ. Note that the thermal conductivities and the dimensions of the evaluated TBS were previously displayed in Table 2.

Table 2. Characterization of the materials used in the partition LSF walls: thicknesses (t) and thermal conductivities (λ).

Material	t [mm]	λ [W/(m·K)]	Ref.
Gypsum plasterboard	12.5	0.175	[29]
Oriented strand board	12.0	0.100	[30]
Mineral wool	90.0	0.035	[31]
Steel studs (C90 × 43 × 15 × 1.5)	90.0	50.000	[32]

2.4. Numerical Simulations

The LSF walls' numerical simulations were performed using the finite element method (FEM) software THERM (version 7.6.1, Lawrence Berkeley National Laboratory: Berkeley, CA, USA) [33]. In the next subsections, the details of the numerical models will be explained, starting with the domain discretization, followed by the boundary conditions and the model accuracy verifications and validation.

2.4.1. Discretization of the Models' Domain

Only a representative bidimensional segment of the LSF walls' cross-sections were modelled, as illustrated in Figure 1, for the reference LSF walls (400 mm width) in order to decrease computation time and effort. The materials' thermal properties were previously presented in Tables 1 and 2. The maximum error accepted on the FEM calculations was set to 2% for the models assessed in this work.

2.4.2. Boundary Conditions

The environment air temperatures and the surface thermal resistances were the two sets of boundary conditions defined for each model. In this parametric study, the exterior air temperature was set to 0 °C, while the interior air temperature was set to 20 °C. Regarding superficial thermal resistances, we used the default values for horizontal heat flow indicated in ISO 6946 [34], i.e., 0.04 and 0.13 $m^2 \cdot K/W$ for external (Rse) and internal resistance (Rsi), respectively.

2.4.3. Verifications of Model Accuracy and Validation

For the accuracy verification of the 2D THERM software [33] models, 3 verifications were performed: (1) the ISO 10211 (Annex C) [35] test cases; (2) the analytical approach following ISO 6946 [34], assuming homogeneous layers, i.e., walls with no steel studs, and; (3) by comparison with the results provided by 3D simulations performed in the ANSYS FEM software [36]. For the model validation, experimental lab measurements were performed, as detailed in the following paragraphs.

(1) ISO 10211 Test Cases Verification

The first accuracy verification was to model two distinct bidimensional test-cases specified in Annex C of standard ISO 10211 [35]. The achieved results were within the tolerance range and are not displayed here for the sake of conciseness. However, they can be found in earlier scientific papers, e.g., [26,37,38] by the same author, confirming the accuracy of this 2D FEM software algorithm and the models.

(2) ISO 6946 Analytical Approach Verification

In this second precision confirmation, simplified LSF walls were assumed by making use of homogeneous layers (i.e., having no steel frame), for which the standard ISO 6946 analytical values were compared. This comparison was performed for a wall similar to the reference LSF partition. The input values, such as thermal conductivities, thicknesses of the layers, and used materials, were previously displayed in Table 1. The adopted superficial thermal resistances were prescribed by the standard ISO 6946 [34] for horizontal heat flow, as presented in Section 2.4.2. The obtained results for the numerical and analytical thermal transmittances are listed in Table 3. As expected, the numerical and analytical results perfectly matched, confirming the accuracy of the implemented THERM models [33].

Table 3. Thermal transmittance computed for the simplified LSF walls assuming homogeneous layers.

LSF Wall Type	*U*-Value [W/(m^2·K)]	
	Numerical (THERM)	**Analytical (ISO 6946)**
Partition	0.328	0.328

(3) 3D FEM Verification

The last accuracy verification was performed by comparing the THERM 2D model results for the LSF partitions with some 3D models implemented in the ANSYS software [36]. The thermal conductivities of the modelled materials are presented in Tables 1 and 2; the LSF partition wall dimensions and geometry are displayed in Figure 1; and for the boundary conditions, the values used in the models are the values presented in Section 2.4.2. Figure 3 displays the distribution of temperature and the computed conductive thermal resistances (*R*-values) for the reference LSF partition wall (Figure 3a) and for an LSF partition with an outer TBS (Figure 3a), using both 2D and 3D models. The modelled TBS was 50 mm wide, 10 mm thick, and had a thermal conductivity equal to 30 mW/(m· K). The obtained 2D and 3D models' *R*-values were very similar (+0.5% and +0.3% difference), highlighting again the excellent THERM [33] model accuracy.

(4) Lab Measurement Validation

To validate the THERM model, experimental measurements were performed under laboratory-controlled conditions (Figure 4) using a test sample of the reference LSF partition wall. This test-sample LSF wall was placed between two climatic chambers, as illustrated in Figure 4a. Electrical thermal resistance was used to heat the hot chamber, while a refrigerator was used to cool the cold chamber. The wall sample perimeter was covered by 80 mm foam insulation made of polyurethane with a thermal conductivity equal to 0.036 W/(m·K). The objective was to mitigate the heat losses across the sample wall's perimeter, as shown in Figure 4a.

Two heat flux meters (model HFP01, from Hukseflux) with an accuracy of ±3% and 6 PFA insulated thermocouples (TCs), Type K (1/0.315), class 1 precision certified, were used on each face (outer and inner) of the LSF wall test-sample (cold and hot), totalling 4 heat flux meters (HFM) and 12 TCs.

On both sides, one HFM was placed over the middle metallic stud, and the other in the central region of the batt insulation, as displayed in Figure 4b, to obtain the measurements of both dissimilar thermal behaviour regions within the partition of the LSF test sample.

The TCs were placed in different locations: two were placed to measure the wall surface temperature near the HFM, and two were placed over the HFM to monitor the air temperature in the vicinity of the wall surface, and, as illustrated for the cold surface in Figure 4b, the remaining two TCs were put inside the climatic chamber to measure the environment air temperature.

The cold and hot chambers were set to preserve a temperature of 5 °C and 40 °C, respectively. To ensure a nearly steady-state heat transfer condition and to mitigate the surrounding heat transmission, these chambers were adequately insulated.

The temperature and heat flux data measurements were registered using two PICO TC-08 data loggers, with an accuracy of ±0.5%; one on each surface of the LSF partition sample (cold and hot). The recorded data were managed by the software PicoLog® (version 6.1.10) [39].

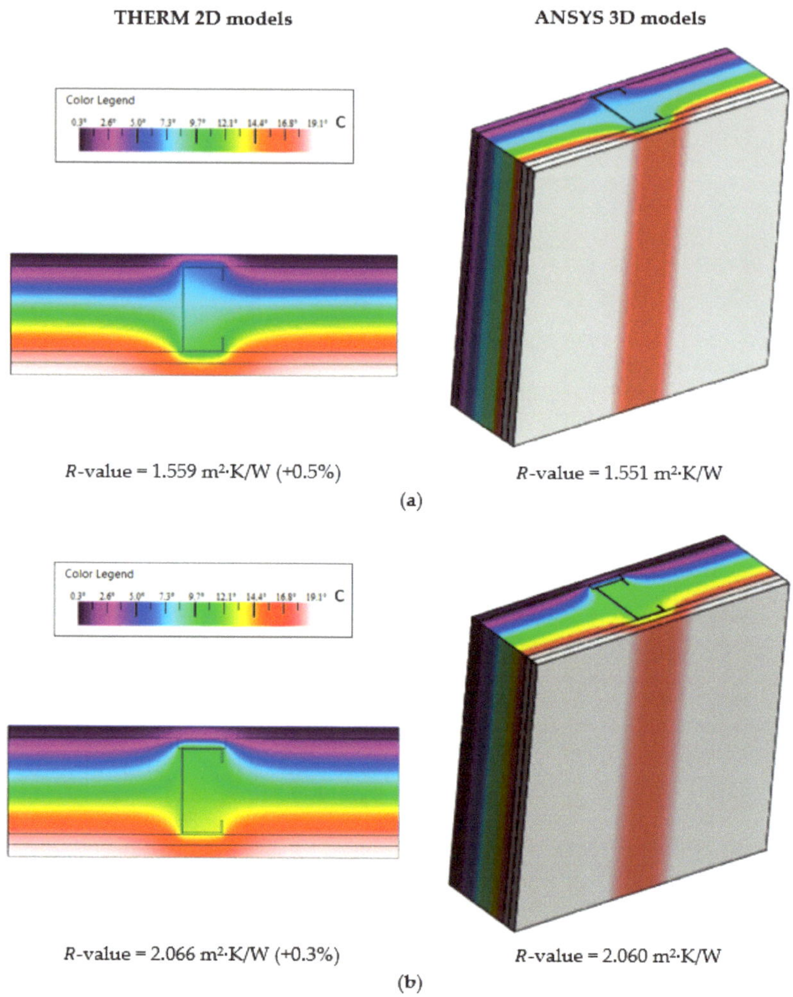

Figure 3. Model accuracy verification of the LSF partitions: predicted temperature distribution and surface-to-surface thermal resistance. (**a**) Reference LSF partition. (**b**) LSF partition with an outer thermal break strip.

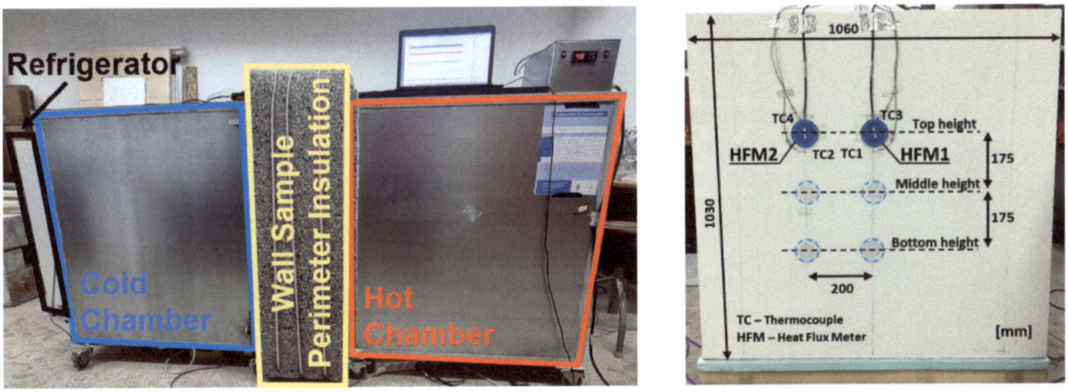

(a) (b)

Figure 4. Experimental apparatus for the *R*-value lab measurements. (**a**) Thermal chambers. (**b**) Sensors on the cold surface.

The LSF partition thermal performance was quantified using the HFM method [40], adjusted for two HFM sensors, one on each wall surface, to increase the measurement accuracy and reduce the time duration of the tests, as suggested by Rasooli and Itard [41].

The lowest time duration for each experimental test was twenty-four hours. Moreover, only the hourly average *R*-values with an absolute change lower than 10% in relation to the previous recorded thermal transmittance were included in the measurements. The "summation technique", as defined in standard ASTM C1155-95 [42], was adopted for the convergence criteria.

The experimental measurements' repeatability was ensured by performing, for each sample LSF partition, a test at three height locations: (1) bottom, (2) middle, and (3) top, as illustrated in Figure 4b.

Table 4 displays the conductive *R*-values measured for the three height positions and the corresponding average conductive thermal resistance for the reference partition LSF wall. The conductive *R*-value predicted by the THERM 2D model is also displayed in this table. The very small differences between the measured and the numerical predicted *R*-values (only +0.1%) indicated the accuracy of the presented THERM models and allowed for their validation.

Table 4. Reference LSF partition wall surface-to-surface *R*-values for both the experimental and numerical approaches. Percentage deviation included.

Test N.	Sensors Position	*R*-Value [$m^2{\cdot}K/W$]
1	Bottom	1.607
2	Middle	1.576
3	Top	1.491
Measurement Average		1.558
Computed in THERM		1.559
Percentage Deviation		+0.1%

3. Results and Discussion

3.1. *Reference Partition LSF Wall*

A representative horizontal transversal section of the reference partition LSF wall, with commercial cold formed studs C90 × 43 × 15 × 1.5 spaced 400 mm apart, was previously displayed in Figure 1. Their predicted conductive *R*-value is 1.559 $m^2{\cdot}K/W$ (see

Figure 3a). When this steel stud spacing is increased to 600 mm, their *R*-value is augmented to 1.851 $m^2 \cdot K/W$ (+19%). As expected, given the reduced steel content by wall area, increasing the metallic studs' spacing improves the LSF partition's thermal performance.

3.2. One Thermal Break Strip

In this subsection, the computed results when using a single thermal break strip, placed in the outer steel stud flange, are presented.

3.2.1. The Influence of TBS Thickness and Conductivity

Figure 5a exhibits the conductive thermal resistances computed for the partition LSF walls with a single TBS of variable thicknesses and 50 mm width when the steel studs were spaced at intervals of 400 mm. Each line corresponds to a specific thermal conductivity of the TBS material, from 7.5 up to 120 mW/m/W. Looking to the *R*-values for the higher thermal conductivity (black line), even the thinner TBS (5 mm) allowed a thermal performance increase from 1.559 $m^2 \cdot K/W$ (reference value) to 1.667 $m^2 \cdot K/W$ (+7%). When the TBS thickness increased up to 15 mm, the *R*-value also increased up to 1.890 $m^2 \cdot K/W$, with a nearly linear variation. As illustrated in the right graph of Figure 5, this *R*-value variation, in relation to the reference wall, increased from +7% (5 mm thick) up to +21% (15 mm thick).

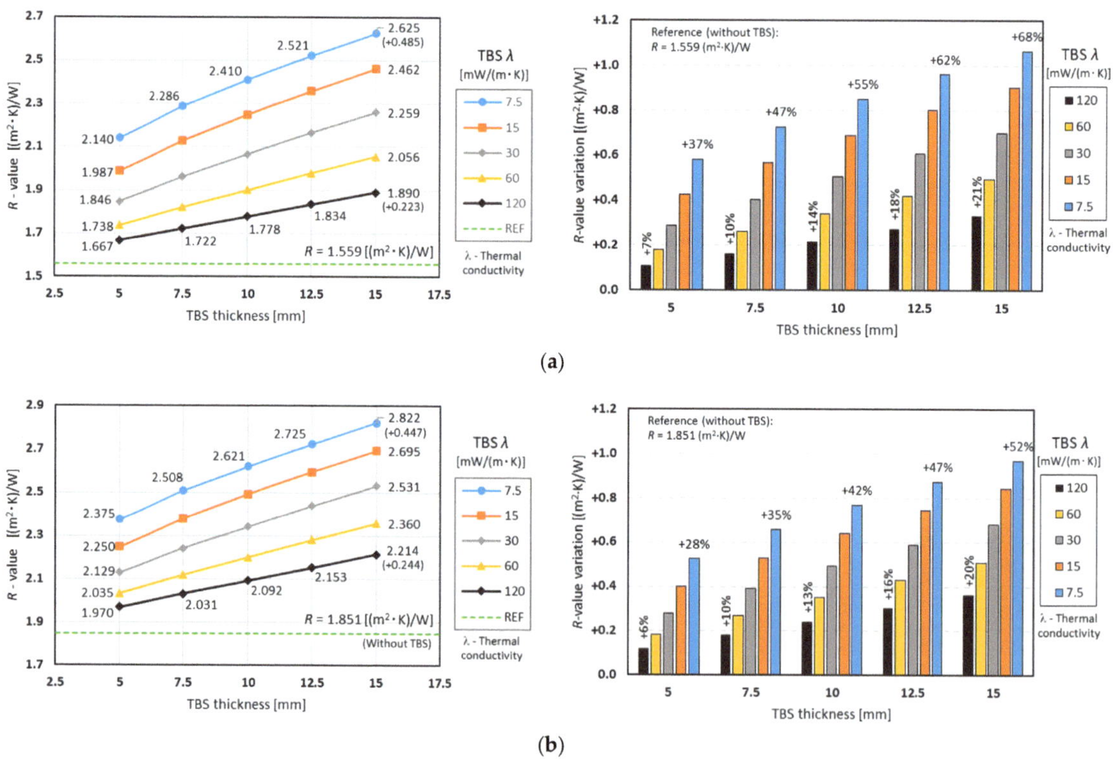

Figure 5. Conductive thermal resistances for partition LSF walls with one thermal break strip: 50 mm wide and variable thickness, for two different steel stud spacings. (**a**) 400 mm stud spacing. (**b**) 600 mm stud spacing.

In terms of the other evaluated TBS thermal conductivities, the slope of the corresponding *R*-value lines also increased with decreasing TBS conductivity. As expected, the major

R-values were achieved for the smaller TBS thermal conductivity (7.5 mW/m/W), ranging from 2.140 $m^2{\cdot}K/W$ (5 mm thick) up to 2.625 $m^2{\cdot}K/W$ (15 mm thick). In percentages, these R-value increments corresponded to +37% up to +68%.

Figure 5b displays two similar charts, but instead of 400 mm steel stud spacing, these R-values were obtained for 600 mm spacing. This plot shows a similar trend, but all the obtained thermal resistance values were higher than the previous ones, including the reference (1.851 $m^2{\cdot}K/W$), as previously mentioned. Another difference is that (Figure 5) for the same TBS thickness, the thermal performance improvement due to the TBS conductivity decrease was smaller in relation to the 400 mm steel stud spacing, Figure 5a. Similarly, the relevance of the TBS thickness increase in the partition LSF wall's thermal performance improvement was greater for higher TBS conductivities (increased black line slope) and smaller for lower TBS conductivities (reduced blue line slope). Obviously, due to a higher reference R-value for the 600 mm stud spacing (Figure 5b), all percentage values became smaller, with this reduction being greater for smaller thermal conductivities.

Regardless of the thermal conductivity of the TBS, increasing its thickness always leads to better performance, as increasing the thickness of the TBS also increases the MW (due to its volumetric expansion).

3.2.2. The Influence of TBS Width and Conductivity

Figure 6a illustrates the conductive thermal resistance values obtained for the partition LSF walls with one attached TBS of 10 mm thickness and variable width, for a 400 mm steel stud spacing, as well as the R-values variation relative to a reference wall (without TBS). The R-values for the higher thermal conductivity (120 mW/m/W), black line, even for a 30 mm width and 10 mm thickness, showed that TBS increases thermal performance from 1.559 $m^2{\cdot}K/W$ (reference value) to 1.855 $m^2{\cdot}K/W$ (+19%).

Surprisingly, when the TBS width increased up to 70 mm, the R-value decreased to 1.750 $m^2{\cdot}K/W$, with a small negative variation (−0.105 $m^2{\cdot}K/W$). Analysing the grey line (thermal conductivity equal to 30 mW/m/W), there was almost no variation of the thermal resistance when increasing the width of the TBS, starting at 2.049 and going up to 2.073 $m^2{\cdot}K/W$. The lower thermal conductivity (7.5 mW/m/W), blue line, presented the largest increase in relation to the reference LSF wall without TBS (1.559 up to 2.207 $m^2{\cdot}K/W$), and also with the increase in width from 30 to 70 mm, with a +0.314 $m^2{\cdot}K/W$ variation.

Figure 6b displays a similar chart, but instead of 400 mm steel stud spacing, these R-values were obtained for 600 mm spacing. This plot has a similar trend, but all the obtained thermal resistance values were higher than the previous ones, including the reference (1.851 $m^2{\cdot}K/W$), as previously mentioned. The thermal resistance increase was reduced for smaller thermal conductivities (e.g., 7.5 mW/(m·K)) and slightly increased for larger thermal conductivities (e.g., 120 mW/(m·K)), as shown on the right-hand graphs.

Comparing Figures 5 and 6, the TBS of 30 mm width and 10 mm thickness (Figure 6) presented higher R-values than a TBS of 50 mm width and 5 mm thickness (Figure 5). However, increasing the TBS thickness (Figure 5) up to 15 mm is more advantageous (higher R-values) than increasing the TBS width (Figure 6) up to 70 mm due to the increase in the wall cavity thickness whenever the TBS thickness increases. Given the MW natural expansion, the thickness of this thermal insulation also increases, which leads to an incremental change of the overall R-value of the LSF partition wall.

Moreover, increasing the TBS thickness always leads to an increased thermal resistance (Figure 5), while the increase in the TBS width for higher thermal conductivity values may lead to a decreased thermal performance (Figure 6). This happens whenever the TBS material has a thermal conductivity that is higher than the batt insulation (in this case, mineral wool with 35 mW/m/K). In Figure 6, the grey line, corresponding to a TBS thermal conductivity of 30 mW/m/K, was the nearest one, where the thermal resistance was slightly increasing. However, as expected, there was an increased R-value diminishment for the yellow (60 mW/m/K) and black (120 mW/m/K) lines, given their higher TBS thermal conductivities.

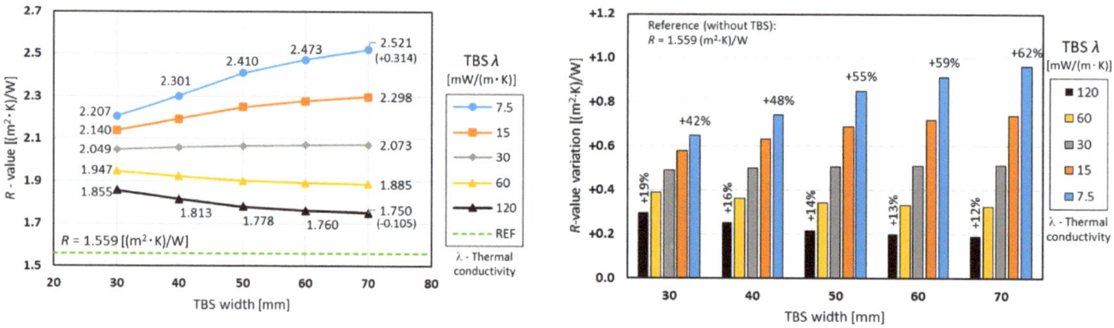

(a)

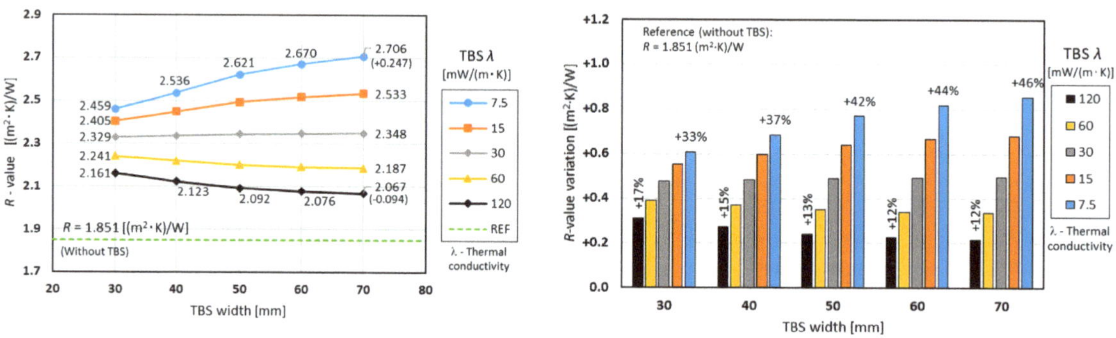

(b)

Figure 6. Surface-to-surface thermal resistance for LSF partitions, with one thermal break strip: 10 mm thick and variable width, for two different steel stud spacings. (**a**) 400 mm stud spacing. (**b**) 600 mm stud spacing.

3.3. *Two Thermal Break Strips*

Here, the results when using two TBS are presented and compared with the previous results (single TBS).

3.3.1. The Influence of TBS Thickness and Conductivity

Figure 7a exhibits the conductive thermal resistances obtained for the partition LSF walls with two TBS of 50 mm width and variable thicknesses when the steel studs were spaced 400 mm. Based on the *R*-values for the higher thermal conductivity (120 mW/m/W), black line, even the thinner TBS (5 mm) allowed a thermal performance increase from 1.559 $m^2 \cdot K/W$ (reference value) to 1.769 $m^2 \cdot K/W$ (+13%). When the TBS thickness increased up to 15 mm, the *R*-value also increased up to 2.204 $m^2 \cdot K/W$ (+41% relative to the reference LSF wall), with a nearly linear variation.

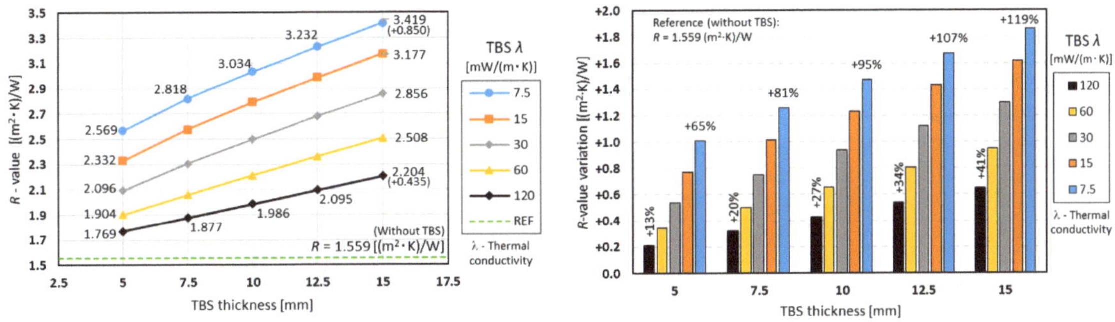

(**a**)

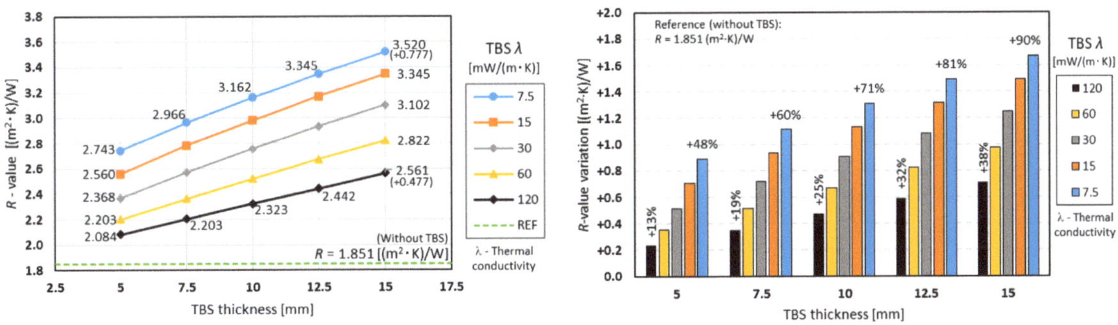

(**b**)

Figure 7. Conductive thermal resistance for partition LSF walls, with two thermal break strips: 50 mm wide and variable thickness. (**a**) 400 mm stud spacing. (**b**) 600 mm stud spacing.

Looking to the other evaluated TBS thermal conductivities, the slope of the corresponding *R*-value lines also increased with the decreasing TBS conductivity. Once again, the major *R*-values were reached for the smaller TBS thermal conductivity (7.5 mW/m/W), ranging from 2.569 $m^2{\cdot}K/W$ (5 mm thick, +65%) up to 3.419 $m^2{\cdot}K/W$ (15 mm thick, +119%).

Figure 7b displays a similar chart, but instead of 400 mm steel stud spacing, these *R*-values were obtained for 600 mm spacing. This plot shows a similar trend, but all obtained thermal resistance values were higher than the previous ones. Another difference is that (Figure 7b) for the same TBS thickness, the thermal performance improvement due to the TBS conductivity decrease was smaller in relation to the previous values for the 400 mm steel stud spacing Figure 7a. As mentioned before, the relevance of the TBS thickness increase to the partition LSF wall thermal performance improvement was smaller for the 600 mm spacing, with this being more visible for the smaller TBS thermal conductivities.

3.3.2. The Influence of TBS Width and Conductivity

Figure 8a exhibits the conductive thermal resistance values obtained for the partition LSF walls with two TBS of 10 mm thickness and variable width when the steel studs were spaced 400 mm apart. The *R*-values for the higher thermal conductivity (120 mW/m/W), black line, showed that two TBS of 30 mm width and 10 mm thickness allowed a thermal performance increase from 1.559 $m^2{\cdot}K/W$ (reference value) up to 2.134 $m^2{\cdot}K/W$ (+37% relative to the reference LSF partition wall). When the TBS width increased up to 70 mm, the *R*-value decreased to 1.928 $m^2{\cdot}K/W$, corresponding to +24% relative to the reference partition (without TBS). Analysing the grey line (thermal conductivity of 30 mW/m/W), once again we observed that there was a very reduced variation of the *R*-value when increasing the width of the TBS (2.470 to 2.510 $m^2{\cdot}K/W$). The lower thermal conductivity

(7.5 mW/m/W), as expected (blue line), presented the largest increase in relation to the reference (1.559 to 2.727 m^2·K/W, +75%), and with the increase in width from 30 to 70 mm, presented a 0.467 m^2·K/W variation.

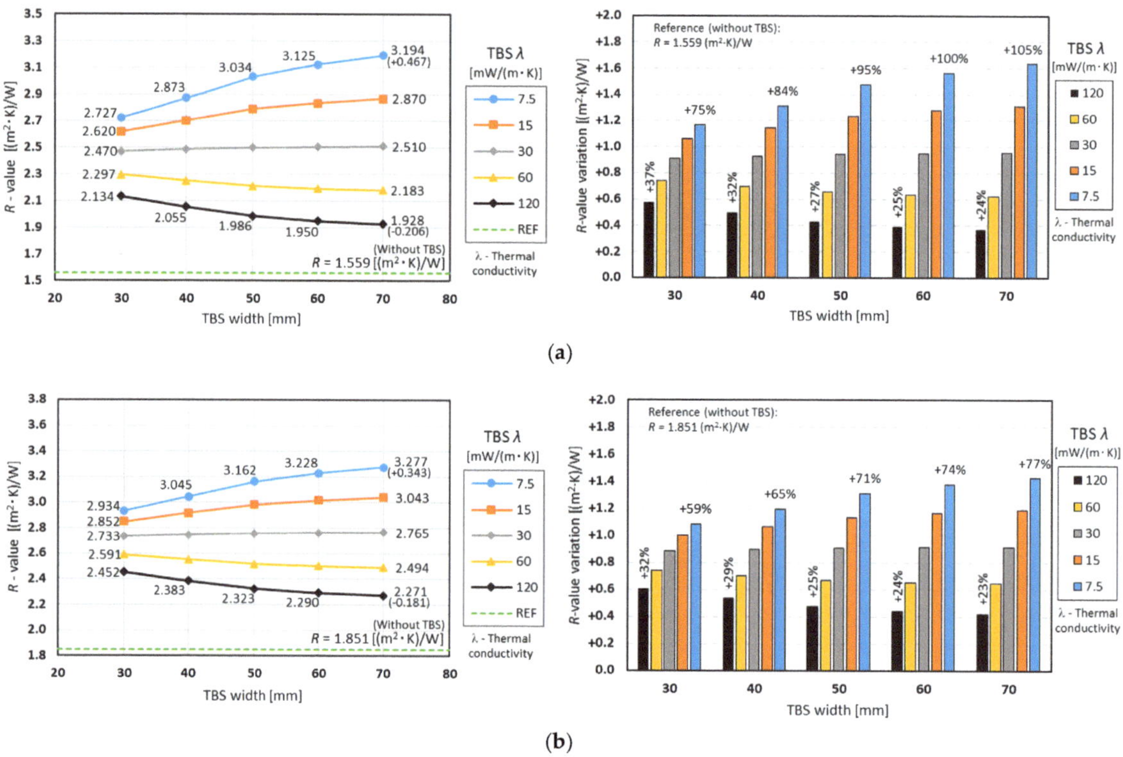

Figure 8. Surface-to-surface thermal resistances for LSF partitions, with two thermal break strips: 10 mm thick and variable width. (**a**) 400 mm stud spacing. (**b**) 600 mm stud spacing.

Figure 6b displays a similar chart, but instead of 400 mm steel stud spacing, these *R*-values were obtained for 600 mm spacing. This plot shows a similar trend, but all the obtained thermal resistance values were higher than the previous values, including the reference (1.851 m^2·K/W), as previously mentioned.

Comparing Figure 7 (TBS thickness relevance) and Figure 8 (TBS width relevance), it can be observed, once again, that a TBS with 30 mm width and 10 mm thickness presented higher *R*-values (Figure 8) than a TBS with 50 mm width and 5 mm thickness (Figure 7). However, the increase in TBS thickness (Figure 7) was more effective at improving the overall partition thermal performance than the increment of TBS width for all evaluated TBS thermal conductivities. Moreover, as seen before (one TBS), an increase in TBS width for higher thermal conductivity materials is counter-productive, i.e., will decrease their *R*-value.

4. Conclusions

In this research, the relevance of TBS to the thermal performance improvement of load-bearing LSF partitions was assessed through a parametric study. The bidimensional numerical models used in this study were previously validated, and their accuracy and reliability ensured. Five different key parameters were changed and their importance evaluated, namely the TBS thickness, width, thermal conductivity, and number, as well as the vertical steel stud spacing.

The key conclusions of this parametric study are as follows:

- As expected, only by increasing the vertical steel stud spacing from 400 mm to 600 mm did we achieve a relevant thermal resistance improvement (+19%).
- Regardless of the TBS thermal conductivity, it is always worth increasing the thickness of the TBS due to the consequent increase in the wall cavity thickness and the resulting expansion of the batt insulation (in this study, mineral wool).
- Moreover, regardless of the evaluated TBS conductivities, the thermal resistances provided by the smaller assessed TBS width (30 mm) were always bigger than the ones provided by the smaller evaluated TBS thickness (5 mm).
- Surprisingly, the increase in the TBS width did not always lead to increased thermal resistance.
- In fact, for higher TBS conductivities, a thermal performance reduction occurred when increasing the width of the TBS.
- The previous happens whenever the thermal conductivity of the TBS is greater than the conductivity of the expansible batt insulation.
- Considering the preceding features, it was concluded that it is more effective to increase the TBS thickness rather than the width.
- The abovementioned features are more relevant for smaller stud spacing (400 mm instead of 600 mm) and when using two TBS instead of a single one.

Regarding the TBS thickness higher effectiveness, comparatively to increasing their width, notice that this conclusion is also valid for other expansible thermal insulation materials, as the TBS thickness increase will directly induce a batt insulation thickness increase and consequent thermal performance improvement. Obviously, this LSF wall overall thermal resistance rise will depend on the thermal conductivity of the adopted batt insulation, and it will be higher for lower conductivities.

The two main limitations of this research were that (1) only LSF partition walls with vertical steel studs were considered, and (2) only one batt insulation material (mineral wool) was used. To overcome the first restriction, 3D numerical models instead of 2D models, as previously illustrated in Figure 3, could be used. Regarding the second limitation, it was assumed that the mineral wool batt insulation was expansible enough to fill the LSL wall cavity.

This research allowed us to better understand, quantify, and compare the effectiveness of TBS for the thermal performance improvement of load-bearing partition LSF walls, which was not previously available in the literature. This knowledge could be useful for designers when they need to decide the cross-section dimensions of the TBS, as well as the material thermal conductivity and the number of TBS.

Author Contributions: Conceptualization, P.S.; Methodology, P.S., P.L. and D.A.; Validation, P.S. and P.L.; Formal analysis, P.L. and D.A.; Investigation, P.S., P.L. and D.A.; Resources, P.S.; Writing—original draft, P.L. and D.A.; Writing—review & editing, P.S.; Visualization, P.S. and D.A.; Supervision, P.S.; Project administration, P.S.; Funding acquisition, P.S. All authors have read and agreed to the published version of the manuscript.

Funding: This research was funded by FEDER funds through the Competitivity Factors Operational Programme–COMPETE, and by national funds through FCT, Foundation for Science and Technology, within the scope of the project POCI-01-0145-FEDER-032061.

Cofinanciado por: POCI-01-0145-FEDER-032061

Acknowledgments: The authors also want to thank the following companies: Pertecno, Gyptec Ibéria, Volcalis, Sotinco, Kronospan, Hulkseflux, Hilti, and Metabo.

Conflicts of Interest: The authors declare no conflict of interest.

References

1. *Energy Efficiency in Buildings*; European Commission, Department Energy: Brussels, Belgium, 2020.
2. Global Alliance of Buildings and Construction. *2021 Global Status Report for Buildings and Construction*; United Nations Environment Programme: Nairobi, Kenya, 2021.
3. EU. Directive 8(EU) 2018/844 of the European Parliament and of the Council on the energy performance of buildings and on energy efficiency. *Off. J. Eur. Union* **2018**, *2018*, 75–91.
4. EU. Directive (EU) 2018/2001 of the European Parliament and of the Council on the promotion of the use of energy from renewable sources. *Off. J. Eur. Union* **2018**, *2018*, 82–209.
5. Zhan, Q.; Xiao, Y.; Musso, F.; Zhang, L. Assessing the hygrothermal performance of typical lightweight steel-framed wall assemblies in hot-humid climate regions by monitoring and numerical analysis. *Build. Environ.* **2020**, *188*, 107512. [CrossRef]
6. Soares, N.; Santos, P.; Gervásio, H.; Costa, J.J.; da Silva, L.S. Energy efficiency and thermal performance of lightweight steel-framed (LSF) construction: A review. *Renew. Sustain. Energy Rev.* **2017**, *78*, 194–209. [CrossRef]
7. Perera, D.; Upasiri, I.; Poologanathan, K.; Gatheeshgar, P.; Sherlock, P.; Hewavitharana, T.; Suntharalingam, T. Energy performance of fire rated LSF walls under UK climate conditions. *J. Build. Eng.* **2021**, *44*, 103293. [CrossRef]
8. Kapoor, D.R.; Peterman, K.D. Quantification and prediction of the thermal performance of cold-formed steel wall assemblies. *Structures* **2021**, *30*, 305–315. [CrossRef]
9. Kempton, L.; Kokogiannakis, G.; Green, A.; Cooper, P. Evaluation of thermal bridging mitigation techniques and impact of calculation methods for lightweight steel frame external wall systems. *J. Build. Eng.* **2021**, *43*, 102893. [CrossRef]
10. Ge, H.; Baba, F. Dynamic effect of thermal bridges on the energy performance of a low-rise residential building. *Energy Build.* **2015**, *105*, 106–118. [CrossRef]
11. Erhorn-Kluttig, H.; Erhorn, H. *Impact of Thermal Bridges on the Energy Performance of Buildings*; Buildings Platform; European Communities: Brussels, Belgium, 2009; pp. 1–8.
12. Al-Sanea, S.A.; Zedan, M. Effect of thermal bridges on transmission loads and thermal resistance of building walls under dynamic conditions. *Appl. Energy* **2012**, *98*, 584–593. [CrossRef]
13. Theodosiou, T.; Papadopoulos, A. The impact of thermal bridges on the energy demand of buildings with double brick wall constructions. *Energy Build.* **2008**, *40*, 2083–2089. [CrossRef]
14. Jedidi, M.; Benjeddou, O. Effect of Thermal Bridges on the Heat Balance of Buildings. *Int. J. Sci. Res. Civ. Eng.* **2018**, *2*, 2456–6667.
15. Santos, P.; Silva, L.; Ungureanu, V. *Energy Efficiency of Lightweight Steel-Framed Buildings*; Technical Committee 14-Sustainability & Eco-Efficiency of Steel Construction; European Convention for Constructional Steelwork (ECCS): Brussels, Belgium, 2012; ISBN 978-92-9147-105-8.
16. Martins, C.; Santos, P.; Simoesdasilva, L. Lightweight steel-framed thermal bridges mitigation strategies: A parametric study. *J. Build. Phys.* **2015**, *39*, 342–372. [CrossRef]
17. Santos, P.; Poologanathan, K. The importance of stud flanges size and shape on the thermal performance of lightweight steel framed walls. *Sustainability* **2021**, *13*, 3970. [CrossRef]
18. Atsonios, I.A.; Mandilaras, I.D.; Kontogeorgos, D.A.; Founti, M.A. Two new methods for the in-situ measurement of the overall thermal transmittance of cold frame lightweight steel-framed walls. *Energy Build.* **2018**, *170*, 183–194. [CrossRef]
19. François, A.; Ibos, L.; Feuillet, V.; Meulemans, J. In situ measurement method for the quantification of the thermal transmittance of a non-homogeneous wall or a thermal bridge using an inverse technique and active infrared thermography. *Energy Build.* **2021**, *233*, 110633. [CrossRef]
20. Santos, P.; Lemes, G.; Mateus, D. Analytical Methods to Estimate the Thermal Transmittance of LSF Walls: Calculation Procedures Review and Accuracy Comparison. *Energies* **2020**, *13*, 840. [CrossRef]
21. Santos, P.; Mateus, D. Experimental assessment of thermal break strips performance in load-bearing and non-load-bearing LSF walls. *J. Build. Eng.* **2020**, *32*, 101693. [CrossRef]
22. Santos, P.; Abrantes, D.; Lopes, P.; Mateus, D. Experimental and Numerical Performance Evaluation of Bio-Based and Recycled Thermal Break Strips in LSF Partition Walls. *Buildings* **2022**, *12*, 1237. [CrossRef]
23. Santos, P.; Mateus, D.; Ferrandez, D.; Verdu, A. Numerical Simulation and Experimental Validation of Thermal Break Strips' Improvement in Facade LSF Walls. *Energies* **2022**, *15*, 8169. [CrossRef]
24. Gomes, A.P.; de Souza, H.A.; Tribess, A. Impact of thermal bridging on the performance of buildings using Light Steel Framing in Brazil. *Appl. Therm. Eng.* **2012**, *52*, 84–89. [CrossRef]
25. Kosny, J.; Christian, J.E. Thermal evaluation of several configurations of insulation and structural materials for some metal stud walls. *Energy Build.* **1995**, *22*, 157–163. [CrossRef]
26. Santos, P.; Gonçalves, M.; Martins, C.; Soares, N.; Costa, J.J. Thermal transmittance of lightweight steel framed walls: Experimental versus numerical and analytical approaches. *J. Build. Eng.* **2019**, *25*, 100776. [CrossRef]
27. Lupan, L.M.; Manea, D.L.; Moga, L.M. Improving Thermal Performance of the Wall Panels Using Slotted Steel Stud Framing. *Procedia Technol.* **2016**, *22*, 351–357. [CrossRef]
28. Henriques, J.; Rosa, N.; Gervasio, H.; Santos, P.; da Silva, L.S. Structural performance of light steel framing panels using screw connections subjected to lateral loading. *Thin Walled Struct.* **2017**, *121*, 67–88. [CrossRef]

29. Ibérica, G. Technical Sheet: Standard Gypsum Plasterboard. 2022. Available online: https://www.gyptec.eu/documentos/Ficha_Tecnica_Gyptec_A.pdf (accessed on 10 February 2022).
30. Kronospan. *Technical Sheet: Kronobuild Materials*; Kronospan GmbH: Steinheim-Sandebeck, Germany, 2013.
31. Volcalis. Technical Sheet: Alpha Mineral Wool. 2022. Available online: https://www.volcalis.pt/categoria_file_docs/fichatecnica_volcalis_alpharollo-386.pdf (accessed on 10 February 2022).
32. Santos, C.; Matias, L. *ITE50—Coeficientes de Transmissão Térmica de Elementos da Envolvente dos Edifícios (in Portuguese)*; LNEC—Laboratório Nacional de Engenharia Civil: Lisboa, Portugal, 2006.
33. THERM. Software Version 7.6.1. Lawrence Berkeley National Laboratory, United States Department of Energy. 2017. Available online: https://windows.lbl.gov/software/therm (accessed on 14 February 2019).
34. *ISO 6946*; Building Components and Building Elements—Thermal Resistance and Thermal Transmittance—Calculation Methods. ISO—International Organization for Standardization: Geneva, Switzerland, 2017.
35. *ISO 10211*; Thermal Bridges in Building Construction-Heat Flows and Surface Temperatures-Detailed. ISO—International Organization for Standardization: Geneva, Switzerland, 2017.
36. *ANSYS Workbench Software*, Version 19.1; ANSYS, Inc.: Canonsburg, PA, USA, 2018. Available online: http://www.ansys.com/products/(accessed on 8 June 2020).
37. Roque, E.; Santos, P. The effectiveness of thermal insulation in lightweight steel-framedwalls with respect to its position. *Buildings* **2017**, *7*, 13. [CrossRef]
38. Santos, P.; Lemes, G.; Mateus, D. Thermal transmittance of internal partition and external facade LSF walls: A parametric study. *Energies* **2019**, *12*, 2671. [CrossRef]
39. Pico Technology. PicoLog 6 Software, Version 6.1.10. 2019. Available online: https://www.picotech.com/downloads (accessed on 1 September 2022).
40. *ISO 9869-1*; Thermal Insulation–Building Elements—In-Situ Measurement of Thermal Resistance and Thermal Transmittance. Part 1: Heat Flow Meter Method. ISO—International Organization for Standardization: Geneva, Switzerland, 2014.
41. Rasooli, A.; Itard, L. In-situ characterization of walls' thermal resistance: An extension to the ISO 9869 standard method. *Energy Build.* **2018**, *179*, 374–383. [CrossRef]
42. *ASTM C1155-95*; Standard Practice for Determining Thermal Resistance of Building Envelope Components from the In-Situ Data. ASTM—American Society for Testing and Materials: Philadelphia, PA, USA, 2013.

energies

Article

Metal-Organic Framework Adsorbent Materials in HVAC Systems: General Survey and Theoretical Assessment

Andrea Rocchetti [1,*], Martina Lippi [1], Luca Socci [1], Paride Gullo [2], Vahid Khorshidi [3] and Lorenzo Talluri [1,*]

[1] Department of Industrial Engineering, University of Florence, 50139 Florence, Italy
[2] Department of Mechanical and Electrical Engineering, University of Southern Denmark, 6400 Sønderborg, Denmark
[3] RAC R&D Innovation Lab, Danfoss A/S, 6430 Nordborg, Denmark
* Correspondence: andrea.rocchetti@unifi.it (A.R.); lorenzo.talluri@unifi.it (L.T.)

Abstract: In this paper, the use of Metal-Organic Framework (MOF) materials as an option for the energy efficiency enhancement of HVAC systems is investigated. In particular, the possibility of using MOFs as dehumidifying materials to reduce the latent load associated with the moisture content of the airflows is studied. A literature review is proposed, highlighting the benefits of using MOFs instead of other adsorbents (e.g., silica-gel) and discussing the unique features (high water uptake capacity and low regeneration temperatures) that make MOFs a preferential desiccant. The possibility to finely tune these properties is also underlined, reporting some explicative examples. A theoretical proposal of a psychrometric transformation, to be performed in a HVAC system equipped with a MOF-Assisted Dehumidifier (MAD), is presented. This transformation is compared with a traditional one (cooling and dehumidification operated by a cooling coil with low temperatures of the coolant). The preliminary numerical simulations, conducted on a reference case study in Florence, Italy, show an estimated energy saving of 30–50%, leading us to consider the use of this technology as a very competitive one in the air-conditioning sector.

Keywords: metal-organic framework; water-uptake; HVAC; dehumidification; energy-saving

Citation: Rocchetti, A.; Lippi, M.; Socci, L.; Gullo, P.; Khorshidi, V.; Talluri, L. Metal-Organic Framework Adsorbent Materials in HVAC Systems: General Survey and Theoretical Assessment. *Energies* **2022**, *15*, 8908. https://doi.org/10.3390/en15238908

Academic Editors: Alessandro Cannavale and Ubaldo Ayr

Received: 4 November 2022
Accepted: 22 November 2022
Published: 25 November 2022

1. Introduction

The energy requirements of many countries for refrigeration and air conditioning (AC) systems are dramatically increasing [1–3]. AC devices account for 20–30% of the worldwide electricity consumption of buildings [4,5], which corresponds to at least 10% of global electricity [4]. Considering these numbers and thinking about the non-postponable necessity to reduce electricity consumption and obtain wide-scale decarbonisation, the efficiency enhancement of the AC sector is crucial. AC systems (Heating Ventilation and Air Conditioning, HVAC), ensure thermo-hygrometric comfort in an indoor environment and operate a dehumidification process over the airflow that they treat. The latent heat of the building load (i.e., the water vapour content of the air treated by the system) could account for a great part of the total load (around 30–40%) [6], with greater value in humid climates or in applications where consistent outdoor airflow is required. In traditional HVAC systems, the dehumidification process is driven by a cooling coil (e.g., the evaporator of a Vapour Compression Refrigeration cycle, VCR). To obtain the dehumidification, the coolant of the coil is kept at a temperature much lower than the dew point temperature of the treated airflow, resulting in an over-cooling of the airflow (energy waste) and with a consequent necessity of re-heating. Moreover, these low temperatures of the coolant lead to a low value of the Energy Efficiency Ratio (EER) of the VCR device, with huge electricity consumption. A strategy to minimise the impact of these problems is the utilisation of materials able to remove water vapour contained in the airstream. Between them, an important place is occupied by the adsorbent (or desiccant) materials [7]. Once put in contact with an airflow, these materials can remove part of the water vapour content from it. In this way, sensible and latent heat could be managed separately [5]. Depending on thermal load and environmental conditions, it is possible to achieve consistent savings. For example, Mazzei et al. showed energy savings of up to 35% using desiccant materials in the HVAC systems of a reference Italian building [8].

Ideally, a sorbent material capable of taking up water from the air should show hydrolytic stability and a relevant adsorption capacity, maintaining its efficient performance over a multitude of uptake-release cycles. Moreover, the regeneration process must require low energy consumption. Desiccant materials act through adsorption processes, taking up water molecules on their surface. This behaviour is well described by the adsorption isotherms, graphs defined by a non-linear relationship between the amount of the adsorbate and the relative humidity measured at defined temperatures. The isotherm trend describes the mass transfer between the aqueous vapour and a porous matter. A suitable material thoughtfully designed for the dehumidification process should be able to adsorb water even at low relative humidity ($0.05 < RH < 0.40$) in order to provide a low energy demanding adsorbent technology. Thereby, a step-function isotherm represents the best adsorption performance.

During the past few decades, the most common adsorbent solids employed for water harvesting have been silica-gels and zeolites [7,9–11]. Concerning the International Union of Pure and Applied Chemistry (IUPAC) classification of physisorption isotherms [12], microporous silica-gels generally present a type II trend, which reflects slow uptake kinetics. On the other hand, although microporous zeolites exhibit a steep uptake tendency at low relative humidity, due to the presence of numerous and strong binding sites for water molecules in their structure, the recovery of the material is energetically costly. Indeed, high temperatures are required to evacuate zeolite porous frameworks from the adsorbed water molecules. This is the main reason for the incomplete degassing process, which translates into low cycling durability for this kind of material [13]. Recently, a zeolite material was proposed as a high-performance water adsorbent, exhibiting fast adsorption kinetic hydrothermal stability, and most importantly, a lower temperature regeneration around 65 °C [14]. Authors in the paper well characterised the adsorption process, describing how water molecules first coordinate the aluminium centres (at $RH = 0.01$) and then start to fill the 1D pores, forming a dense H-bonding network once the vapour pressure increases; however, defining a mechanism characterized by a low enthalpy value can lead to a low-cost regeneration process. This study explains how the water affinity of the adsorbers drives the adsorption process, maintaining high uptake efficiencies, without increasing the energy consumption for the water desorption. This major challenge can be addressed via a targeted structural design of the material. Consequently, many studies are focusing on the development of tunable materials.

In this regard, very promising adsorbents that are successfully emerging for this purpose have been identified in Metal-Organic Framework (MOF), a class of hybrid organic-inorganic crystalline materials built upon metal-containing units linked to organic ligands through coordination bonds extending within a regular and porous framework. Their robustness, porosity and uncommonly high specific surface area (up to 6000 m^2/g) have made MOFs an attractive proposal in adsorption applications [6,15–17]. Nowadays, theoretical and experimental studies on MOF adsorption abilities have demonstrated great loading and selectivity performances for specific gasses, such as CO_2 and many volatile organic compounds [18–20]. In this sense, MOFs have gained great consideration for gas-cleaning applications.

MOFs benefit from the possibility of modulating their final properties by a careful design of their reticular structures, thus broadening their range of applications. This representative feature differentiates MOFs from other materials such as their inorganic analogues (i.e., zeolites). The tunability of MOF structures can be exploited to provide suitable adsorption profiles for a high-efficient water capture system (as HVAC airflows dehumidification) [21]. Therefore, owing to their large specific surface area, low framework density and wide-reaching tunable porosity, MOFs can exceed potential traditional microporous adsorbent materials in terms of water adsorption capability. The water adsorption process in a MOF is mediated by chemisorption occurring on the open metal sites, physisorption characterized by the formation of weak interactions with the framework and capillary condensation, which occurs when large pores are present. Since the regeneration of material is strongly connected to these events, tuning the structural properties of the MOF is a valuable pathway to control the desired adsorption and desorption qualities, with the aim of achieving exceptional water harvester systems [22]. Yagi et al. gave efforts to the progress that these functional materials have shown as harvesters for water, mostly in terms of their hydrolytic resilience and their tunable porosity [23]. The Authors underlined the correlation between water uptake and pore volume, emphasizing the need to overcome the energy barrier during the regeneration process of MOF structures displaying large pores. In these cases, hysteresis phenomena are present in water adsorption isotherms, due to irreversible capillarity condensation, responsible for the high-temperature demand for water release. Thus, the pore size should be designed in terms of both high moisture capture capacity and relatively easy material regeneration, for example, by modulating the length of the organic linker or by introducing chemical functionalization during post-synthetic modification [24].

It is crucial to understand the water uptake mechanism by evaluating how the water molecules populate their binding sites within the MOF structure. In this sense, X-ray diffraction in combination with Density-Functional Theory calculations can be appropriate tools, as demonstrated by Hanikel et al. with the studies performed on MOF-303, which exhibits water adsorption with a steep trend at very low RH [25]. Understanding the hierarchically filling of the pores was demonstrated to be essential in order to induce a more adequate water uptake behaviour. In particular, they proposed a multivariate approach for the modification of the architecture of the pores, finding the most appropriate material design in compliance with the water adsorption enthalpy values, the limiting desorption temperature, the stability and the water capacity. In addition, the condition of the adsorption mechanism can be varied to understand the mechanism of water harvesting as was reported by Yanagita et al. [26]. In detail, they measured the adsorption/desorption isotherms of chromium terephthalate MIL-101 and the time trend of the amount of adsorbed water by stepwise modification of the relative humidity, describing how the porosity and the hydrophobic-hydrophilic structural composition of the MOFs drive the uptake/release kinetic of the water. Based on these all considerations, it is evident that a proper design of the material is a reasonable strategy to allow a successful material development for water uptake applications.

In the literature, it is possible to find few works that deal with the utilisation of MOF materials as desiccants in HVAC applications. Some of them propose using the traditional desiccant wheels coated with MOF materials. Bareschino et al. performed a numerical simulation of a desiccant wheel with MOF MIL-101 [27], including a gas-side resistance model to define the behaviour of MOF material. Dehumidification effectiveness of 30% better than the silica-gel wheel is demonstrated. The Authors estimated a reduction of 20.5% in CO_2 emissions compared to the HVAC system with silica-gel. Shahvari et al. explained the functioning of a MOF desiccant wheel [15], proposing a complete first-principle analysis validated with experimental tests. They showed that MOF-based wheels are regenerable at lower temperatures (40–60 °C) compared to silica-gel (80–140 °C), with consistent energy saving for regeneration (10–50%, depending on environmental conditions). Shahvari obtained promising results for a system that integrates MOF-assisted dehumidification with indirect evaporative cooling [28]. Again, the low regeneration temperatures (40–75 °C) have been confirmed, with consequent gain in the efficiency of the MOF system with respect to the silica-gel system (MOF system has an efficiency 2.7–6 times higher). The idea of using MOF-coated desiccant wheels has been well-reviewed and studied, as in the publication [29] and in the work of Wang [30], where the MOF materials are cited as good desiccant options due to their low regeneration temperatures (30–60 °C). Cui et al. proposed to coat the surface of the coils of a VCR cycle with MIL-100 (Fe) (Figure 1), [6].

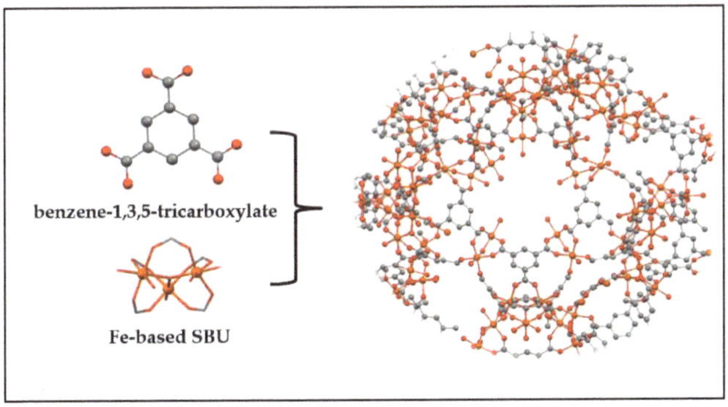

Figure 1. Crystal structure of MIL100(Fe) from the reaction between the benzene-1,3,5-tricarboxylate and the iron(III)-based cluster.

In adsorption mode, the evaporator acts as a cooler and dehumidifier while the waste heat of the condenser regenerates the wet MOF. The MOF-coated evaporator removes at the same time as the sensible and latent heat of the airflow, operating an isothermal-dehumidification. In the paper, the low regeneration temperature of the material (about 50 °C) is highlighted. With a theoretical EER of 7.9 in typical European summer conditions, the estimated energy-saving is 36.0% compared to traditional dehumidification with a cooling coil. The idea of coating a heat exchanger with desiccant material has been reported by review papers of Saeed et al. [31] and Venegas et al. [32], which considered MOFs as material to be used for the purpose. Even if this solution is not directly applied in the HVAC system, it is important to cite in the context of the control of thermo-hygrometric parameters of indoor environments and the energy-enhancement of HVACs, which was proposed by Feng et al. [16]. They proposed, in accordance with the results of a lumped-model simulation, to install a MIL-100 (Fe) in an indoor environment with the role of a moisture buffer; the removal of indoor humidity accounts for 73.4% of latent heat, leading to minor efforts of the HVAC system.

In this paper, the Authors theoretically compare a HVAC system that operates traditional dehumidification with a HVAC system equipped with a MOF-Assisted Dehumidifier (MAD). First of all, the reference psychrometric transformations are shown. Then, the mathematical model used in the simulations is presented. The two kinds of dehumidification are compared by considering a case study, based on real boundary conditions. The benefits, in terms of energy consumption, that characterise MOF-assisted dehumidification are quantified, also highlighting better results obtainable with MOF materials compared to other desiccant materials, in terms of consumption for the regeneration process. The obtained results lead the Authors to consider this technology as a very competitive option in the HVAC sector, and it inspires us to further research this topic.

2. Traditional Dehumidification vs. MOF-Assisted Dehumidification: Psychrometric Transformations

The proposed MOF-assisted dehumidification is compared here, in terms of reference psychrometric transformations, with a traditional dehumidification process operated by a cooling coil. In both cases, the setpoint indoor conditions are reached from the supply conditions, ensured by the HVAC system, considering the sensible and latent thermal loads.

2.1. HVAC with Traditional Dehumidification

In a HVAC system with traditional dehumidification (Figure 2), the dehumidification process is operated by a cooling coil.

- Pre-cooling of the outdoor air with an air-to-air heat exchanger, where the exhaust airflow coming from indoors is used as a pre-cooling medium (transformation 1–2).

- Cooling and dehumidification with a cooling coil (with a low temperature of the coolant). The air at the outlet of the coil is saturated, and the temperature is equal to the dew point temperature of the supply conditions (2–3). Coolant temperature must be lower than this limit. Additionally, in HVAC applications, a reference temperature for the coolant entering the coil is 7 °C.
- Heating with a post-heating coil (e.g., the desuperheater and/or condenser of the VCR, or other thermal sources) to reach supply conditions (3–4).

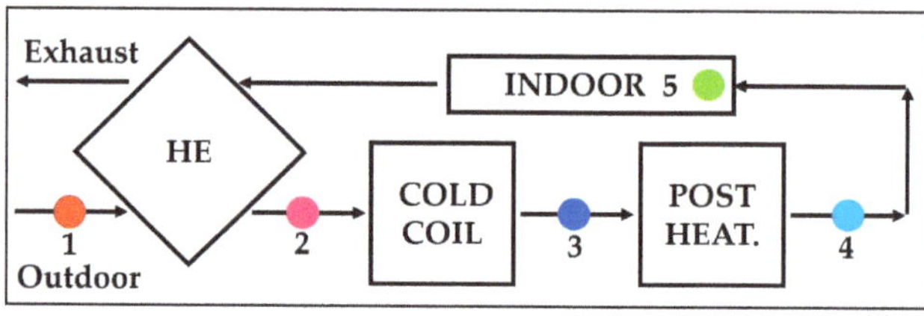

Figure 2. Schematic representation of a HVAC system with traditional dehumidification.

A generic complete process is represented in Figure 3.

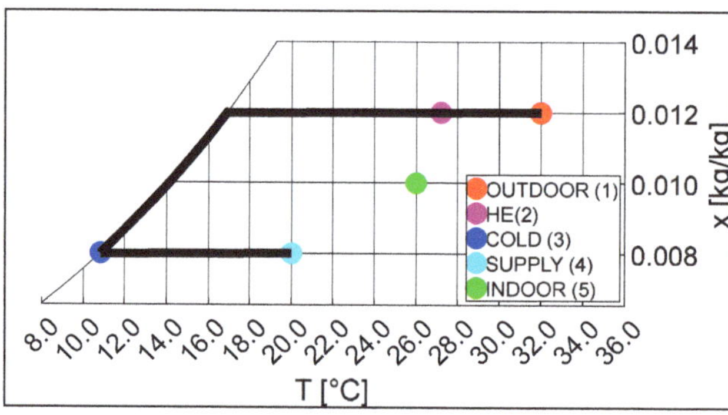

Figure 3. Transformations in an HVAC system with traditional dehumidification.

2.2. HVAC with MOF-Assisted Dehumidification

In a HVAC system with MOF-assisted dehumidification (Figure 4), the dehumidification process is operated by a device that connects the outdoor airflow with the MOF material, ensuring an isothermal transformation. In the system, the outdoor airflow undergoes the following transformations:

- Isothermal dehumidification with a MOF-Assisted Dehumidifier (MAD). The final point of the dehumidification process has a humidity ratio equal to the ones of supply conditions (transformations).
- Sensible cooling with the cooling coil, with a higher temperature of the coolant. In this case the coolant temperature must be lower than the air at the coil outlet, but these values are significantly higher than in the traditional cooling/dehumidification process.

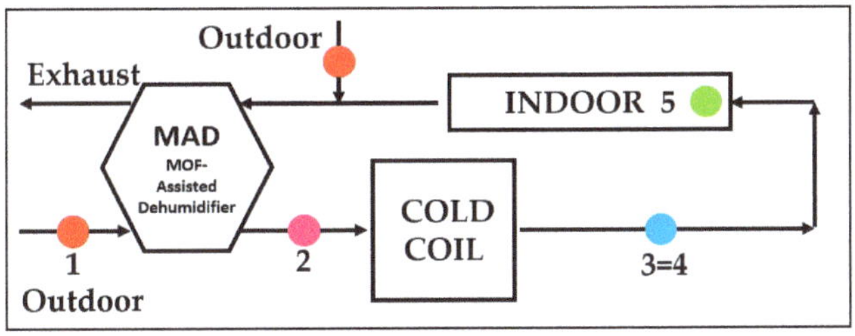

Figure 4. Schematic representation of a HVAC system with MOF-assisted dehumidification.

A generic complete process is represented in Figure 5.

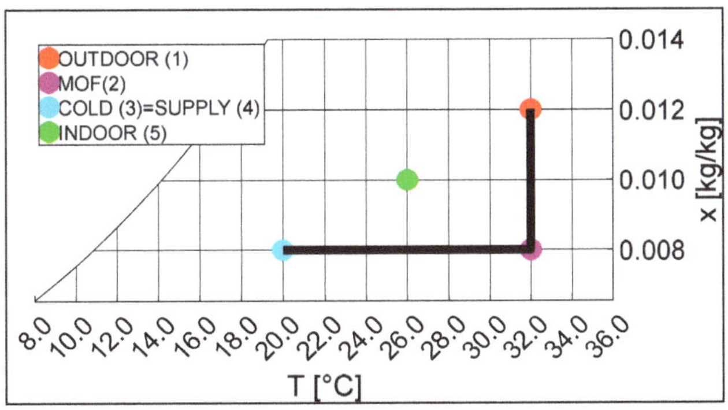

Figure 5. Transformations in a HVAC system with MOF-assisted dehumidification.

This description refers to the adsorption process. This process generates heat, which must be removed to obtain isothermal dehumidification of the airflow. Otherwise, the airflow will undergo an isenthalpic transformation, with a hypothetical consistent rise of its temperature (typical psychrometric transformation of desiccant wheels). Therefore, a cooling medium is necessary to avoid the heating of the system. For this purpose, it should use an airflow resulting from a mix between the exhausted airflow extracted from the indoor environment and an additional outdoor airflow. This mix will have a temperature within the same range as points one and five (Figures 3 and 5).

For the regeneration (that is not described in this paper), it is supposed that the utilisation of an outdoor airflow will be heated by the condenser of the VCR (e.g., at temperature T_{II}, Figure 6).

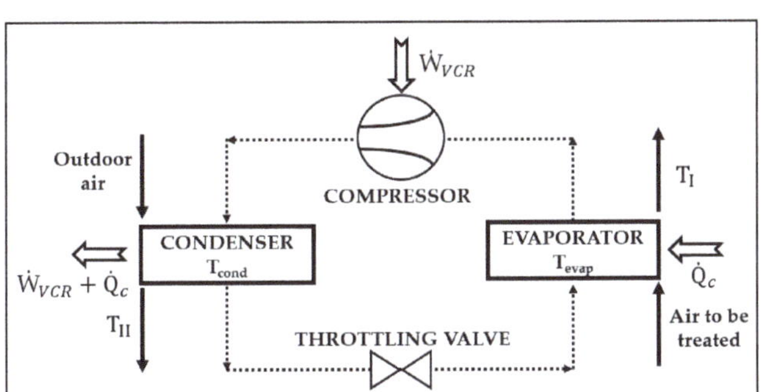

Figure 6. Reference scheme for the VCR cycle.

3. Mathematical Modelling

In the following, a simplified mathematical model used for the evaluation of the different psychrometric transformations is shown. The supply airflow is formed completely by outdoor air. The supply conditions, in terms of temperature and humidity ratio, are defined as follows:

$$T_s = T_i - \frac{\dot{Q}_{sens}}{c * m} \tag{1}$$

$$x_s = x_i - \frac{\dot{Q}_{lat}}{r * \dot{m}} \tag{2}$$

where the constant pressure specific heat of air c is assumed equal to 1.0 kJ/(kg∗K) and the latent heat of vaporisation of water r is assumed equal to 2500.0 kJ/kg.

In the different scenarios object of the analysis, the repartition between sensible and latent heat has been changed according to the Sensible Heat Ratio (SHR) parameter:

$$SHR = \frac{\dot{Q}_{sens}}{\dot{Q}_{sens} + \dot{Q}_{lat}} \tag{3}$$

3.1. HVAC with Traditional Dehumidification

The air-to-air heat exchanger is modelled with the Kays and London efficiency:

$$eff_{HE} = \frac{T_o - T_{he}}{T_o - T_e} \tag{4}$$

where the outdoor airflow, that must be treated, is cooled by the same quantity of exhaust airflow (characterised by the same thermo-hygrometric conditions as the indoor environment).

The condition at the outlet of the cooling coil corresponds to a saturation condition, corresponding to the dew point state of the supply air:

$$T_c = T_{dp,s} \tag{5}$$

$$x_c = x_s \tag{6}$$

The cooling power provided by the cooling coil (e.g., the cooling power provided by the evaporator of VCR in this context) is equal to:

$$\dot{Q}_c = \dot{m} * (j_{he} - j_c) \tag{7}$$

To evaluate the electric power consumption of the VCR cycle, associated with the activation of the cooling coil, the temperature levels of the coolant must be considered. They are set by the air temperature conditions at the cooling coil (evaporator) and at the condenser, and by opportune temperature differences between coolant and air (Figure 6). For the evaporator, the outlet air temperature is known (T_I, equal to T_c in this context), consequently the evaporation temperature; for the condenser, an outlet condensing air temperature is fixed (T_{II}), consequently the condensation temperature. The temperature differences at the evaporator and the condenser (DT_{evap} and DT_{cond}) are defined, taking into account common efficiencies for these kinds of heat exchangers:

$$T_{evap} = T_I - DT_{evap} \tag{8}$$

$$T_{cond} = T_{II} + DT_{cond} \tag{9}$$

Once the refrigerant levels are defined, it is possible to evaluate the EER of the VCR with parametric functions of T_{evap} and T_{cond}, obtained by analyzing some commercial models [33]. Then, the power consumption of the VCR cycle is:

$$\dot{W}_{VCR} = \frac{\dot{Q}_c}{EER} \quad (10)$$

The necessary heating power, to bring the air to the supply conditions, is:

$$\dot{Q}_h = \dot{m} * \left(j_h - j_c\right) \quad (11)$$

3.2. HVAC with MOF-Assisted Dehumidification

At the MOF device, the outdoor air undergoes isothermal dehumidification down to the supply humidity ratio.

The condition at the outlet of the cooling coil corresponds to the supply conditions.

The cooling power provided by the cooling coil (e.g., evaporator of VCR) is:

$$\dot{Q}_c = \dot{m} * \left(j_{mof} - j_s\right) \quad (12)$$

The power consumption of the VCR cycle is calculated as previously explained. It is possible to observe from Figure 5 that, in this case, the temperature of refrigerant can be kept higher, resulting in favourable conditions for the VCR performance.

In this case, no heating power is required to bring the airflow to supply conditions.

4. Case Study

The selected case study is based on a reference building located in Florence, Italy, served by a HVAC system. Locale climate conditions for typical summer days have been considered (Table 1, derived from ASHRAE [34] and Italian Standards [35] indications). The indoor setpoint conditions, the number of people present in the building (which is relevant to define the repartition between sensible and latent thermal load) and the necessary outdoor air have been chosen according to the Italian technical standards for thermo-hygrometric comfort and building management (Table 2) [35].

Table 1. Outdoor air conditions assumed in the simulations.

CASE	T_o [°C]	RH_o [%]
A	35.0	30.0
B	33.5	45.0

Table 2. Boundary conditions assumed in the simulations (part 1).

T_i [°C]	26.0
RH_i [%]	50.0
n	80
$\dot{m}_p$ [kg/s/s]	0.0125
$\dot{m}$ [kg/s]	1.0
$\dot{Q}_{sens,p}$ [W]	50.0
$\dot{Q}_{lat,p}$ [W]	50.0

Three different scenarios of thermal load, with different supply air conditions (Equations (1) and (2)), have been considered (Table 3). In the following table, the latent load is related to the presence of people, and the sensible one has been varied according to some SHRs. Thus, the sensible and latent loads do not refer to the loads that derive from ventilation.

Table 3. Boundary conditions used in the simulations (part 2).

SCENARIO	SHR	$\dot{Q}_{sens}$ [kW]	$\dot{Q}_{lat}$ [kW]	T_s [°C]	x_s [kg/kg]
1	0.50	4.0	4.0	22.0	0.009
2	0.60	6.0	4.0	20.0	0.009
3	0.67	8.0	4.0	18.0	0.009

5. Results and Discussion

In the following, MOF-assisted dehumidification is compared with the dehumidification operated by a traditional HVAC system. Then, some comparisons between the performances of MOFs as desiccants and traditional desiccant materials (i.e., silica-gel) are proposed.

5.1. Comparison of Psychrometric Transformations

The following figures show the psychrometric transformations (for outdoor conditions of case B) that the airflow undergoes in the HVAC with traditional dehumidification and in the HVAC with MOF-assisted dehumidification.

From these figures, apart from the energy savings that will be discussed in the following parts, a crucial point emerges. The lowest temperatures of the airflow are set as follows: the supply dew point temperature in the traditional dehumidification is about 12.0 °C, Figure 7, and the supply temperature in the MOF-assisted dehumidification is 22.0 °C, Figure 8. This fact has important consequences: there is a temperature difference of 10.0 °C in the evaporation temperature of the VCR in the two systems (assuming the same temperature difference evaporating coolant-outlet air), with a general gain in EER for the second configuration. In any case, in the following calculations, the gain in EER is partially penalized by the necessity to have slightly hotter air exiting the condenser (T_{II}, Figure 6) to enhance the regeneration process. This assumption has been made to obtain more reliable results, taking into consideration possible parasitic phenomena in the regeneration process, even if, according to the literature [6], no higher condensing coolant temperature, with respect to the traditional HVAC, is needed.

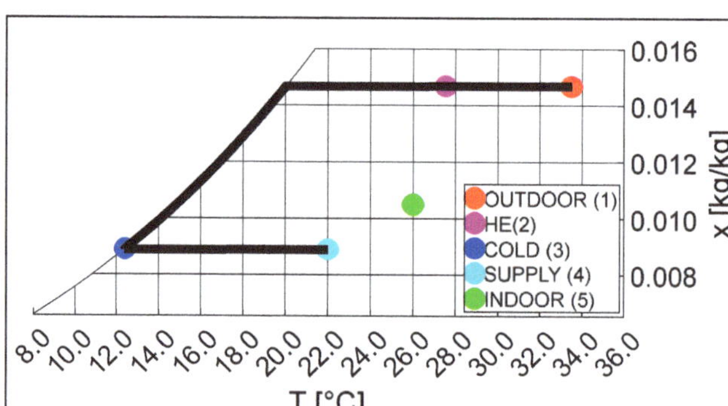

Figure 7. Transformations for HVAC with traditional dehumidification (case B, scenario 1).

5.2. Comparison between Traditional and MOF-Assisted Dehumidification

It is possible to compare the different configurations in terms of cooling power provided by the cooling coil, consequently in terms of electrical power consumption of the VCR (Table 4). The cooling power required in the different scenarios is always the same for the HVAC with traditional dehumidification, because the cooling coil must cool down the inlet air to the dew point temperature of the supply air. In the last column, the savings compared to the traditional configuration are reported.

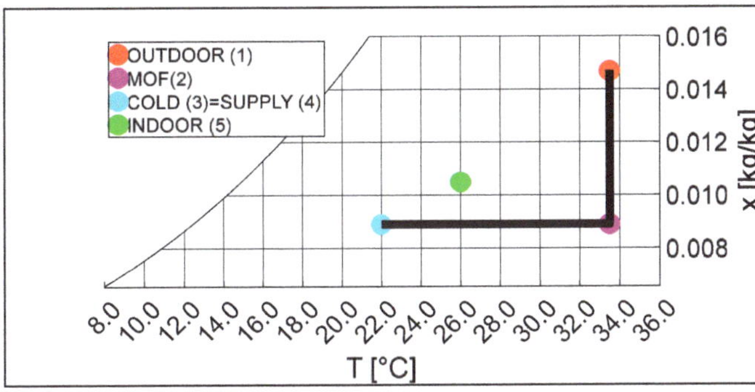

Figure 8. Transformations for HVAC with MOF-assisted dehumidification (case B, scenario 1).

Table 4. Cooling power and electric power consumption for the two HVAC configurations.

CASE	SCENARIO	$\dot{Q}_c$ [kW] TRAD.	$\dot{W}_{VCR}$ [kW] TRAD.	$\dot{Q}_c$ [kW] MOF	$\dot{W}_{VCR}$ [kW] MOF	Savings [%]
	1	19.9	6.2	13.2	3.6	41.9
A	2	19.9	6.2	15.2	4.3	30.6
	3	19.9	6.2	17.2	5.2	16.1
	1	30.2	9.6	11.7	3.3	65.6
B	2	30.2	9.6	13.7	4.2	56.3
	3	30.2	9.6	15.7	5.1	46.9

Comparing the respective scenarios for both cases, and mediating the values, it is possible to assert that the energy savings with MOF dehumidification is in the range of 30–50%.

5.3. Comparison between MOF and Other Desiccants

As already explained in the introduction, isotherm profiles analysis can be a good approach to show the benefits of MOFs with respect to other common desiccants, in terms of adsorbate/adsorbent mass ratio and temperature required for regeneration. In the following figures, the isotherms of good examples belonging to the three main classes of desiccants are reported:

- A common silica-gel [36].
- A recently reported zeolite [14].
- One of the best performant MOFs, e.g., MIL-100 (Fe) [6].

In the graphs, the water uptake (adsorbed water mass/adsorbent mass) is reported as a function of vapour pressure (instead of relative humidity, as described in the introduction), in order to clearly show the adsorption behaviour at different temperatures.

In each graph (Figures 9–11), the isotherms corresponding to the adsorption (T_o = 33.5 °C of case B) and typical regeneration temperatures are reported. For silica-gel and zeolite the isotherm associated with the temperature of the air exiting the condenser of VCR (T_{II}) is shown, to make an easy comparison with MOF material.

Considering the boundary conditions of case B (T_o = 33.5 °C, RH_o = 45.0%) and the regeneration temperature indicated in [6,14,36], the obtained results are resumed in Table 5:

Table 5. Uptake and regeneration characteristics of the material under comparison.

MATERIAL	u [kg/kg]	T_{reg} [°C]
Silica-gel	0.24	90.0
Zeolite	0.30	65.0
MOF MIL-100	0.48	50.0

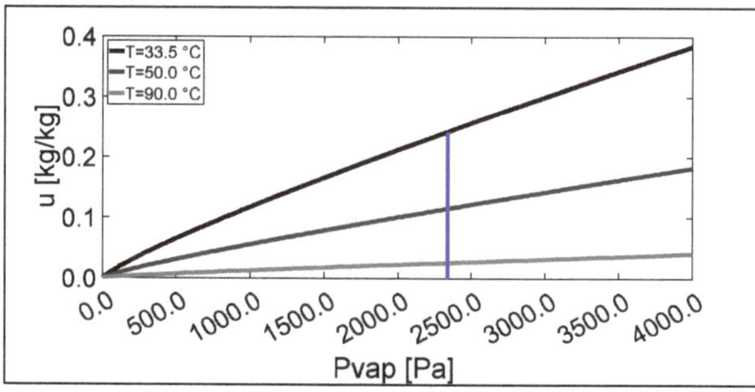

Figure 9. Isotherms of silica-gel under analysis [36].

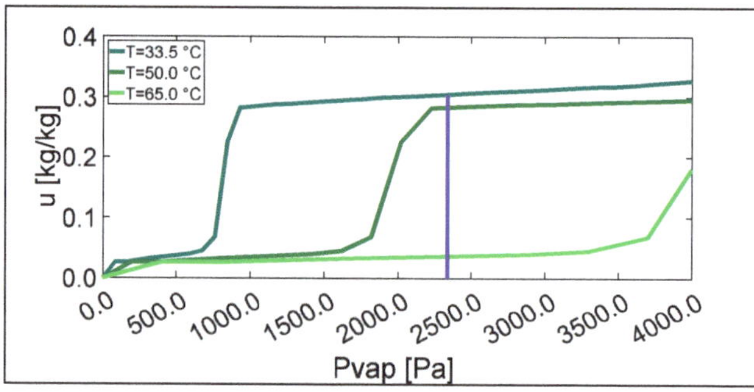

Figure 10. Isotherms of zeolite [14].

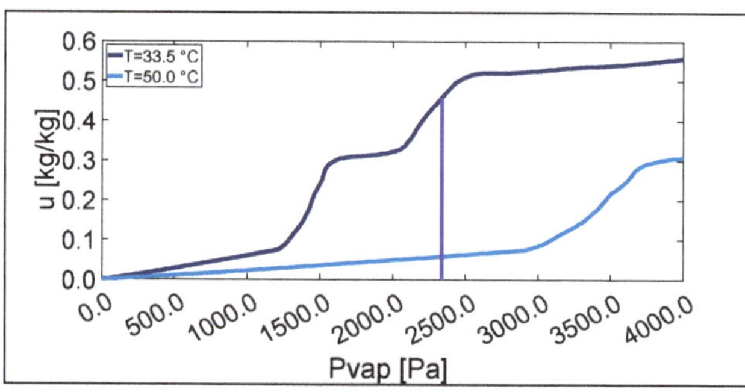

Figure 11. Isotherms of MOF MIL-100 (Fe) [6].

It is possible to note that MOF material shows the best performance of water uptake at these conditions (0.48 kg/kg against 0.24 kg/kg of silica-gel and 0.30 kg/kg of zeolite). This means that, for the same used quantity of adsorbent material and the same boundary conditions, MOF material achieves a better dehumidification result (i.e., amount of water vapour adsorbed). These results are representative of a typical situation in which air dehumidification is required. Generally speaking, MOF materials have a favourable isotherm, so they are strongly effective, in the range of sufficiently high values of RH that otherwise would request an overload on the cooling device (e.g., low coolant temperatures) for the dehumidification process. Regarding the regeneration capacity, it is useful to compare in the graphs the temperatures required to obtain a value of uptake close to zero. MIL-100 presents, compared to the other options, a low regeneration temperature to remove the uptake water content from the material and restore its adsorption capacity. It is vital to remark that this temperature is well coupled with the conditions of the VCR condenser (low-grade thermal energy), while additional heat sources are requested for the other desiccants (as it is possible to observe from Figures 9–11).

6. Conclusions

In this paper, the importance of using desiccant materials for the energy enhancement of HVAC systems has been discussed. Desiccant materials can dehumidify the airflows, removing the latent heat, with better performances of cooling devices and consequent energy savings. Among the many available materials, attention has been addressed to Metal-Organic Framework (MOF) materials, which have unique adsorption characteristics (high water uptake, low regeneration temperatures) compared to other known desiccants (i.e., silica-gel).

In the first part of the paper, a general overview of MOFs is presented, in particular emphasising the possibility to control the design of their structures, aimed at the desired properties. Moreover, the application of MOFs in the HVAC context has been reviewed. The literature analysis shows that the big advantage of MOFs over silica-gel is the low regeneration temperature (40–60 °C against 90–100 °C). The proposed solutions are the desiccant wheel and exchangers coated with MOFs and cooled by a refrigerant.

Then, a comparison between a traditional HVAC (with dehumidification operated by a cooling coil) and a HVAC equipped with a MOF-Assisted Dehumidifier (MAD) has been discussed. The mathematical model of the two HVAC alternatives is applied to a real case study. The energy savings achievable with MOF dehumidification are in the range of 30–50%, depending on the outdoor air conditions and the repartition between sensible and latent heat. Comparing the MOF material with other desiccants (silica-gel and a new zeolite), the MOF-driven dehumidification shows the highest water uptake and the lowest regeneration temperature. The obtained results lead us to consider the utilisation of MOF-assisted dehumidification as a very competitive strategy to reduce the energy consumption of the HVAC sector, making decarbonization closer, and inspiring the Authors to make deeper studies on this topic.

Author Contributions: Conceptualization, A.R., L.S. and L.T.; Methodology, M.L. and L.S.; Software, L.S.; Validation, L.S.; Formal analysis, M.L. and L.S.; Investigation, M.L. and L.S.; Resources, A.R.; Data curation, M.L. and L.S.; Writing—original draft, L.S.; Writing—review & editing, A.R., M.L., L.S. and L.T.; Supervision, A.R., P.G., V.K. and L.T.; Project administration, P.G. and L.T.; Funding acquisition, A.R., P.G., V.K. and L.T. All authors have read and agreed to the published version of the manuscript.

Funding: This research received no external funding.

Conflicts of Interest: The authors declare no conflict of interest.

Abbreviations

AC	Air Conditioning
c	Constant pressure specific heat [kJ/kg/K]
DT	Temperature difference [°C]
EER	Energy Efficiency Ratio
eff	Efficiency
HVAC	Heating Ventilation Air Conditioning
j	Specific enthalpy [kJ/kg]
$\dot{m}$	Air mass flow rate [kg/s]
MAD	MOF-Assisted Dehumidifier
MOF	Metal-Organic Framework
n	Number of people

$\dot{Q}$	Thermal load [kW]
r	Latent heat of vaporisation [kJ/kg]
RH	Relative Humidity [%]
SHR	Sensible Heat Ratio
T	Temperature [°C]
u	Water uptake [kg water/kg adsorbent]
VCR	Vapour Compression Refrigeration
$\dot{W}$	Power [kW]
x	Humidity ratio [kg/kg]
Subscripts	
c	Cooling coil
dp	Dew point
e	Exhaust airflow (from indoor environment)
h	Heating coil
he	Air-to-air heat exchanger
i	Indoor air
lat	Latent
mof	MOF-Assisted Dehumidifier
o	Outdoor air
p	Person
s	Supply air
sens	Sensible
evap	Evaporator of the VCR cycle
cond	Condenser of the VCR cycle
I	Air at the outlet of the evaporator
II	Air at the outlet of the condenser

References

1. Grazzini, G.; Milazzo, A. *Tecnica Del Freddo*; Società Editrice Esculapio: Bologna, Italy, 2017; ISBN 9788874889969.
2. IEA. World Energy Outlook 2022. IEA Report. 2022. Available online: https://www.iea.org/reports/world-energy-outlook-2022 (accessed on 21 November 2022).
3. IEA. Net Zero by 2050. IEA Report. 2021. Available online: https://www.iea.org/reports/net-zero-by-2050 (accessed on 21 November 2022).
4. IEA. The future of cooling. IEA Report. 2018. Available online: https://www.iea.org/reports/the-future-of-cooling (accessed on 21 November 2022).
5. Zaki, O.M.; Mohammed, R.H.; Abdelaziz, O. Separate sensible and latent cooling technologies: A comprehensive review. *Energy Convers. Manag.* **2022**, *256*, 115380. [CrossRef]
6. Cui, S.; Qin, M.; Marandi, A.; Steggles, V.; Wang, S.; Feng, X.; Nouar, F.; Serre, C. Metal-Organic Frameworks as advanced moisture sorbents for energy-efficient high temperature cooling. *Sci. Rep.* **2018**, *8*, 15284. [CrossRef] [PubMed]
7. Sultan, M.; El-Sharkawy, I.I.; Miyazaki, T.; Saha, B.B.; Koyama, S. An overview of solid desiccant dehumidification and air conditioning systems. *Renew. Sustain. Energy Rev.* **2015**, *46*, 16–29. [CrossRef]
8. Mazzei, P.; Minichiello, F.; Palma, D. Desiccant HVAC systems for commercial buildings. *Appl. Therm. Eng.* **2002**, *22*, 545–560. [CrossRef]
9. Ng, E.P.; Mintova, S. Nanoporous materials with enhanced hydrophilicity and high water sorption capacity. *Microporous Mesoporous Mater.* **2008**, *114*, 1–26. [CrossRef]
10. LaPotin, A.; Zhong, Y.; Zhang, L.; Zhao, L.; Leroy, A.; Kim, H.; Rao, S.R.; Wang, E.N. Dual-Stage Atmospheric Water Harvesting Device for Scalable Solar-Driven Water Production. *Joule* **2021**, *5*, 166–182. [CrossRef]
11. Goldsworthy, M.J. Measurements of water vapour sorption isotherms for RD silica gel, AQSOA-Z01, AQSOA-Z02, AQSOA-Z05 and CECA zeolite 3A. *Microporous Mesoporous Mater.* **2014**, *196*, 59–67. [CrossRef]
12. Thommes, M.; Kaneko, K.; Neimark, A.V.; Olivier, J.P.; Rodriguez-Reinoso, F.; Rouquerol, J.; Sing, K.S.W. Physisorption of gases, with special reference to the evaluation of surface area and pore size distribution (IUPAC Technical Report). *Pure Appl. Chem.* **2015**, *87*, 1051–1069. [CrossRef]
13. Castillo, J.M.; Silvestre-Albero, J.; Rodriguez-Reinoso, F.; Vlugt, T.J.H.; Calero, S. Water adsorption in hydrophilic zeolites: Experiment and simulation. *Phys. Chem. Chem. Phys.* **2013**, *15*, 17374–17382. [CrossRef]
14. Liu, Z.; Xu, J.; Xu, M.; Huang, C.; Wang, R.; Li, T.; Huai, X. Ultralow-temperature-driven water-based sorption refrigeration enabled by low-cost zeolite-like porous aluminophosphate. *Nat. Commun.* **2022**, *13*, 193. [CrossRef]
15. Shahvari, S.Z.; Kalkhorani, V.A.; Wade, C.R.; Clark, J.D. Benefits of metal–organic frameworks sorbents for sorbent wheels used in air conditioning systems. *Appl. Therm. Eng.* **2022**, *210*, 118407. [CrossRef]

16. Feng, X.; Qin, M.; Cui, S.; Rode, C. Metal-organic framework MIL-100(Fe) as a novel moisture buffer material for energy-efficient indoor humidity control. *Build. Environ.* **2018**, *145*, 234–242. [CrossRef]
17. Ming, Y.; Kumar, N.; Siegel, D.J. Water Adsorption and Insertion in MOF-5. *ACS Omega* **2017**, *2*, 4921–4928. [CrossRef] [PubMed]
18. Hu, Z.; Wang, Y.; Shah, B.B.; Zhao, D. CO_2 Capture in Metal-Organic Framework Adsorbents: An Engineering Perspective. *Adv. Sustain. Syst.* **2019**, *3*, 1800080. [CrossRef]
19. González-Zamora, E.; Ibarra, I.A. CO_2 capture under humid conditions in metal-organic frameworks. *Mater. Chem. Front.* **2017**, *1*, 1471–1484. [CrossRef]
20. Li, H.-Y.; Zhao, S.-N.; Zang, S.-Q.; Li, J. Functional metal-organic frameworks as effective sensors of gases and volatile compounds. *Chem. Soc. Rev.* **2020**, *49*, 6364–6401. [CrossRef]
21. LaPotin, A.; Kim, H.; Rao, S.R.; Wang, E.N. Adsorption-Based Atmospheric Water Harvesting: Impact of Material and Component Properties on System-Level Performance. *Acc. Chem. Res.* **2019**, *52*, 1588–1597. [CrossRef]
22. Zhou, X.; Lu, H.; Zhao, F.; Yu, G. Atmospheric Water Harvesting: A Review of Material and Structural Designs. *ACS Mater. Lett.* **2020**, *2*, 671–684. [CrossRef]
23. Hanikel, N.; Prévot, M.S.; Yaghi, O.M. MOF water harvesters. *Nat. Nanotechnol.* **2020**, *15*, 348–355. [CrossRef]
24. Khutia, A.; Rammelberg, H.U.; Schmidt, T.; Henninger, S.; Janiak, C. Water sorption cycle measurements on functionalized MIL-101Cr for heat transformation application. *Chem. Mater.* **2013**, *25*, 790–798. [CrossRef]
25. Hanikel, N.; Pei, X.; Chheda, S.; Lyu, H.; Jeong, W.; Sauer, J.; Gagliardi, L.; Yaghi, O.M. Evolution of water structures in metal-organic frameworks for improved atmospheric water harvesting. *Science* **2021**, *374*, 454–459. [CrossRef]
26. Yanagita, K.; Hwang, J.; Shamim, J.A.; Hsu, W.-L.; Matsuda, R.; Endo, A.; Delaunay, J.-J.; Daiguji, H. Kinetics of Water Vapor Adsorption and Desorption in MIL-101 Metal-Organic Frameworks. *J. Phys. Chem. C* **2019**, *123*, 387–398. [CrossRef]
27. Bareschino, P.; Diglio, G.; Pepe, F.; Angrisani, G.; Roselli, C.; Sasso, M. Numerical study of a MIL101 metal organic framework based desiccant cooling system for air conditioning applications. *Appl. Therm. Eng.* **2017**, *124*, 641–651. [CrossRef]
28. Shahvari, S.Z.; Kalkhorani, V.A.; Clark, J.D. Performance evaluation of a metal organic frameworks based combined dehumidification and indirect evaporative cooling system in different climates. *Int. J. Refrig.* **2022**, *140*, 186–197. [CrossRef]
29. Affordable and Sustainable Cooling Using Metal-Organic Frameworks. Catalyzing Commercialization. Available online: https://www.aiche.org/sites/default/files/cep/20200916.pdf (accessed on 21 November 2022).
30. Wang, W.; Wu, L.; Li, Z.; Fang, Y.; Ding, J.; Xiao, J. An Overview of Adsorbents in the Rotary Desiccant Dehumidifier for Air Dehumidification. *Drying Technol.* **2013**, *31*, 1334–1345. [CrossRef]
31. Saeed, A.; Al-Alili, A. A review on desiccant coated heat exchangers. *Sci. Technol. Built Environ.* **2017**, *23*, 136–150. [CrossRef]
32. Venegas, T.; Qu, M.; Nawaz, K.; Wang, L. Critical review and future prospects for desiccant coated heat exchangers: Materials, design, and manufacturing. *Renew. Sustain. Energy Rev.* **2021**, *151*, 111531. [CrossRef]
33. Danfoss. CoolSelector Software. Available online: https://www.danfoss.com/en-gb/service-and-support/downloads/dcs/coolselector-2/ (accessed on 21 November 2022).
34. ASHRAE. *ASHRAE Handbook-Fundamentals*; ASHRAE: Atlanta, GA, USA, 2021.
35. *Standard UNI 10339:1995*; Air-conditioning systems for thermal comfort in buildings. General, classification and requirements. Offer, order and supply specifications. UNI Ente Nazionale di Normazione: Milano, Italy, 1995.
36. Mohammed, R.H.; Mesalhy, O.; Elsayed, M.L.; Su, M.; Chow, L.C. Revisiting the adsorption equilibrium equations of silica-gel/water for adsorption cooling applications. *Int. J. Refrig.* **2018**, *86*, 40–47. [CrossRef]

energies

MDPI

Article

Aging of Polymeric Insulators under Various Conditions and Environments: Another Look

Xinhan Qiao [1,2], Yue Ming [3], Ke Xu [4], Ning Yi [5] and Raji Sundararajan [6,*]

1 School of Electrical Engineering, China University of Mining and Technology, Xuzhou 221116, China
2 Sunten Electric Equipment Co., Ltd., Foshan 528300, China
3 School of Mechanical & Electrical Engineering, Xuzhou University of Technology, Xuzhou 221018, China
4 State Key Laboratory of Environmental Adaptability for Industrial Products, China National Electric Apparatus Research Institute Co., Ltd., Guangzhou 510663, China
5 State Grid Zhenjiang Power Supply Company, Zhenjiang 212000, China
6 School of Engineering Technology, Purdue University, West Lafayette, IN 47907, USA
* Correspondence: raji@purdue.edu; Tel.: +1-765-494-6912

Abstract: Polymeric insulators have lightweight, excellent hydrophobicity and convenient transportation and installation. They are widely used in the external insulation for distribution and transmission lines. However, due to the long-term effects of pollution, ultraviolet radiation, discharge, temperature, humidity, altitude and other natural and complex environmental and service factors, the silicone rubber and other materials of polymeric insulators gradually age and lose their hydrophobicity and electrical insulation characteristics. The operability is significantly reduced, which seriously affects the safety and reliability of the power system. Hence, there is a need for assessing and evaluating the long-term aging and degradation of polymeric insulators under various operating conditions and environments. In this review, the various aging and characterization techniques of the polymeric insulators and their aging performance under the action of multiple factors are discussed. To enhance the performance of polymeric insulators, nano-coating, surface treatment and other techniques are also indicated. In addition, future potential fields that should be explored from a high-voltage electrical insulation perspective are also presented.

Keywords: polymeric insulators; silicone rubber; EPDM; aging; nano-coating

Citation: Qiao, X.; Ming, Y.; Xu, K.; Yi, N.; Sundararajan, R. Aging of Polymeric Insulators under Various Conditions and Environments: Another Look. *Energies* **2022**, *15*, 8809. https://doi.org/10.3390/en15238809

Academic Editors: Alessandro Cannavale and Ubaldo Ayr

Received: 20 September 2022
Accepted: 11 November 2022
Published: 22 November 2022

1. Introduction

Polymeric insulators have the advantages of being lightweight, excellent hydrophobicity and convenient for transportation and installation [1–4]. They are widely used in external insulation for distribution and transmission lines [5,6].

However, in the long term, they can fail. The numerous failures of polymeric insulators around the world drew attention to the shortcomings of the material and design and the need for diagnostic techniques to determine their state [7]. The damage and failure mechanisms of polymeric insulators differ significantly from ceramic and glass insulators.

The failures of polymeric insulators can be mainly divided into two categories. One is the mechanical fault of the insulator core rod breaking caused by typhoons and other factors, which is less likely to occur. The second is the electrical failure caused by the aging phenomenon [8–10], such as the decline of the hydrophobic performance of polymeric insulators. Such incidents will cause the surface flashover of polymeric insulators and a large-scale power outage.

Figure 1 shows various examples of the aging and degradation of polymeric insulators [11]. Table 1 shows a list of the various failures, such as brittle fractures that are experienced by polymeric insulators, as identified by the Electric Power Research Institute (EPRI) [7].

Figure 1. Illustrations of the damage, degradation and aging of polymeric insulators [11].

Table 1. List of the various types of failures of polymeric insulators identified by EPRI [7].

Failure Type	% of Failure Recorded (221)
Brittle fracture	51.1%
Flashunder	24.9%
Mechanical failure: rod failure	10.4%
Rod destruction by discharge activity	8.1%
End fitting pullout	0.5%

The aging of polymeric insulators is studied extensively [10,12–14]. The research on aging mainly focuses on the aging mechanisms and characterization methods. With the development of advanced material engineering technology, the performance of polymeric insulator materials can be improved in various ways to enhance their long-term use [10,15–18].

This article reviews the various aging methods of polymeric insulators and the analysis of the aging mechanisms in combination with environmental factors. The application of nano and micro fillers, nano-coating and other state-of-the-art techniques are discussed, alongside which technologies can enhance the material's performance. For this purpose, a literature survey was conducted using the keywords polymeric insulator aging, silicone rubber insulator aging, hydrophobicity loss and aging in standard scholarly databases and Google Scholar.

2. Characterization Techniques of Electrical Performance and Aging Characteristics of Polymeric Insulators

2.1. Electrical Performance

After being in service for several years, the power operation department conducts various withstand study tests, such as a salt-fog flashover [19], a natural pollution flashover, an artificial pollution flashover [4,20–25] and a 50% lightning impulse flashover [26,27] to evaluate the degree of deterioration for polymeric insulator performance according to the flashover voltage level. Different flashover voltage tests simulate the electrical performance of insulators under different climatic and operating conditions of the system. This is the most direct method to judge the operating characteristics of polymeric insulators, but it is a destructive test. The insulator needs to be removed from the transmission line for the test, so this method has limitations.

The trap characteristic test [28] can also be used to identify the aging of polymeric insulators. According to reports, when the corona time increases, the sample's trap energy level gradually increases and the trapped charge rises in general [29].

2.2. Physical Appearance

(1) Visual inspection.

Various obvious and serious aging problems can be identified through a manual visual inspection of the appearance of the insulators (Figure 2.), to be replaced over time [29]. However, the main disadvantage of this method is that it is difficult to identify any internal defects or early aging signs in time. Visual inspections can be conducted by a manual patrol inspection and unmanned aerial vehicle patrol inspections, which are a non-contact inspection method [30].

Figure 2. Deteriorated insulators that can be found by visual inspection [11].

(2) Hydrophobicity classification (HC) test.

Hydrophobicity [31–33] is a crucial factor in determining the level of deterioration of silicone rubber and other polymeric materials. With an increase in aging, the hydrophobicity of polymeric insulator surface decreases or is lost completely, leading to the decrease in flashover voltages.

To comprehensively evaluate the hydrophobicity changes of the silicone rubber and other surfaces of polymeric insulators in time, the HC water spray classification method and static contact angle method can be used [32]. Table 2 and Figure 3 illustrate the HC obtained in the 345 kV EPDM insulators, in-service aged at a coastal environment for 5 years [34]. Here, the middle was the mid part of the 2.862 m insulator. The top surface of the shed was white, which was exposed to UV radiation for 5 years and, therefore, aged and discolored, while dark was the original color of the bottom surface of the shed, which was unexposed to UV radiation. Figure 3 shows how, for the same shed, the side protected from the sun is dark in color with an HC class of 2–3, compared to the other exposed (white) side, with HC 5–7.

Table 2. HC of the 345 kV EPDM insulator [34].

Weathershed	Surface	HC
High voltage end	White (aged/discolored)	HC5
	Dark	HC2–3
Middle	White (aged/discolored)	HC4
	Dark	HC2
Low voltage end	White (aged/discolored)	HC5
	Dark	HC3

Figure 3. The hydrophobicity classification of the white and dark surfaces.

The static and dynamic contact angle methods are the two most common techniques used to study the HC conditions. If the HC level is less than four and the static contact angle is larger than ninety degrees, the hydrophobicity of the sample surface is good. The dynamic contact angle error increases with the water drop volume [35]. Thus, the water drop velocity [36] could be a measure of the hydrophobicity change. The above are direct methods used to evaluate hydrophobicity. Other non-contact indirect methods can also be used to evaluate hydrophobicity. The non-contact methods include the DC discharge-induced acoustic wave [32] and the PD-induced electromagnetic wave [37]. These can be used to assess the hydrophobicity of polymeric insulators by recognizing the surface discharge characteristics.

(3) Mechanical properties.

The mechanical properties, such as hardness, tensile strength and elongation at break, are important parameters to characterize the aging of silicone rubber and other polymeric insulators. A Shore hardness tester can be used to characterize the hardness of silicone rubber, which increases with aging [38]. A new approach is reported in ref [39]. The findings suggest that laser-induced breakdown spectroscopy (LIBS) can be a reliable source of hardness information for polymeric materials, which is useful for ensuring the reliability of power lines. However, the tensile strength and elongation at break (tested by a tension machine with a load sensor) decrease with aging [40].

(4) Surface morphology study using scanning electron microscope.

A scanning electron microscope (SEM) [41,42] is useful to visually explore the micromorphology changes on the surfaces of polymeric insulators and identify the defects or deformation, such as roughness change, that are not easily seen by the naked eye. This helps to provide a general overview of the micromorphology properties of the aged polymeric insulator and establish whether the laboratory-accelerated aging test and the in-service aging of actual samples are comparable.

Figure 4 illustrates the SEM surface morphology changes of new, 3, 6 and 9-year-old in-service insulators [43]. Figure 5 indicates the SEM images of the surface morphologies of 345 kV EPDM insulators that were installed in 1995 and removed from service in 2000, in a coastal environment in the USA [34]. Here, bulk indicates the SEM morphology of the bulk of the material (equivalent to new/unaged). The white surface indicates the discoloration caused by UV exposure and the black surface indicates the original surface—all from the high-voltage end of the insulator.

Figure 4. SEM images of the surface morphology changes of a (**a**) new (unaged) sample, (**b**) 3 years, (**c**) 6 years and (**d**) 9 years of in-service polymeric insulators [43].

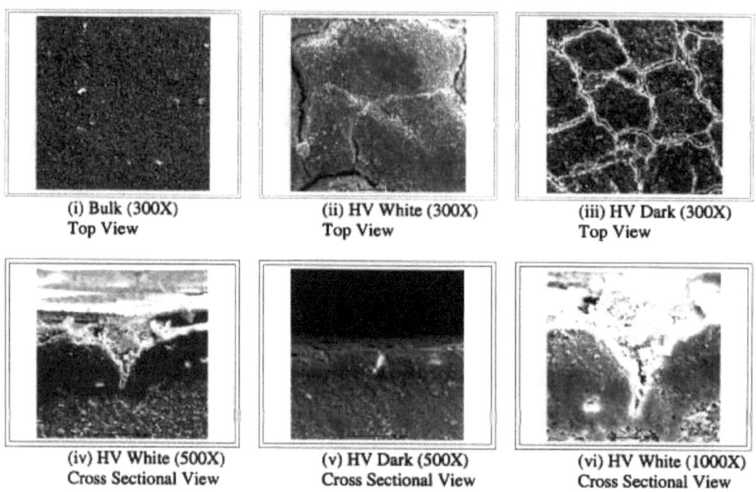

Figure 5. SEM images of the surface morphology changes of 345 kV EPDM insulators removed from service after 5 years in a coastal environment [34].

It was also verified that the aging evaluation of silicone rubber insulators could be realized through a portable digital microscope-B011 [30], as shown in Figure 6. The unmanned-aerial-vehicle-mounted electron microscope is expected to be used to conduct the non-contact evaluation of the aging of composite insulators in transmission lines.

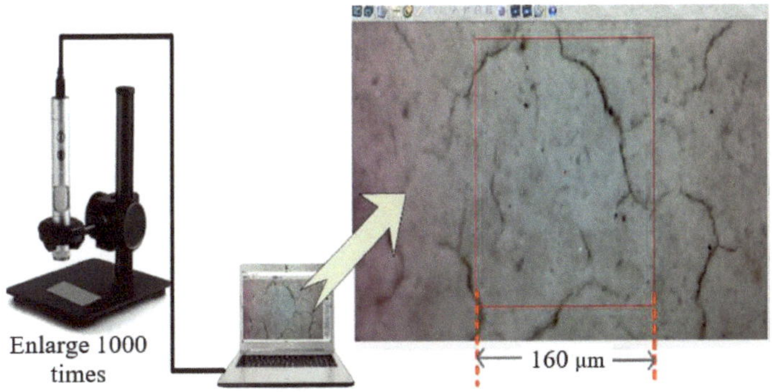

Figure 6. Characterization of the aging surface seen through the portable digital microscope-B011 [30].

2.3. Chemical Performance

FTIR spectroscopy and EDX analyses [44–46] are the two most common methods used to analyze the material contents of polymeric insulators. The analyses of these polymeric insulators using EDX and FTIR demonstrated their superiority over other methods [47].

FTIR is useful to identify the number of typical functional groups, and the aging degree can be analyzed based on this. For silicone rubber materials used for polymeric insulator skirts, the corresponding bands of the functional groups in the absorption band diagram are shown in Table 3 [44–46]. The height of the absorption peak associated with each functional group is proportional to the functional group's composition. The greater the absorbance at the absorption peak in a particular band, the greater the absorption spectrum capacity and the greater the number of functional groups that correspond to that band. The lower the absorption peak or the smaller the absorption area of Si–$(CH_3)_2$ and Si–O–Si functional groups, the more serious the molecular chain fracture and the more serious the aging degree. Similar results were also reported for 345 kV in-service aged EPDM insulators [34].

Table 3. Characteristic peaks of silicon rubber in IR analysis.

Characteristic Group	Wave Number/cm^{-1}
O–H	3700–3200
CH_3(C–H)	2960
C–H	1440–1410
Si–CH_3(C–H)	1270–1255
Si–O–Si(Si–O)	1100–1000
O–Si$(CH_3)_2$–O(Si–O)	840–790
Si$(CH_3)_3$	800–700

EDX analyses are used to obtain the surface element percentage of the polymeric insulators. The working principle of EDX is that the incident beam scans the surface line by line. The X-rays generated by the electron beam and material interaction are collected and classified according to the characteristic X-ray energy spectrum. Since each element has characteristic X-rays, the chemical composition of a given sample can be found according to the energy scale of the abscissa. With an increase in aging, the contents of Si and C usually decrease, while the content of oxygen increases [44–46].

The dielectric constant [48] is used to indicate the relative ability of a dielectric to store electrostatic energy in an electric field. It can also indicate the degree of polarization of the dielectric. The smaller the dielectric constant, the weaker the ability of the dielectric to store static electricity, which, to some extent, means the better the insulation performance. Therefore, the dielectric constant measured by the broadband dielectric spectrometer can also be used as a parameter to characterize the aging of polymeric insulators.

In addition, a variety of thermal analysis techniques, including differential scanning calorimetry (DSC) and thermal gravity analysis (TGA), were utilized in order to investigate the transformation of organic components in polymeric insulators [49]. According to the findings of the tests described in ref [49], thermal analysis techniques have the potential to significantly contribute to the performance evaluation of polymeric insulators.

To summarize, there are many methods to characterize the aging of polymeric insulators, and they can be divided into two main categories from the technical means. One is the non-contact characterization method. The insulator in operation is suitable for non-contact characterization. The other is to take samples and return them to the laboratory for testing. This method usually requires a power cut and more time and human resources. In addition, the polymeric insulator has hydrophobic recovery properties [50,51]. When the sample is returned to the laboratory, the characteristics of the polymeric could change. Therefore, the non-contact aging characterization method is preferred to evaluate the aging of polymeric insulators.

3. Aging Mechanisms and the Performance of Polymeric Insulators under Various Operating Conditions and Environments

3.1. Aging under UV, Acidic and High Field Environments

Ultraviolet is one of the dominant factors that influence the aging of polymeric insulators. Although the atmospheric ultraviolet intensity is not enough to interrupt the main chain of silicone rubber (the shortest wavelength of ultraviolet transmitted to the ground is 290 nm, and the energy is 396.3 kJ/mol), it can oxidize the methyl group of the side chain by interacting with other factors, resulting in the aging of the material surface. The Si—O bond energy in the main chain of silicone rubber is 444 kJ/mol. As shown in Figure 7, under ultraviolet light, the C–Si bond or the C–H bond on the silicone rubber surfaces can break and form free radicals. These free radicals have high energy and are prone to cross-linking reactions. Generally, oxygen in the air will react with free radicals to form hydrophilic OH and other polar groups on the silicone rubber surface, generating methane and other gases [52] and leading to long-term aging and degradation.

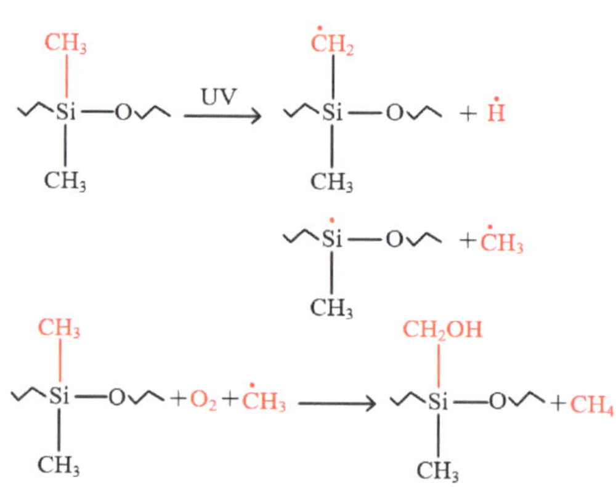

Figure 7. Chemical reactions in silicone rubber under UV irradiation [52].

The other factors that cause the chemical aging of silicone rubber include acid and alkali, ozone and nitrogen oxides where the nitrogen oxides will be converted into nitric acid by moisture absorption. Previous studies have shown that the surface of the composite insulators will be seriously damaged when they are in a strong acidic environment for a long time [11,53–57]. The acidic substance will cause the polar, Si–O bond to break on the main chain of silicone rubber and generate the polar Si–OH bond, as shown in Figure 8. This is different from the breaking of the Si–C bond and the C–H bond on the surface of silicone rubber. The breaking of the main chain will greatly damage the silicone rubber material and the generated Si–OH bond will produce the silicone rubber hydrophilic. Table 4 shows the various compounds studied in ref [57] for enhancing the acidic stability of silicone rubber insulators.

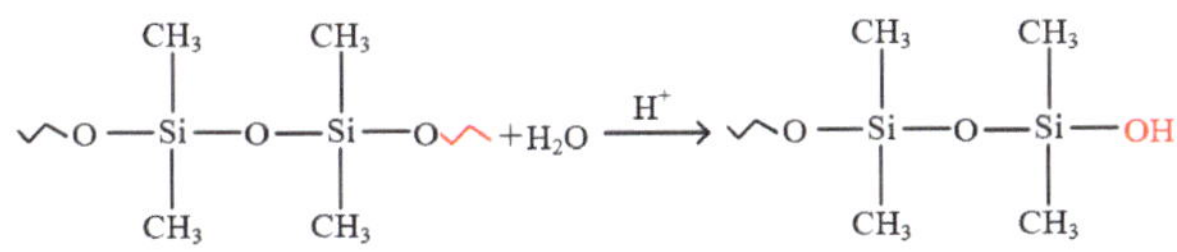

Figure 8. The formation of Si–OH bonds due to the reaction between the acid and silicone rubber [52].

Table 4. Silicone rubber compounds tested for enhancing acid stability [57].

Silicone Rubber Compounds	ATH Parts Per Hundred Rubber/Weight%	Inert Parts Per Hundred Rubber/Weight%
52 wt% untreated ATH	110/52	-
52 wt% precoated ATH	110/52	-
52 wt% insitu coated ATH vinyl silane	110/52	-
ATH + 25 wt% SiO_2	55/25	55/25
ATH + 9 wt% $BaSO_4$	100/9	20/9
ATH + 13.5 wt% $BaSO_4$	90/13.5	30/13.5
ATH + 18 wt% $BaSO_4$	80/18	40/18
ATH + 36 wt% $BaSO_4$	40/36	80/36
54 wt% $BaSO_4$	0/0	120/54

Silicone rubber composite insulators are not only affected by natural conditions but also by high electric fields [28,58,59]. The aging effect of the electric field on silicone rubber is more severe and faster. During the operation of silicone rubber, the impact of charged particles generated by corona or arc discharge, such as the impact of ions and electrons, causes the main chain of silicone rubber to break and the molecules to depolymerize. At the same time, the energy generated by the discharge decomposes and reacts with oxygen and other gases in the surrounding air to produce highly oxidizing substances, such as active oxygen atoms, ozone and nitrogen-oxygen compounds, which can also cause the molecular chain of silicone rubber to break and damage its performance. Therefore, the electrical aging process is the most complex, often accompanied by physical and chemical aging, significantly impacting the silicone rubber's performance [60].

3.2. Aging under Various Environments

According to the aging mechanism of polymeric insulators analyzed above, the focus was on the effects of the polymeric insulators in tropical, coastal high humidity, plateau strong ultraviolet, acid-based, electric stress, low temperature, icing and high altitude environments [53,58].

(1) Tropical environment.

To determine if insulators can withstand electrical stress beyond the prescribed limits in polluted and clean tropical regions, a field and lab investigation was conducted [61]. During the field experiment, two sets of insulators were erected and activated in coastal and inland Sri Lanka. None of the tested field SiR insulators flashed over. However, a tree-like surface discoloration was most likely caused by some discharge activity. Thus, biological growth was identified on the installed insulators. Another study [62] summarizes fieldwork. Microbiological development is unlikely to cause substantial degradation on non-ceramic insulator rubber housings, according to the collected data. Growth appears to have no effect on insulator performance.

Algae have been observed on polymeric and RTV-coated insulators in southwest and southern China. The research results showed that the algae enhanced salt dispersion and decreased insulator hydrophobicity. When algal covering was less than 20%, the effects on insulators were limited [63].

(2) Salt-fog environment.

In ref [64], salt fog was used to simulate the high humidity environment with salt in a coastal area, and it was found that the sample's performance deteriorated after the salt-fog treatment. The salt-fog-treated samples became rough and porous under electric and thermal stress. The hydrophobic absorption peak decreased, indicating that the SR molecular chain was broken, and the filler was consumed, lowering the sample's arc resistance. Moisture absorption affected the insulating performance and electrical strength. The physicochemical deterioration will reduce the electrical strength. Thus, in salt-fog environments, samples with higher conductivity are more deteriorated, showing that salt fog accelerates silicone rubber aging [65].

(3) Radiation environment.

Polymeric materials exposed to gamma rays in a radiation environment [59] have two impacts. First, the bond scission reduces the molecular weight by cross-linking. Second, the surface oxygenation forms oxygenated molecules. As irradiation doses increase, the transmitted energy and oxidation index lead to polymeric cross-linking. Similar conclusions have been obtained from the experiments conducted in central Saudi Arabia where the UV radiation level is high [50].

The hydrophobicity transfer was also affected by the contaminated species, pollution layer thickness, pollution level, temperature, UV radiation and corona activity. The study's key conclusions [66] note that UV radiation can accelerate the migration of LMW molecules from the SiR bulk to the contamination layer surface and the pollution and layer thickness can increase the hydrophobicity transfer time.

(4) Electric stress.

Further, the hydrophobicity of the SiR surface worsens with increasing the corona discharge voltage and treatment time. The vertical wind can accelerate the hydrophobicity loss of SiR, while the parallel wind inhibits the hydrophobicity loss. The wind with higher speed has greater influence [67]. Thus, positive polarity has a greater deleterious effect on aging than negative polarity [68].

(5) Low temperature icing environment.

The electrical characteristic test of the aging composite insulator in a low-temperature, icing environment was conducted in ref [6]. The test environment is shown in Figure 9. Three kinds of ice were observed (Figure 10A–C), including rime ice (at below 0 °C), glazed ice (at below 0 °C) and continuous water film due to the loss of hydrophobicity (at above 0 °C).

Figure 9. Iced test for transmission lines and insulators [6].

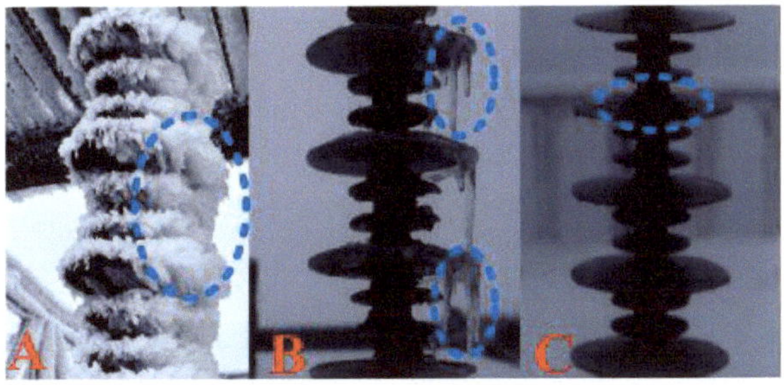

Figure 10. Iced insulators with various types of ice coverage, such as (**A**) rime rice (at below 0 °C), (**B**) glazed-ice (at below 0 °C) and (**C**) continuous water film due to loss of hydrophobicity (at above 0 °C) [6].

The rime ice and the glazed ice greatly reduced the flashover voltage of the insulators. In low-temperature and dense-fog environments, the hydrophobicity of aging composite insulators is more likely to be lost, as shown in Figure 10C.

(6) High altitude.

There are a few studies on the influence of altitude (air pressure) on the aging of composite insulators, and in ref [69,70], the influence of altitude on the flashover voltage of insulators were investigated. Polymer insulators have a 48.72% larger flashover voltage gradient than porcelain and a 72.35% higher voltage than glass at 59 kPa (corresponding to 4484 m height) [70].

To summarize, the aging of silicone rubber can be divided into three main types: physical, chemical and electrical. The factors causing physical aging include ultraviolet irradiation and local high temperature. The factors causing chemical aging include acid and alkali, ozone and nitrogen oxides. The electrical aging process often accompanies physical and chemical aging. However, the aging effect on silicone rubber insulators is more serious and faster. Polymeric insulators are most prone to aging with corona discharge under high humidity, especially in salt-fog, acid-based and strong radiation environments. Therefore, in the above environments, more attention should be paid to online monitoring of polymeric insulators and improving the anti-aging performance of silicone rubber materials.

4. Enhanced Performance of Polymeric Insulators

4.1. Enhancement Methods

The nano-modified coating is considered to be an effective method to increase the long-term durability of polymeric insulators. A multifunctional nano-coating based on SiO_2/PDMS/EP are proposed in ref [71]. SEM images of nano-coated SiR samples at various magnifications are shown in Figure 11. Nanoparticles and aggregates form hierarchical patterns on the covered surface.

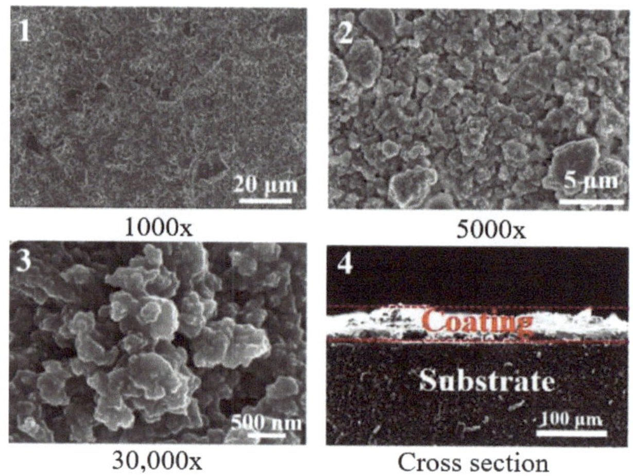

Figure 11. SEM images of nano-coated SiR samples at various magnifications (1–3) and the cross-sectional morphology [71].

The multiscale structure and low surface energy modification of this coating can improve water repellency. Coated silicone rubber has a large contact angle (>160) and a tiny rolling angle (around 0), enabling self-cleaning. Using this nano-coating, the electric field distortion caused by residual water droplets on the surface can be reduced, and the aging of the silicone rubber insulator caused by the corona can also be slowed. Different test samples coated with ATH and SiO_2 were subjected to multi-stress aging under both DC voltage polarities [68]. Considering the combined effect of numerous aging characteristics, a hybrid composite is the most age-resistant. Furthermore, the results of ref [72] showed that the micro and nano SiR compounds had better dielectric properties than plain SiR in terms of the dielectric constant and the dielectric loss. A typical polymer nanocomposite composed of three primary constituents—the polymer matrix, the nanofiller and the interaction zone— is claimed to play a significant role in the improvement of polymer nanocomposite properties [73], as shown in Figure 12.

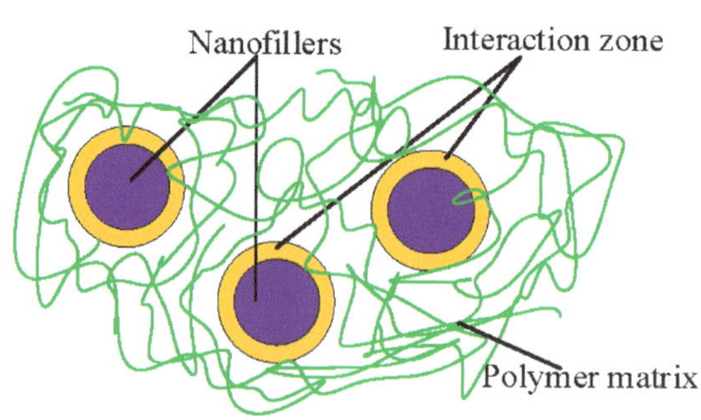

Figure 12. Simple diagram demonstrating the components of polymer nanocomposites [73].

The surface properties of polymeric insulators can also be improved by surface treatment techniques, such as plasma. For example, an atmospheric plasma jet can rapidly increase the hydrophobicity of dirty silicone rubber. According to study findings in [74], plasma can accelerate the transfer of the hydrophobicity of silicone rubber coated in a wet pollution layer. However, the hydrophobicity recovery of contaminated silicone rubber following the plasma treatment is reduced by the high humidity environment. In contrast, it is accelerated by prolonged plasma therapy. According to the findings in [75], the crosslinking density of polymeric materials increased by 8.8 times after a modest dose of electron-beam irradiation (40 kGy) treatment. The crosslinking density of low-irradiated (40 kGy) polymeric material is still six times greater than that of nontreated polymeric material after 1000 h of aging treatment.

Additionally, in particular settings such as acidic situations, inert fillers may be used. The silicone rubber's acid stability can be improved by the inert fillers. Barium-sulfate-loaded silicone elastomers have better acid-aging resistance [57]. The reduced mass loss in concentrated nitric acid decreased the fracture development and improved the mechanical stability with comparable erosion resistance. A comparison of the various surface treatments are shown in Table 5 [71].

Table 5. Comparison of the various surface treatments [71].

Treatment	Contact Angle, °	Self-Cleaning	Surface Charge Optimization	Flashover Improvement
PDMS + ZnO coating	162	Yes	Yes	16.7% increase in dry flashover
Liquid silicone rubber + SiO_2 coating	161.8	Not mentioned	Not mentioned	10.5% increase in wet flashover
DBD plasma treatments	~100	Yes	Yes	Not mentioned
Raisin-based primer + SiO_2 topcoat	161	Not mentioned	Not mentioned	29% increase in pollution flashover
PDMS + ZnO + MWCNT coating	152	Yes	Yes	28.8% increase in dry flashover
Pico second laser-ablated template + PDMS/SiO_2 coating	150.3	Yes	Yes	Not mentioned
Al_2O_3 + CNT + Polyamide mesh	160	Yes	Yes	30% increase in dry flashover
EP + PDMS + grafted SiO_2	162	Yes	Yes	60% increase in wet flashover

The above research shows that the surface properties of the polymeric insulators can be enhanced by nano-coating, surface treatment technology and inert fillers. However, in the case of a corona caused by the electric field distortion, the effect of these methods will be greatly weakened or even ineffective. Therefore, these treatment methods to improve the surface characteristics of polymeric insulators should be combined with electric field control methods to effectively enhance the long-term use of polymeric insulators.

For example, by combining nano-modified coatings with functionally gradient materials, their performance could be enhanced. The design of functionally gradient materials [76–78] can be realized using a finite-element electric field simulation before fabrication. In this way, the polymeric insulator will have better surface hydrophobicity and a lower surface electric field, improving the material's performance and reducing the effect of electrical stress.

4.2. Future Research Directions

(1) The non-contact aging characterization method is preferred for studying the aging of polymeric insulators. However, the research on non-contact characterization methods is usually based on a single method, such as the infrared thermometer, ultraviolet imager or unmanned aerial vehicle patrol inspection. These non-contact monitoring methods are greatly affected by environmental factors and are more error-prone. Therefore, in the future, online monitoring can be carried out through multi non-contact data fusion, and the change characteristics of monitoring data with environmental parameters can be studied.

(2) The surface properties of polymeric insulators can be enhanced by nano-coating, surface treatment technology and inert fillers. It will be beneficial to combine these treatment methods with electric field control techniques to effectively increase the long-term use of polymeric insulators by enhancing their surface characteristics. However, the aging mechanisms of the various environments are different, and different application environments should adopt different material optimization methods. Therefore, the anti-aging improvement methods for different environments should be studied in the future.

5. Conclusions

The aging performance of polymeric insulators under various conditions and environments is reviewed, and the following conclusions are derived.

(1) The aging characteristics of polymer insulators can be expressed as electrical, physical and chemical characteristics. There are different detection methods for different characteristics, but to facilitate the field application, a non-contact monitoring method is preferable.

(2) Polymeric insulators are most prone to aging with a corona discharge under high humidity, especially in salt-fog, acid-based and strong radiation environments. Therefore, in the above environments, more attention should be paid to online monitoring and improving their anti-aging performance.

(3) Combining surface treatment methods, such as nano-coating, surface treatment technology and inert fillers combined with electric field control methods, will help enhance the long-term performance of polymeric insulators effectively.

(4) More online monitoring should be carried out through multi non-contact data fusion, and the change characteristics of monitoring data with environmental parameters. However, different application environments should adopt different material optimization methods. Therefore, the anti-aging improvement methods for different environments should be studied in the future.

Author Contributions: Conceptualization, X.Q., Y.M., R.S., N.Y. and K.X.; methodology, X.Q. and Y.M.; investigation, N.Y. and K.X.; resources, K.X.; writing—original draft preparation, X.Q. and Y.M.; writing—review and editing, R.S.; software, X.Q.; validation, K.X.; supervision, R.S. All authors have read and agreed to the published version of the manuscript.

Funding: This research received no external funding.

Institutional Review Board Statement: Not applicable.

Informed Consent Statement: Not applicable.

Data Availability Statement: Not applicable.

Acknowledgments: The authors would like to express their gratitude to external insulation research group of Chongqing University for their assistance with this research.

Conflicts of Interest: The authors declare no conflict of interest.

References

1. Nan, J.; Li, H.; Wan, X.; Huo, F.; Lin, F. Pollution Flashover Characteristics of Composite Crossarm Insulator with a Large Diameter. *Energies* **2021**, *14*, 20. [CrossRef]
2. Qiao, X.; Zhang, Z.; Sundararajan, R.; Jiang, X.; Hu, J.; Fang, Z. AC Breakdown Characteristics of Polluted 10-kV Post Insulator With Concentric Externally Gapped Line Arrester. *IEEE Trans. Power Deliv.* **2022**, *37*, 59–66. [CrossRef]
3. Zhao, X.; Yang, X.; Hu, J.; Wang, H.; Yang, H.; Li, Q.; He, J.; Xu, Z.; Li, X. Grading of electric field distribution of AC polymeric outdoor insulators using field grading material. *IEEE Trans. Dielectr. Electr. Insul.* **2019**, *26*, 1253–1260. [CrossRef]
4. Ming, L.; Yangyang, L.; Jianlin, H. Influence of Sheds Damage on the AC Pollution Flashover Performance of Different Voltage Class Composite Insulators. *IEEE Access* **2020**, *8*, 84713–84719. [CrossRef]
5. Qiao, X.; Zhang, Z.; Sundararajan, R.; Jiang, X.; Hu, J.; Fang, Z. The failure arc paths of the novel device combining an arrester and an insulator under different pollution levels. *Int. J. Electr. Power Energy Syst.* **2021**, *125*, 106549. [CrossRef]
6. Qiao, X.; Zhang, Z.; Jiang, X.; Sundararajan, R.; Ma, X.; Li, X. AC failure voltage of iced and contaminated composite insulators in different natural environments. *Int. J. Electr. Power Energy Syst.* **2020**, *120*, 105993. [CrossRef]
7. Georgiev, G.; Ivanova, M.; Dimitrova, R.; Rangelov, Y. In Polymeric Composite Insulators for Overhead Power Lines. A Review of In-Service Damages and Diagnostic Approaches. In Proceedings of the 2021 17th Conference on Electrical Machines, Drives and Power Systems (ELMA), Sofia, Bulgaria, 1–4 July 2021; pp. 1–6.
8. Kokkinaki, O.; Klini, A.; Polychronaki, M.; Mavrikakis, N.C.; Siderakis, K.G.; Koudoumas, E.; Pylarinos, D.; Thalassinakis, E.; Kalpouzos, K.; Anglos, D. Assessing the type and quality of high voltage composite outdoor insulators by remote laser-induced breakdown spectroscopy analysis: A feasibility study. *Spectrochim. Acta Part B* **2020**, *165*, 105768. [CrossRef]
9. Jiang, H.; Li, B.; Zhao, B.; Wang, Y. Aging characterization of 500-kV field-serviced silicone rubber composite insulators with self-normalized photothermal radiometry. *Infrared Phys. Technol.* **2021**, *116*, 103763. [CrossRef]

10. Chen, L.; Guo, F.; Yang, T.; Hu, T.; Bennett, P.; Yang, Q.; Liu, D. Aging characteristics and self-healing properties of laser-textured superhydrophobic silicone rubber for composite insulators. *Polym. Degrad. Stab.* **2021**, *192*, 109693. [CrossRef]
11. Learning from Service Experience with Composite Line Insulators; INMR: 2020. Available online: http://www.inmr.com/learning-from-service-experience-with-composite-line-insulators-2/ (accessed on 10 November 2022).
12. Jiao, J.; Li, L.; Wu, B.; He, C. Novel capacitive proximity sensors for assessing the aging of composite insulators. *Sens. Actuators A* **2017**, *253*, 75–84. [CrossRef]
13. Homma, T.; Kumada, A.; Fujii, T.; Homma, H.; Oishi, Y. Depth profiling of surface degradation of silicone rubber composite insulators by remote laser-induced breakdown spectroscopy. *Spectrochim. Acta Part B* **2021**, *180*, 106206. [CrossRef]
14. Francis, U.U.; Philip, A.N. Degradation of epoxy reinforced Banana fibers and eggshell particles hybrid composite high-voltage insulators via accelerated UV aging processes. *Chem. Data Collect.* **2022**, *38*, 100842. [CrossRef]
15. Ullah, I.; Akbar, M. Anti-aging characteristics of RTV-SiR aided HV insulator coatings: Impact of DC polarity and fillers. *Mater. Chem. Phys.* **2022**, *278*, 125634. [CrossRef]
16. Taghvaei, M.; Sedighizadeh, M.; NayebPashaee, N.; Fini, A.S. Reliability assessment of RTV and nano-RTV-coated insulators concerning contamination severity. *Electr. Power Syst. Res.* **2021**, *191*, 106892. [CrossRef]
17. Shen, L.; Zou, L.; Ding, M.; Zhao, T.; Zhang, L.; Li, Q. Investigation of physical properties of epoxy-functionalized graphene nanoplatelets composite coatings on DC-GIL insulators by molecular dynamics simulation. *Appl. Surf. Sci.* **2020**, *505*, 144197. [CrossRef]
18. Liang, M.; Wong, K.L. Improving the Long-term Performance of Composite Insulators Use Nanocomposite: A Review. *Energy Procedia* **2017**, *110*, 168–173. [CrossRef]
19. Gutman, I.; Hartings, R.; Matsuoka, R.; Kondo, K. Experience with IEC 1109 1000 h salt fog ageing test for composite insulators. *IEEE Electr. Insul. Mag.* **1997**, *13*, 36–39. [CrossRef]
20. Sun, Y.; Tu, Y.; Wang, C.; Wang, S.; Cheng, Y.; Peng, Q.; Chen, X. In Contamination and AC pollution flashover characteristics of insulators under fog-haze environment. In Proceedings of the 2015 IEEE 11th International Conference on the Properties and Applications of Dielectric Materials (ICPADM), Sydney, Australia, 19–22 July 2015; pp. 596–599.
21. Saadati, H.; Werle, P.; Gockenbach, E.; Borsi, H.; Seifert, J.M. In Flashover performance of polluted composite insulators under AC and hybrid AC/DC field stress. In Proceedings of the 2016 IEEE Electrical Insulation Conference (EIC), Montreal, QC, Canada, 19–22 June 2016; pp. 170–173.
22. Wang, W.; Lu, H.; Li, C.; Deng, Y.; Lan, L.; Wen, X.; Luo, B.; Xiao, W. Study on the relationship between the characteristics of water droplets on the shed of composite insulators and the pollution flashover voltage. *Energy Rep.* **2022**, *8*, 1071–1077. [CrossRef]
23. Qiao, X.; Zhang, Z.; Jiang, X.; Sundararajan, R.; You, J. DC pollution flashover performance of HVDC composite insulator under different non-uniform pollution conditions. *Electr. Power Syst. Res.* **2020**, *185*, 106351. [CrossRef]
24. Mohammadi Savadkoohi, E.; Mirzaie, M.; Seyyedbarzegar, S.; Mohammadi, M.; Khodsuz, M.; Ghorbani Pashakolae, M.; Biazar Ghadikolaei, M. Experimental investigation on composite insulators AC flashover performance with fan-shaped non-uniform pollution under electro-thermal stress. *Int. J. Electr. Power Energy Syst.* **2020**, *121*, 106142. [CrossRef]
25. Dadashizadeh Samakosh, J.; Mirzaie, M. Investigation and analysis of AC flashover voltage of SiR insulators under longitudinal and fan-shaped non-uniform pollutions. *Int. J. Electr. Power Energy Syst.* **2019**, *108*, 382–391. [CrossRef]
26. Rahman, M.S.A.; Kadir, M.Z.A.A.; Ab-Rahman, M.S.; Osman, M.; Nor, S.F.M. In Lightning Impulse Strength of 275 kV and 132 kV Tower with Composite Crossarm. In Proceedings of the 2019 11th Asia-Pacific International Conference on Lightning (APL), Hong Kong, China, 12–14 June 2019; pp. 1–6.
27. Ilhan, S.; Ozdemir, A. In Flashover performance of 380 kV V-strings with composite insulators under lightning and switching impulses. In Proceedings of the 2009 IEEE Bucharest PowerTech, Bucharest, Romania, 28 June–2 July 2009; pp. 1–6.
28. Zhang, H.; Tu, Y.; Lu, Y.; Chen, J.; Chen, C.; Xie, L. In Influence of the electric field on TSC characteristics of 110kV silicone rubber insulator sheds in service. In Proceedings of the 2012 IEEE International Symposium on Electrical Insulation, San Juan, Puerto R, USA, 10–13 June 2012; pp. 469–472.
29. Zhang, G.; Zhao, L.; Zhou, R.; Shen, W.; Liang, X. Review on aging characterization and evaluation of silicon rubber composite insulator. *High Volt. Appar.* **2016**, *52*, 1–15.
30. Qiao, X.; Zhang, Z.; Jiang, X.; Li, X.; He, Y. A New Evaluation Method of Aging Properties for Silicon Rubber Material Based on Microscopic Images. *IEEE Access* **2019**, *7*, 15162–15169. [CrossRef]
31. Li, S.; Liang, X.; Gao, Y.; Liu, Y.; Yin, Y.; Li, Z. A modified 5000 h test procedure for silicone rubber insulator based on contamination and hydrophobicity change simulation. *IEEE Trans. Dielectr. Electr. Insul.* **2017**, *24*, 1818–1828. [CrossRef]
32. Du, B.X.; Li, Y.P.; Liu, Y. Hydrophobicity evaluation of silicone rubber insulator using DC discharge induced acoustic wave. *IEEE Trans. Dielectr. Electr. Insul.* **2013**, *20*, 1836–1844. [CrossRef]
33. Cao, H.; Yan, D.; Han, J.; Ren, H.; Lu, M.; Lv, Z.; Guo, H. Investigation and corroboration of a novel method to estimate the hydrophobicity of composite insulators. *IEEE Trans. Dielectr. Electr. Insul.* **2012**, *19*, 2029–2036.
34. Sundararajan, R.; Mohammed, A.; Chaipanit, N.; Karcher, T.; Liu, Z. In-service aging and degradation of 345 kV EPDM transmission line insulators in a coastal environment. *IEEE Trans. Dielectr. Electr. Insul.* **2004**, *11*, 348–361. [CrossRef]
35. Xu, Z. An inclined plane based dynamic contact angle algorithm and its validity in application of hydrophobicity measurement for insulating materials. *IEEE Trans. Dielectr. Electr. Insul.* **2013**, *20*, 1832–1835. [CrossRef]

36. Miyoshi, M.; Homma, H.; Goshima, H. In Effect of Local Arc Discharge on Hydrophobicity Decrease of Silicone Rubber Surface: -Investigation on Hydrophobicity Change by Measuring Water Drop Velocity and Surface Analysis. In Proceedings of the 2020 International Symposium on Electrical Insulating Materials (ISEIM), Virtual, 13–17 September 2020; pp. 206–209.
37. Du, B.X.; Ma, Z.L.; Cheng, X.X.; Liu, Y. Hydrophobicity evaluation of silicone rubber insulator using PD-induced electromagnetic wave. *IEEE Trans. Dielectr. Electr. Insul.* **2012**, *19*, 1060–1067. [CrossRef]
38. Valle, P.M.d.; Álvarez, R.E.; Catalano, L.J.; Calo, E.; Mayora, H. In Overhead line transmission composite insulators. Evaluation of their technical condition in the laboratory after their removal from service. In Proceedings of the 2021 IEEE Electrical Insulation Conference (EIC), Virtual, 7–28 June 2021; pp. 238–241.
39. Wang, X.; Hong, X.; Chen, P.; Zhao, C.; Jia, Z.; Wang, L.; Zou, L. Surface Hardness Analysis of Aged Composite Insulators via Laser-Induced Plasma Spectra Characterization. *IEEE Trans. Plasma Sci.* **2019**, *47*, 387–394. [CrossRef]
40. Jing, P.; Zhiqiang, L.; Peng, W.; Qian, J.; Chunjie, W.; Yuxin, L. In Research on Ultraviolet Aging Properties of Modified Silicone Rubber for Composite Insulator. In Proceedings of the 2020 International Symposium on Electrical Insulating Materials (ISEIM), Virtual, 13–17 September 2020; pp. 202–205.
41. Shen, H.; Deng, Y.; Zhou, J.; Wang, J.; Lei, L.; Xu, J.; Wang, X. In The Application of SEM in the Decay-Like Aging of Silicon Rubber Composite Insulator. In Proceedings of the 2020 IEEE International Conference on High Voltage Engineering and Application (ICHVE), Beijing, China, 6–10 September 2020; pp. 1–4.
42. Subba Reddy, B.; Ramamurthy, P.C. Analysis of in-service composite insulators used in overhead railway traction. *Eng. Fail. Anal.* **2020**, *108*, 104227. [CrossRef]
43. Ghosh, D.; Khastgir, D. Degradation and stability of polymeric high-voltage insulators and prediction of their service life through environmental and accelerated aging processes. *ACS Omega* **2018**, *3*, 11317–11330. [CrossRef]
44. Rowland, S.M.; Robertson, J.; Xiong, Y.; Day, R.J. Electrical and material characterization of field-aged 400 kV silicone rubber composite insulators. *IEEE Trans. Dielectr. Electr. Insul.* **2010**, *17*, 375–383. [CrossRef]
45. Mavrikakis, N.C.; Mikropoulos, P.N.; Siderakis, K. Evaluation of field-ageing effects on insulating materials of composite suspension insulators. *IEEE Trans. Dielectr. Electr. Insul.* **2017**, *24*, 490–498. [CrossRef]
46. Gao, Y.; Liang, X.; Bao, W.; Li, S.; Wu, C. Failure analysis of a field brittle fracture composite insulator: Characterization by FTIR analysis and fractography. *IEEE Trans. Dielectr. Electr. Insul.* **2018**, *25*, 919–927. [CrossRef]
47. Xiong, Y.; Rowland, S.M.; Robertson, J.; Hoffmann, S. In Characterization of field-aged 400 kV silicone rubber composite insulators. In Proceedings of the 2006 IEEE Conference on Electrical Insulation and Dielectric Phenomena, Kansas City, MI, USA, 15–18 October 2006; pp. 417–420.
48. Ehsani, M.; Borsi, H.; Gockenbach, E.; Morshedian, J.; Bakhshandeh, G.R.; Shayegani, A.A. In Effect of aging on dielectric behavior of outdoor polymeric insulators. In Proceedings of the 2004 IEEE International Conference on Solid Dielectrics, ICSD 2004, Virtual, 22–26 March 2004; Volume 1, pp. 312–315.
49. Li, Y.; Chen, C.; Jia, Z.; Ye, W. In Thermal stability and organic component analysis of HTV silicone rubber composite insulator. In Proceedings of the 2015 IEEE Conference on Electrical Insulation and Dielectric Phenomena (CEIDP), Arbor, MI, USA, 18–20 October 2015; pp. 721–724.
50. Khan, Y.; Al-Arainy, A.A.; Malik, N.H.; Qureshi, M.I.; Al-Ammar, A.E. In Loss and Recovery of Hydrophobicity of EPDM Insulators in Simulated Arid Desert Environment. In Proceedings of the 2010 Asia-Pacific Power and Energy Engineering Conference, Chengdu, China, 28–31 March 2010; pp. 1–4.
51. Hernández, R.; Vallejo, G.; Montoya, G.; Ramirez, I. In Performance of hydrophobicity of different polymeric insulators in an accelerated ageing test. In Proceedings of the 2012 Annual Report Conference on Electrical Insulation and Dielectric Phenomena, Montreal, QC, Canada, 14–17 October; 2012; pp. 753–756.
52. Peng, X.; Lin, H.; Wang, R.; Wang, Z.; Fang, P. Review on Aging Mechanism and Service Life Prediction of Silicone Rubber Composite Insulator. *High Volt. Appar.* **2018**, *54*, 1–8.
53. Verma, A.R.; Reddy, B.S. Tracking and erosion resistance of LSR and HTV silicon rubber samples under acid rain conditions. *IEEE Trans. Dielectr. Electr. Insul.* **2018**, *25*, 46–52. [CrossRef]
54. Verma, A.R.; Reddy, B.S. In Tracking and erosion resistance of Liquid Silicone Rubber under normal and acidic environment. In Proceedings of the 2017 4th IEEE Uttar Pradesh Section International Conference on Electrical, Computer and Electronics (UPCON), Mathura, India, 26–28 October 2017; pp. 223–227.
55. Sun, H.; Delun, M.; Boveri-Yibo, Z.; Wu, F.; Chen, J.; Hillborg, H. In Performance of liquid silicone rubber exposed to acid fog under DC stress. In Proceedings of the 2015 IEEE Conference on Electrical Insulation and Dielectric Phenomena (CEIDP), Sydney, Australia, 22 July 2015; pp. 229–232.
56. Ma, X.; Zhang, Z.; Ma, X.; Huang, H.; Jiang, X. In Study on Corrosion Failure Characteristics of Silicone Rubber in Acidic Environment. In Proceedings of the 2021 International Conference on Electrical Materials and Power Equipment (ICEMPE), Chongqing, China, 11–15 April 2021; pp. 1–4.
57. Lehretz, F.; Kreßler, J.; Nase, M.; Rennert, M.; Schulte-Fischedick, J.; Seifert, J.M. Enhancement of Acid Stability of Silicone Elastomers by Using Inert Fillers. *IEEE Trans. Dielectr. Electr. Insul.* **2020**, *27*, 2188–2194. [CrossRef]
58. Sarathi, R.; Mishra, P.; Gautam, R.; Vinu, R. Understanding the influence of water droplet initiated discharges on damage caused to corona-aged silicone rubber. *IEEE Trans. Dielectr. Electr. Insul.* **2017**, *24*, 2421–2431. [CrossRef]

59. Rajini, V.; Udayakumar, K. Degradation of Silicone Rubber under AC or DC Voltages in Radiation Environment. *IEEE Trans. Dielectr. Electr. Insul.* **2009**, *16*, 834–841. [CrossRef]
60. Gustavsson, T.G.; Gubanski, S.M.; Hillborg, H.; Karlsson, S.; Gedde, U.W. Aging of silicone rubber under ac or dc voltages in a coastal environment. *IEEE Trans. Dielectr. Electr. Insul.* **2001**, *8*, 1029–1039. [CrossRef]
61. Fernando, M.A.R.M.; Gubanski, S.M. Ageing of silicone rubber insulators in coastal and inland tropical environment. *IEEE Trans. Dielectr. Electr. Insul.* **2010**, *17*, 326–333. [CrossRef]
62. Gubanski, S.M.; Karlsson, S.; Fernando, M.A.R.M. In Performance of Biologically Contaminated High Voltage Insulators. In Proceedings of the First International Conference on Industrial and Information Systems, Tirtayasa, Indonesia, 8–11 August 2006; pp. 30–35.
63. Ouyang, X.; Jia, Z.; Yang, S.; Shang, X.; Wang, X.; Chen, H.; Zhou, D.; Liu, R. Influence of algae growth on the external insulation performance of HVDC insulators. *IEEE Trans. Dielectr. Electr. Insul.* **2018**, *25*, 263–271. [CrossRef]
64. Zhang, Z.; Liang, T.; Li, C.; Jiang, X.; Wu, J.; Wu, B. Electrical Strength and Physicochemical Performances of HTV Silicone Rubber under Salt-Fog Environment with DC Energized. *Polymers* **2020**, *12*, 324. [CrossRef] [PubMed]
65. Zhijin, Z.; Tian, L.; Xingliang, J.; Chen, L.; Shenghuan, Y.; Yi, Z. Characterization of Silicone Rubber Degradation Under Salt-Fog Environment With AC Test Voltage. *IEEE Access* **2019**, *7*, 66714–66724. [CrossRef]
66. Zhu, Y.; Zhang, X.; Fang, J. In Influence of environmental factor on hydrophobicity transfer of silicone rubber used for outdoor insulation. In Proceedings of the 2017 International Symposium on Electrical Insulating Materials (ISEIM), Toyohashi, Japan, 11–15 September 2017; pp. 32–35.
67. Du, B.X.; Xu, H.; Liu, Y. Effects of wind condition on hydrophobicity behavior of silicone rubber in corona discharge environment. *IEEE Trans. Dielectr. Electr. Insul.* **2016**, *23*, 385–393. [CrossRef]
68. Ullah, I.; Akbar, M.; Khan, H.A. Degradation analysis of RTV-SiR based composites under both polarities DC voltage for insulators coating. *Mater. Today Commun.* **2021**, *29*, 102890. [CrossRef]
69. Sundararajan, R.; Nowlin, R.W. In Effect of altitude on the flashover voltage of contaminated insulators. In Proceedings of the Conference on Electrical Insulation and Dielectric Phenomena-CEIDP '96, Denver, CO, USA, 23 October 1996; Volume 2, pp. 433–436.
70. Guo, Y.; Jiang, X.; Sree, V.G.; Sundararajan, R. In AC flashover performance of porcelain, glass and polymeric insulators at high altitudes. In Proceedings of the 2015 IEEE 11th International Conference on the Properties and Applications of Dielectric Materials (ICPADM), Sydney, Australia, 19–22 July 2015; pp. 484–487.
71. Liu, S.; Liu, S.; Wang, Q.; Zuo, Z.; Wei, L.; Chen, Z.; Liang, X. Improving surface performance of silicone rubber for composite insulators by multifunctional Nano-coating. *Chem. Eng. J.* **2023**, *451*, 138679. [CrossRef]
72. Faiza; Khattak, A.; Rehman, A.U.; Ali, A.; Mahmood, A.; Imran, K.; Ulasyar, A.; Sheh Zad, H.; Ullah, N.; Khan, A. Multi-Stressed Nano and Micro-Silica/Silicone Rubber Composites with Improved Dielectric and High-Voltage Insulation Properties. *Polymers* **2021**, *13*, 1400. [CrossRef]
73. Lau, K.Y.; Piah, M. Polymer nanocomposites in high voltage electrical insulation perspective: A review. *Malays. Polym. J.* **2011**, *6*, 58–69.
74. Li, S.; Li, J.; Zhang, R. In Hydrophobicity Improvement of Polluted Silicone Rubber by Plasma Jet in High Humidity Environment. In Proceedings of the 2021 International Conference on Electrical Materials and Power Equipment (ICEMPE), Chongqing, China, 11–15 April 2021; pp. 1–4.
75. Ning, K.; Lu, J.; Xie, P.; Hu, J.; Huang, J.; Sheng, K. Study on surface modification of silicone rubber for composite insulator by electron beam irradiation. *Nucl. Instrum. Methods Phys. Res. Sect. B* **2021**, *499*, 7–16. [CrossRef]
76. Wang, B.; Zhu, D.; Li, C.; Chen, J. Performance of Full Compositional W/Cu Functionally Gradient Materials Under Quasi-Steady-State Heat Loads. *IEEE Trans. Plasma Sci.* **2018**, *46*, 1551–1555. [CrossRef]
77. Jie, X.; Xinhua, Z.; Zhongyan, M. Effect of the interdiffusion reaction on the compatibility in PZT/PNN functionally gradient piezoelectric materials. *IEEE Trans. Compon. Packag. Technol.* **1999**, *22*, 11–16. [CrossRef]
78. Cao, X.; Shi, J.; Jin, F. Effect of gradient dielectric coefficient in a functionally graded material (FGM) substrate on the propagation behavior of love waves in an FGM-piezoelectric layered structure. *IEEE Trans. Ultrason. Ferroelectr. Freq. Control* **2012**, *59*, 1253–1257. [CrossRef]

energies

MDPI

Article

Solid–Gas Sorption System for Ammonia Storage and Delivery Driven by Engine Waste Heat for NOx Reduction of Diesel Engine

Chenjue Wang [1], Shi Chen [1], Long Jiang [2], Yaodong Wang [1] and Huashan Bao [1,*]

[1] Department of Engineering, Durham University, Durham DH1 3LE, UK
[2] Institute of Refrigeration and Cryogenics, Zhejiang University, Hangzhou 310027, China
* Correspondence: huashan.bao@durham.ac.uk

Abstract: A new design of a sorption-selective catalytic reduction (SCR) system is proposed to improve ammonia storage density and meet the ammonia demand for high NOx conversion efficiency at a relatively lower temperature (<100 °C) compared to urea-SCR systems. The major components are a main unit and a start-up unit that each contain a metal halide ammine as the sorbent. The start-up unit can operate without any external heat source, but spontaneously releases ammonia at the ambient temperature and is only used when the main unit is being warmed up for action. The selection criteria for the metal halide ammine for each unit is discussed. The working pair of $SrCl_2$ as the main ammine and NH_4Cl as the start-up ammine is further analyzed as an example to be used in the sorption-SCR system for a diesel engine, the NOx emissions of which were experimentally measured in different operation modes. Based on the experimental data of engine emissions and kinetic models of the chemisorption between ammines and ammonia, the dynamic performance of the sorption system with a total capacity of 180 L sorbent composite in different layouts was investigated and compared. It was found that the achievable desorption conversion degree was lower in smaller reactors and was more sensitive to operating conditions in smaller reactors compared to larger reactors. This suggests that a system using a small reactor layout requires some extra volume to completely meet the required capacity compared to a larger reactor layout. However, because systems with large reactors tend to respond slowly, as they have more thermal mass and take a longer time for preparation, there is a design trade-off required to have optimal performance and balance between the main unit and the start-up unit. In the case studied in this work, a system using three rechargeable reactors with a volume of 60 L each was found to be the preferable layout; it could have about a 90% desorption conversion degree and required around 10 min of warm-up time. Meanwhile, the coupled start-up unit should have a capacity of around 165 mL at least.

Keywords: sorption; NOx emissions reduction; SCR; ammonia storage; dynamic performance; metal halide ammine

Citation: Wang, C.; Chen, S.; Jiang, L.; Wang, Y.; Bao, H. Solid–Gas Sorption System for Ammonia Storage and Delivery Driven by Engine Waste Heat for NOx Reduction of Diesel Engine. *Energies* **2022**, *15*, 8376. https://doi.org/10.3390/en15228376

Academic Editors: Alessandro Cannavale and Ubaldo Ayr

Received: 9 October 2022
Accepted: 6 November 2022
Published: 9 November 2022

1. Introduction

Increasingly stringent regulations on the reduction of nitrogen oxide (NOx) emissions have been driving technology development in both combustion and exhaust after-treatment; in particular, the removal of NOx from diesel engine exhaust without compromising fuel economy has received tremendous attention [1–3]. Currently, either a lean NOx trap (LNT) or selective catalytic reduction (SCR) technology is used to control NOx emissions from the engine exhaust, depending on engine size and other factors. For instance, SCR is favored and adopted worldwide for heavy-duty diesel engines as the more cost-effective solution, with higher achievable NOx conversion compared to LNT. More than 99% of new heavy-duty vehicles are diesel in the European Union; therefore, there is an appreciable goal to meet tightening Euro emissions regulations by improving SCR technology.

The SCR technology converts NOx to nitrogen over a catalyst using ammonia as the reductant. Liquid absorption is one of the most common methods for ammonia separation and storage [4]. As a low-cost and reversible absorbent, urea has been used in commercial de-NOx products, such as AdBlue, a eutectic 32.5 wt% urea-water that can be injected into the exhaust gas flow for NOx conversion [5]. In addition, ionic liquids (ILs) and deep eutectic solvents (DESs) are researched as alternative green solvents for NH_3 absorption, as they have adjustable structures, higher ammonia storage capacity, and lower volatility than traditional water-based urea solutions [4,6,7]. However, rather than ammonia storage and delivery for SCR reaction, more research focuses on using ILs and DESs to absorb and convert NOx directly [8,9]. There is a dilemma in designing ILs and DESs for ammonia capture. The solubility of ammonia improves with the increasing acidity of solvents, while the increase of acidity will increase the difficulty of the regeneration and storage of solvents [10]. This dilemma limits the performance of ILs and DESs technologies in vehicle applications that require a safe and stable cycle system. A number of drawbacks related to the use of liquid absorption of ammonia, especially urea, encourage the development of alternative solutions for ammonia storage and delivery and are mentioned below [2,3,11,12]:

- ➢ The safety issue of liquid-based systems onboard a vehicle;
- ➢ The incomplete conversion of urea solution to ammonia at 200 °C forms deposits and even deactivates the SCR catalyst, leading to a sharp reduction of NOx conversion efficiency;
- ➢ Urea solution is corrosive, and the freezing point of −11 °C is too high to avoid freezing in many areas of the world reliably;
- ➢ An injector should be carefully designed to disperse fine droplets for evaporation into the exhaust. However, the evaporation of water decreases the exhaust gas temperature, which greatly impacts the activity, especially in the low-temperature region. Additionally, it was found that particles were formed in this injection, which could clog the nozzle [13];
- ➢ Mixers and careful design are needed to obtain uniform ammonia concentration at the SCR catalyst face;
- ➢ The ammonia storage potential of AdBlue only amounts to 0.201 kg/L (0.184 kg/kg) NH_3/solution, one-third of that of liquefied ammonia.

Some solid compounds have been discussed as alternative means of ammonia storage, with evident improvement in terms of storage density, freezing, and storage stability. These include (1) ammonium salts, for example, ammonium carbamate and ammonium carbonate, which have an ammonia storage density of 0.698 kg/L (0.436 kg/kg) and 0.531 kg/L (0.354 kg NH_3/kg), respectively [3,14–16]; and (2) ammine compounds of metal salts, such as $MgCl_2{\cdot}6NH_3$, $CaCl_2{\cdot}8NH_3$, $SrCl_2{\cdot}8NH_3$. etc., which typically can store at least 2.5 times the ammonia of AdBlue on a volume basis if the compound density is around 1.1~1.3 kg/m^3 [3,11,17–21]. Many of these solid compounds can thermally decompose and effectively release ammonia at a relatively low temperature (potentially down to 50~60 °C, considering the atmosphere as the back pressure), and direct gaseous ammonia dosing from these compounds also significantly simplifies the operation of the ammonia delivery system. Systems based on both technologies have been tested for commercial applications, and the use of ammonium carbamate was protected by solid SCR [15,16]. However, unlike the metal ammine compound that releases pure ammonia, the desorption of ammonium salts also yields CO_2 and/or H_2O. Since these are reversible reactions, there is a risk that these resultants recombine to generate solid ammonium carbamate on cold surfaces, which could clog and inhibit dosing systems. Hence, this work focuses only on ammine compounds to explore the feasibility of a solid–gas sorption-SCR system.

A solid sorption cycle based on the reversible chemisorption reaction between metal ammine compounds and ammonia has been researched and developed for decades. With interest mainly in thermal applications such as heat pumps and thermal energy storage. Apart from the aforementioned superiority of ammonia storage compared to AdBlue, it is also advantageous for its higher vapor pressure at a low temperature, so energy consumption for heating the materials is not prohibitive. On the other hand, it has a much lower vapor pressure at ambient temperature compared to liquid ammonia, which is desirable for the safe handling of materials. Ammonia-based chemisorption has the feature of monovariant behavior. This indicates that the kinetics can be controlled on demand by adjusting one thermodynamic parameter, either temperature or pressure. Therefore, the release of ammonia can be tuned to match the NOx emissions by controlling the desorption temperature and/or the backpressure.

Elmøe et al. [12] tested a small reactor containing about 260 g of $MgCl_2{\cdot}6NH_3$ with a volume of 785 mL and demonstrated the desorption of $MgCl_2{\cdot}6NH_3$, which required heating to at least 184.4 °C to initiate the first phase of desorption when the set-point buffer pressure was maintained at 5 bar. A simple ON–OFF controller with the buffer pressure as the control variable was employed to generate an oscillating varying profile of ammonia out-flow around the set-point pressure. By applying such a controlling strategy under the studied operating conditions, it was found that approximately 99% of the ammonia stored in the salt could be delivered for NOx conversion. Johannessen et al. [11,19,20] designed and studied an ammonia storage and delivery system (ASDS/AdAmmine) with two main cartridges containing the sorbent of chloride ammine compound and a start-up unit, and feasibility and great competitiveness over the urea-SCR system were demonstrated for urban driving conditions. AdAmmine, with a dosing temperature of 100 °C, reduced tailpipe NOx emissions by half compared to the urea-SCR system dosing at 180 °C. An electric heater was integrated within the AdAmmine as the only heat source, and the total usage of electric power could be tuned according to the system size and vehicle power. The start-up unit initially consumed 550 W in high-duty engine applications and 250 W in light-duty engine applications; the main units received 150 W to warm up; afterwards, more power was required [19] for the main units. Johannessen et al. also suggested that a high uniformity (>0.98) of ammonia concentration in a shorter mixing length could be achieved by using an ammonia inlet tube with multiple entry points, which releases ammonia at several positions in the exhaust cross-section to give adequate mixing. A low-pressure drop obstacle can be used to provide efficient turbulent mixing of ammonia vapor with the exhaust gas. It is interesting to note that the early generation of AdAmmine developed by Johannessen et al. was a pure metal compound powder that is mechanically compressed into dense bulk (around 1250 kg/m^3), with an ammonia storage capacity of around 600 g/L [11]; in their later works, a new version of AdAmmine was produced from a veiled special formulation with a usable capacity of 450 g/L [21]. It is well-known that the repeatability of the cyclic performance of dense pure-chemical bulk is questionable due to the expansion and agglomeration of chemical materials within a confined space [22,23].

Jiang et al. [24] compared the theoretical ammonia storage capacity of different metal salts mixed with expanded graphite treated with sulfuric acid (ENG-TSA). The ENG-TSA was used as a matrix to enhance heat transfer properties and to avoid the potential performance degradation caused by swelling and agglomeration associated with pure chemicals [25,26]. It was found that the NH_4Cl-ENG-TSA composite slightly excelled over other salts studied under ideal equilibrium conditions, as it only required one-third of the mass and 80% of volume required by AdBlue to satisfy the same NOx reduction, indicating an ammonia storage capacity of around 250 g/L. The addition of a porous host matrix compromises the volumetric sorption capacity of the bulk sorbents. On the other hand, although NH_4Cl has a higher ammonia storage capacity, its high working pressure will require a thicker reactor wall, resulting in a heavier reaction unit [27]. Thus, Wang et al. [27] concluded that $CaCl_2$ and $SrCl_2$ are better ammoniate choices after comprehensive consideration of ammonia sorption capacity and working pressure.

Zhang et al. [28] proposed four types of the sorption-SCR system based on multi-functional multi-stage ammonia production cycles. The first type used a single ASDS unit to provide ammonia. The second type added an ammonia transfer (AT) unit filled with low-temperature ammoniate to the ASDS unit for a lower starting temperature. Then, an ammonia warning (AW) unit containing high-temperature ammoniate was integrated with the second type to provide NH_3 for NOx conversion when the ammonia output from the main ASDS unit was insufficient. The fourth type combined an ASDS unit and an AW unit into one reactor for a more compact system. The test results of using NH_4Cl, NaBr, $CaCl_2$, $SrCl_2$, and $MnCl_2$ as candidate ammoniates show that adding the AT unit is valuable, as it can ensure the starting temperature at 29–38.2 °C. Although adding the AW unit made the lowest ammonia storage density in the fourth-type system (570 g/L) higher than the lowest ammonia storage density in the second-type system (420 g/L), the highest ammonia storage density was in the second-type system. When using NH_4Cl-ENG-TSA in the AT unit, and $CaCl_2$ -ENG-TSA in the ASDS unit, the ammonia storage density of the second-type sorption-SCR system can reach 710 g/L, and its minimum required mass of sorbent accounts is 28% the needs of AdBlue to obtain the same amount of ammonia.

There is another influential factor that also should be taken into account, but has been barely discussed in previous research for determining the real size of sorption-based SCR system: the reaction kinetics of the ammine compound. Based on intrinsic kinetics, it is essential to strategically control the thermodynamic conditions, heating conditions, and/or working pressure for the ammonia desorption to meet the demanding mass flow rate of ammonia. However, there is very limited literature on this topic, nor detailed in-depth investigation of dynamic performance.

The present paper proposes a new solid–gas sorption-SCR system design by integrating a secondary reactor as a start-up unit with the main sorption reactors. This small start-up unit should operate spontaneously under cold start conditions to completely eliminate the need for an electric heater, while the main reactor uses the existing waste heat onboard the vehicle, either engine coolant or engine exhaust. Moreover, to obtain more insight into optimal reactor design, this paper investigates the interrelation between reactor sizing and dynamic dosing performance by examining experimental data of NOx emissions from a heavy-duty diesel engine and the chemisorption kinetics of the ammonia–metal ammine compound reaction.

2. Working Principle

The proposed sorption-based SCR system schematic is illustrated in Figure 1, which outlines the integrated design mainly comprised of main reactor(s), one start-up unit, dosing valves, temperature and pressure sensors, and a dosing control unit (DCU). The main reactor(s) is filled with a ammonia sorbent and constructed with sufficient heat exchange space for engine coolant or engine exhaust to flow through for efficient heat transfer. The start-up unit with a comparatively smaller volume contains a secondary ammonia sorbent, which has a lower equilibrium temperature than the main sorbent at the same working pressure. Namely, the desorption of this secondary sorbent occurs at a lower temperature so that it can extract heat from ambient surroundings to carry out endothermic desorption of ammonia. Ideally, the equilibrium pressure corresponding to the ambient temperature should be higher than the engine exhaust pressure (≈1 bar) to have the proper equilibrium drop as the driving force. The start-up unit should be thermally isolated from the main reactor to avoid heat exchange.

The *P–T* diagram shown in Figure 2 exemplifies the state points of synergistic operation between the main and secondary sorbents. The start-up unit has a pressure higher than the exhaust pressure when at ambient temperature (Point 1 in Figure 2); consequently, when Valve 1 in Figure 1 opens as the engine starts, the desorption in the start-up unit is triggered by the pressure difference and the ammonia is released into the exhaust line. In the meantime, the inside pressure of the main reactor(s) climbs from Point 2 as it is heated by engine coolant, for example. Once the pressure in the main reactor reaches Point 3, where the equilibrium pressure is higher than both the start-up pressure and the exhaust pressure, Valve 2 opens to release the ammonia from the main reactor for NOx conversion. At the same time, the start-up reactor is also receiving ammonia from the main reactor to refill it for the next cold-start operation. A certain amount of equilibrium pressure difference ($P_{ms} > P_{ss} > P_{exh}$), as shown in Figure 2, is required to activate the desorption and drive the process at the desired reaction rate.

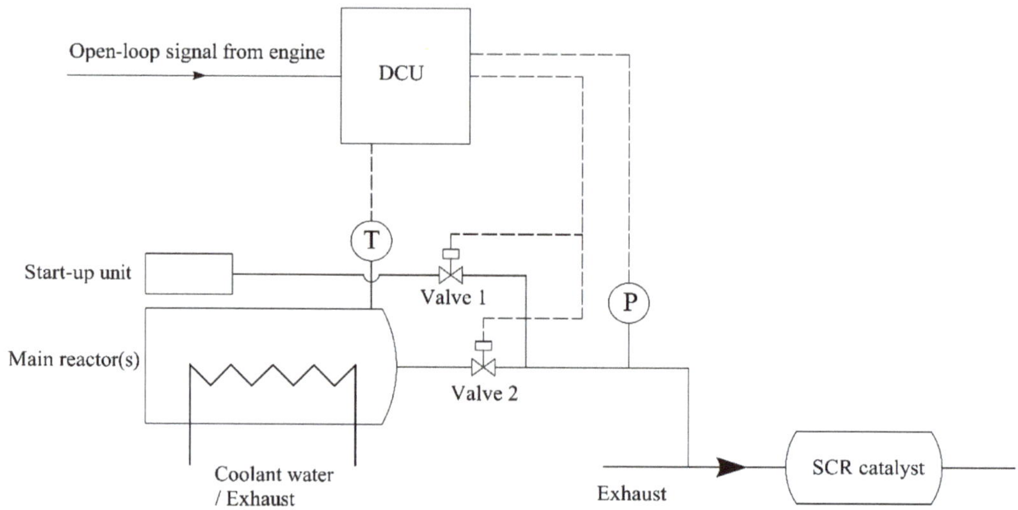

Figure 1. Schematics of sorption-SRC system with a start-up unit and main reactors.

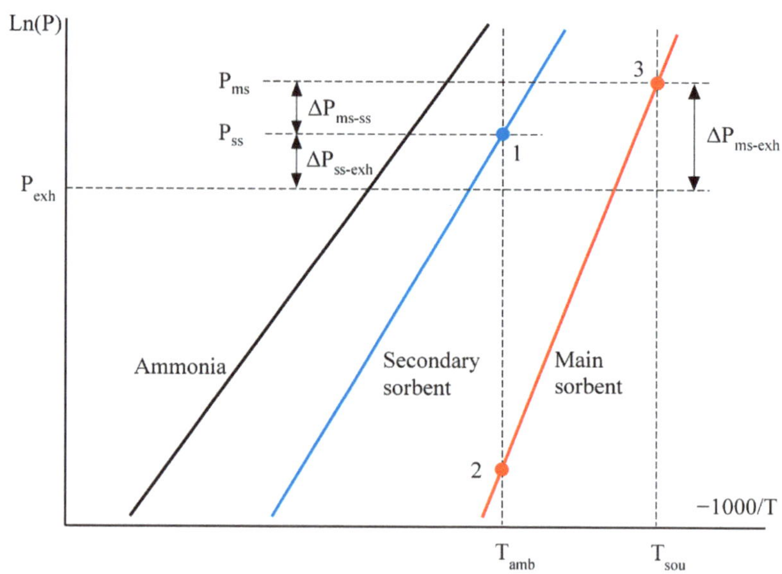

Figure 2. Clausius–Clapeyron diagram of the sorption-SCR system.

Temperature and pressure sensors are used in a feedback control algorithm conducted by the DCU to control the backpressure of desorption to ensure sufficient ammonia dosing into the exhaust line at the desired mass flow rates. The DCU communicates with the engine control unit to acquire information on ammonia demand and combines the temperature/pressure feedback to work out the control algorithm and tune the dosing valve for ammonia release.

3. Materials

Composites that consist of metal halide ammines (MHA) and porous matrix have been developed and studied for sorption systems. They benefit system performance with improved thermal conductivity, a supporting porous structure for sufficient mass transfer, and by preventing the storage capacity from degrading over multiple cycles due to swelling and agglomeration of chemicals [25]. In this work, expanded natural graphite treated with sulfuric acid (ENG-TSA) can be used as a supporting matrix. The thermal conductivity of the ENG-TSA composites was reported to be 44 times higher than that of ENG-based consolidate and 400 times higher compared to ordinary granular pure salt, while the permeability is potentially improved by more than 100 times if compared to pure granular salt after adsorption [29].

There is a great number of MHAs that can react with ammonia in a wide temperature range (50~350 °C) [30]. In a complete cycle, they undertake endothermic desorption with ammonia release in one-half cycle and experience exothermic adsorption in the other half cycle when returning to the initial state. The thermodynamic equilibrium of the most commonly studied MHAs with a desorption temperature below 150 °C at atmospheric pressure is plotted in the *P–T* Clausius–Clapeyron diagram in Figure 3. The properties of each MHA composite in terms of specific ammonia concentration, ammonia storage density, volume ratio of MHA composite to the AdBlue with the same ammonia content, and equilibrium temperature/pressure of the MHA-NH_3 reaction at some key points are summarized in Table 1, where the bulk density of the composites is assumed at 800 kg/m^3, with a mass ratio between MHAs and ENG-TSA of 5:1.

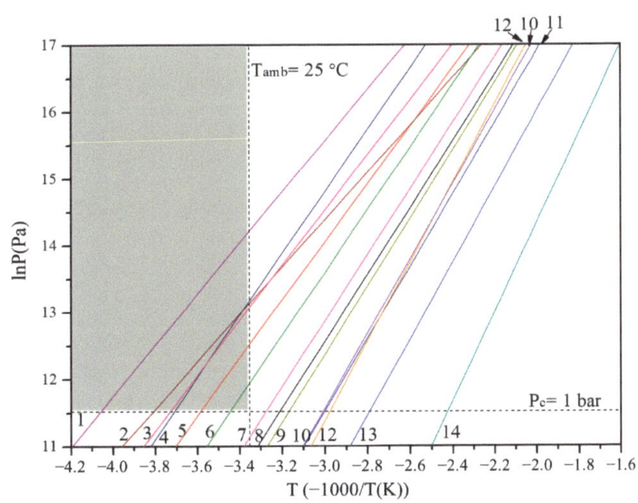

Figure 3. Clausius–Clapeyron diagram of typical metal halide ammines in equilibrium reaction with ammonia.

The first criterion for the candidate MHAs of the start-up unit is that the equilibrium pressure at the ambient temperature should be higher than its backpressure in the exhaust line that is close to atmospheric pressure. Namely, the equilibrium line should be located in the grey-shadowed zone in Figure 3, with the assumed ambient temperature at 25 °C. In this instance, only MHAs No. 1–6 are shortlisted. However, only three of them, i.e., NH_4Cl, NaBr, and $BaCl_2$, can provide appealing improvement in ammonia storage density compared to AdBlue, as they reduce the volume to less than half of the AdBlue, while the $PbCl_2$ and $SnCl_2$ composites are radically ruled out since they have smaller ammonia storage density than AdBlue. Considering the mass transfer issue within the densely compressed composites, the $BaCl_2$ will be screened out because its equilibrium pressure at ambient temperature is lower than 4 bar, which is a recommended threshold of working pressure without the concern of mass transfer limitation [31]. Therefore, NH_4Cl and NaBr are the final candidates for the secondary sorbent in the start-up unit. These two sorbents have the lowest working temperatures at −10.8 °C and −3.7 °C, respectively, at a backpressure of 1 bar, making these two candidates adoptable in cold winter. The sorptions of these two sorbents are formulated as follows:

$$NH_4Cl + 3NH_3 \leftrightarrow NH_4Cl{\cdot}3NH_3 + 3\Delta H_r \quad (1)$$

$$NaBr + 5.25NH_3 \leftrightarrow NaBr{\cdot}5.25NH_3 + 5.25\Delta H_r \quad (2)$$

Table 1. The properties of different reactant salts for the composite sorbents [32].

No.	Salt	Molecular Weight (g/mol)	Initial Mole of NH_3	Final Mole of NH_3	kg NH_3 Per kg Composite	NH_3 g Per Liter Composite	Volume Factor (Norm to AdBlue)	Desorption Temp. (°C, at 1 Bar)	NH_3 Vapour Pressure (Bar, at 25 °C)
1	$SnCl_2$	189.7	4	9	0.330	264	0.76	−26.9	14.91
2	NH_4Cl	53.5	0	3	0.953	763	0.26	−10.8	5.05
3	$PbCl_2$	278.2	3.25	8	0.242	194	1.04	−5.1	4.74
4	NaBr	102.97	0	5.25	0.867	693	0.29	−3.7	5.24
5	$BaCl_2$	208.23	0	8	0.653	522	0.38	5.7	2.74
6	$SnCl_2$	189.7	2.5	4	0.110	88	2.28	17.1	1.54
7	$CaCl_2$	111.68	4	8	0.378	303	0.66	32.0	0.68
8	$SrCl_2$	158.52	1	8	0.678	542	0.37	38.2	0.49
9	$CaCl_2$	111.68	2	4	0.233	187	1.07	42.3	0.39
10	$ZnCl_2$	136.69	2	6	0.200	160	1.26	59.0	0.15
11	$ZnCl_2$	136.69	4	6	0.166	133	1.51	59.7	0.15
12	LiCl	42.44	1	3	0.572	458	0.44	62.5	0.11
13	$MnCl_2$	125.84	2	6	0.425	340	0.59	85.1	0.04
14	$MgCl_2$	95.3	2	6	0.526	421	0.48	139.4	0.002

Likewise, the potential MHA candidates for the main reactors could be $SrCl_2$ and LiCl. Although $MgCl_2$ has a volume factor of less than 0.5, it requires the demanding condition of high temperature for desorption. The chemical formulas of these two main MHA candidates are expressed in Equations (3) and (4). The salts are specified in Table 1.

$$SrCl_2{\cdot}NH_3 + 7NH_3 \leftrightarrow SrCl_2{\cdot}8NH_3 + 7\Delta H_r \quad (3)$$

$$LiCl{\cdot}NH_3 + 2NH_3 \leftrightarrow LiCl{\cdot}3NH_3 + 2\Delta H_r \quad (4)$$

4. Analytical Methods

Limited information about the kinetics of solid–gas sorption between MHAs and ammonia could be found. The authors only found published data for $SrCl_2$, but unfortunately, none for LiCl. This work will take the analysis and discussion of the potential performance of a solid–gas sorption-SCR system using $SrCl_2$ as a case study.

The dynamic evaluation of ammonia desorption from sorbent highly depends on the kinetic models used. However, the kinetic models determined by different methods deviate from each other mainly due to the different heat and mass transfer conditions of the used experimental apparatus. To obtain more insights for reactor and system design, the current study used two different kinetic models to analyze the proposed sorption-SCR system. One represents the ideal performance of the sorbent and reactor with extremely good heat and mass transfer, while the other one represents a relatively practical scenario with limited heat and mass transfer.

The kinetic model developed by Veselovskaya and Tokarev is based on the large temperature jump (LTJ) method [33]. A monolayer of sorbent weighing 0.833 g is placed on flat metal support in the LTJ test unit so that the ammonia adsorption and desorption can process with negligible influence of heat and mass transfer. The LTJ method emulates real operation conditions where the adsorption/desorption is driven by a temperature jump at isobaric conditions. The time-dependent variation of conversion degree, $x(t)$, is expressed by the following equation:

$$x(t) = \exp(-t{\cdot}k) \quad (5)$$

where x is the conversion degree from 0 to 1; the rate constant k is believed to have a linear relationship with the temperature equilibrium drop, ΔT_{eq}, as expressed in Equation (6).

$$k = c\Delta T_{eq} = \frac{\alpha \cdot S}{\Delta H_r \cdot \Delta N} \cdot \Delta T_{eq} \tag{6}$$

where S denotes the surface area of the heat exchanger, equal to the surface of the metal plate (1.96×10^{-3} m^2 in [24]), α is the heat transfer coefficient between the grains and the metal plate, which was calculated as 90 ± 10 W/(m^{2}K) in [33], ΔH_r is the enthalpy of the chemical reaction, and ΔN is the maximal amount of ammonia exchanged; hence, the multiplication of $\Delta H_r \cdot \Delta N$ in Equation (6) represents the total enthalpy change during the adsorption/desorption of the tested composite. The value of c can be calculated to be 0.000333 K^{-1} s^{-1} based on the parameters given in [33]. ΔT_{eq} is the main driving force for the chemical reaction in the isobaric sorption process, and can be calculated by Equation (7):

$$\Delta T_{eq} = T_{eq}(P_c) - T_c \tag{7}$$

where T_{eq} is the equilibrium temperature of the sorbent corresponding to the working pressure P_c; T_c is the constrain temperature, which is equal to the heat-exchanging fluid temperature and approximates the sorbent temperature in the test. Although this model was developed based on $BaCl_2$ test data, considering the similarity of MHA reaction activity and heat and mass transfer performance on the grain scale, it is reasonable to apply it to other MHAs under similar conditions. The value of c for $SrCl_2$ in the current work was calculated to be 0.000129 K^{-1} s^{-1} using the enthalpy change of $SrCl_2$-ammonia sorption.

Another kinetic model of $SrCl_2$-NH_3 sorption expressed in Equation (8) was reported by Huang et al. [34] using a test rig that consisted of a cylindrical reactor and an evaporator/condenser. The cylindrical sorbent had a diameter of 150 mm and length of 100 mm, with an 8 mm hole in the center for gas diffusion; the volume of the sorbent was around 1.8 L, and the mass amount was in the magnitude of hundreds of grams.

$$\frac{dx}{dt} = s_d \exp\left(\frac{-E_d}{RT}\right)(1-x)^{m_d} \frac{P_c - P_{eq}(T)}{P_{eq}(T)} \tag{8}$$

where T is the sorbent temperature and P_{eq} is the equilibrium pressure corresponding to the sorbent temperature. The kinetic parameters s_d, E_d, and m_d were determined as 0.125 s^{-1}, 9000 J/mol and 3.02, respectively, for desorption by fitting to experimental results. This model is intended to be a local model; only the local degree of conversion can be referred to and the use of an experimental global degree of conversion to fit the kinetic model must be avoided. Huang et al. built heat and mass transfer equations together with a kinetic model for small elements to determine those parameters. Hence, theoretically, Huang et al.'s model depresses the influence of heat and mass transfer on kinetics. However, to some extent, the model still reflects the relatively practical performance of the sorbent and reactor with limited heat and mass transfer.

With the calculated desorption rate, the released ammonia flow rate $\dot{m}_{NH3}$ can be calculated by the following equation:

$$\dot{m}_{NH3} = \frac{dx}{dt} \frac{\rho V f}{M_s} \gamma M_{NH3} \tag{9}$$

where ρ is the bulk density of the sorbent, V is the sorbent volume, f is the salt mass fraction in the sorbent, γ is the stoichiometric coefficient of the sorption, and M_s and M_{NH3} are the molar mass of the salt and ammonia, respectively.

5. NOx Emissions of a Diesel Engine

A heavy-duty diesel engine (Cummins ISBe 5) was experimentally tested to obtain NOx emissions data under different operating conditions. The schematic diagram of the diesel engine test bench is shown in Figure 4. The specifications of the tested engine are shown in Table 2. All exhausts were measured by a HORIBA© MEXA 1600D gas analyzer, and solid particle data were collected by a HORIBA© MEXA 1000 SPCS. The diesel engine ran for more than 30 min for warm-up prior to it being considered to have steady-state performance. A more detailed description of this engine test can be found in the previous work [24].

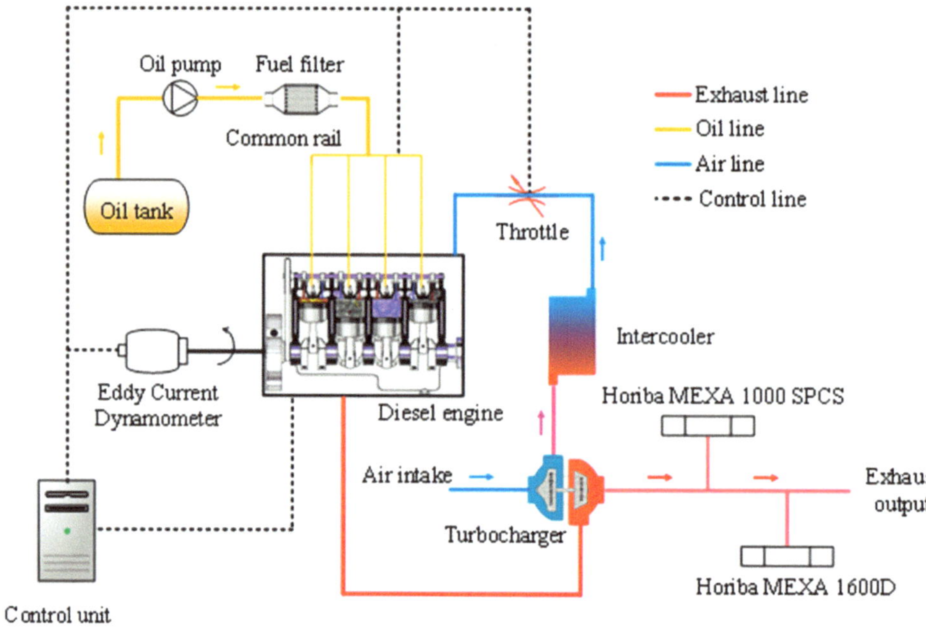

Figure 4. Experimental test rig of NOx emissions of diesel engine [24].

Table 2. Specification of the tested diesel engine [24].

Specification	Values
Engine type	Vertical-inline, turbo-charged, water cooling, 4-stroke
Cylinder	4
Chamber	ω type
Displacement	4.5 L
Bore × Stroke	107 mm × 124 mm
Max. power	152 kW@2300 rpm
Max. torque	760 Nm@1400 rpm
Compression ratio	17.3
Emissions standard	Euro V

According to the European stationary cycle, 12 different operation modes with three different rotation speeds (1680 rpm, 1810 rpm, and 1940 rpm) and four load levels ranging from 25% to 100% were tested. The specification of the operation modes is shown in Table 3. The real-time NOx emissions of 10 modes are presented in Figure 5 (the other two curves overlap and tangle with other curves, so they are not shown for the sake of clear presentation). The highest average NOx emission of 196 mg/s occurred in Mode 8 with a 25% load and the highest exhaust flow rate; the lowest average NOx emission of 112 mg/s happened in Mode 6 with the same load, but the lowest rotation speed. The emission data of operation Mode 6 and 8, respectively, representing two extreme cases, will be considered in the performance evaluation and comparison for the proposed sorption-SCR system. In each operation mode, the instantaneous NOx emission appeared in a slightly wavy profile, but was almost stable with a small amplitude. Therefore, it is sensible to use the average NOx emission for stable engine operation. The average emissions are also presented in Table 3.

Table 3. European stationary cycle test modes and corresponding average NOx emission [24].

Mode	Engine Speed (rpm)	Load (%)	Weight (%)	NOx Emission (mg/s)	NH_3 Flow Rate Required (mg/s)
1	1680	100	8	132.26	72.34
2	1810	50	10	157.55	86.17
3	1810	75	10	152.06	83.17
4	1680	50	5	159.82	87.41
5	1680	75	5	145.02	79.32
6	1680	25	5	111.67	61.08
7	1810	100	9	142.77	78.09
8	1810	25	10	196.08	107.25
9	1940	100	8	135.04	73.86
10	1940	25	5	172.93	94.59
11	1940	75	5	131.10	71.71
12	1940	50	5	182.10	99.60

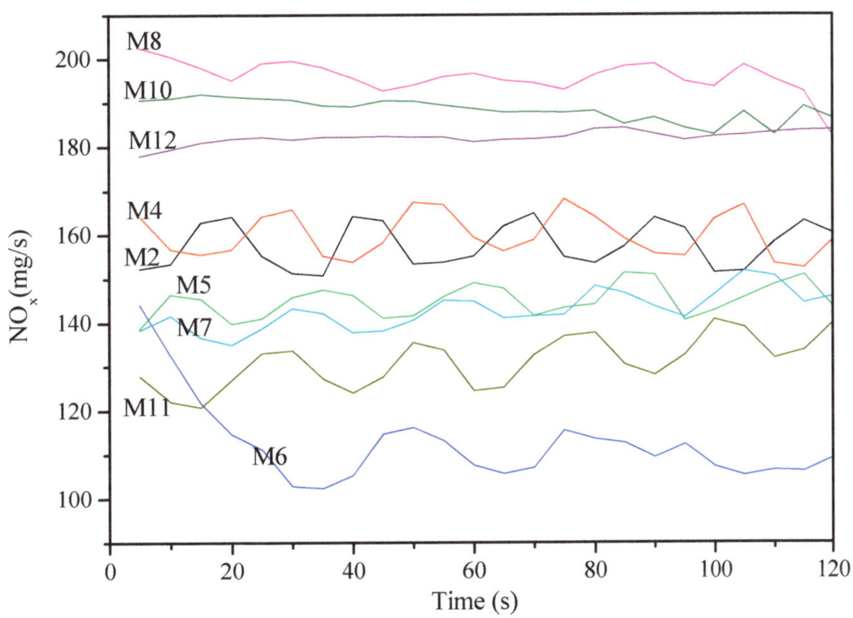

Figure 5. Real-time NOx emission in different operation modes, specifications of the operation modes are given in Table 3 [24].

According to the reactions between NOx and NH_3 described by Equations (10) and (11), to fully converse NOx to nitrogen and water, the required molar ratio of ammonia to NOx is 1:1.

$$4NH_3 + 4NO + O_2 \leftrightarrow 4N_2 + 6H_2O \tag{10}$$

$$2NH_3 + NO + NO_2 \leftrightarrow 2N_2 + 3H_2O \tag{11}$$

6. Results and Discussion

The ammonia charging of the start-up unit occurs simultaneously when the SCR chamber receives ammonia dominantly from the main reactors. This requires a pressure difference between the main reactor and the start-up unit. For example, if NH_4Cl, which has an equilibrium pressure of 5 bar at the ambient temperature (assumed as 25 °C in this work), is used as the start-up MHA, the desorption pressure of the main MHA has to be higher than 5 bar to expel ammonia to the start-up unit and SCR chamber at the same time. This therefore sets a threshold of desorption temperature for the main MHA. This work evaluated the performance of the sorption-SCR system using the $SrCl_2$ (main)/NH_4Cl (start-up) pair based on two different kinetic models, as explained in Section 4. The desorption temperature of the main MHA was set as 80 °C, 90 °C, and 100 °C, with an allowance of more than 1 bar pressure equilibrium drop for the adsorption of the NH_4Cl ammine at the ambient temperature.

In general, a full AdBlue tank is designed to enable the average user to continue driving for one year. The size of the AdBlue tank depends on many factors, including the diesel vehicle's engine and mileage, as well as several additional factors, such as the type of driving, the ambient temperature, or the type of journey. The estimated average use of AdBlue vs. diesel for trucks is between 4% and 8%, so approximately 500 L AdBlue is needed for local distribution in one year [35]. According to the volume factor of $SrCl_2$ ammine/ENG-TSA composite shown in Table 1, it can be estimated that a sorption ammonia storage system using such a composite needs a main unit with an overall 180 L capacity for one-year operation. Different layouts using different sizes of replaceable reactors (from 10 L to 180 L) were discussed and compared; for example, a possible layout could be 9 × 20 L, 6 × 30 L, 3 × 60 L, 2 × 90 L, or 1 × 180 L.

6.1. LTJ Model and Huang et al.' Model

Comparing the dynamic desorption based on two different kinetic models, as shown in Figure 6, the LTJ model represents the more desirable kinetics with a faster reaction rate. In contrast, the curves of Huang's model change rapidly before the desorption conversion degree reaches 0.4. Afterward, the reaction rate slows down quickly, and the ammonia conversion degree hardly changes once it goes beyond 0.8. It is not shown in Figure 6, but the ammonia conversion degree is still lower than 0.9 after 3 h of reaction. The LTJ model reflects relatively pure chemical reaction kinetics in a microscopic grain scale, as the equilibrium drop dominates the evolution and almost eliminates all other possible influences. These excluded influences, such as the heat conduction at the pellet level, should have been coupled into heat transfer property and have a prominent impact on the reaction rate of bulk reactant. Huang et al.'s model was developed by fitting the experimental results on a reactor containing a few hundred grams of bulk reactant. At such a testing scale, besides the dependence of pressure and temperature, the kinetics also involve a function of conversion degree, $f(x)$, which is the term representing the reactant reactivity as the progression of the reaction and the changes in the physical structure of the reactant [36,37]. The reaction rate is directly related to the conversion degree, as shown in Equation (8). The higher the conversion degree is, the lower the reaction rate. The reaction rate drastically decreases when the desorption is halfway through. This implies that the quantity/scale of the bulk MHA directly impacts the reaction rate and the final conversion degree, and therefore the actual capacity of ammonia storage and delivery.

6.2. The Main Reactor

As described in Section 5, when the diesel engine operates at a stable condition, the NOx emissions can be considered at a constant rate; therefore, a stable ammonia dosing rate is required, as given in Table 3. However, it is the nature of the MHA-ammonia chemical reaction that the desorption rate of MHA is an extremely unstable process if it is subject to a fixed constraint pressure and temperature. The desorption is violent within the first few minutes and then slows down quickly. Therefore, the current study proposed using a pressure valve to adjust the backpressure to achieve stable ammonia dosing. This valve is located downstream of the reactor, but upstream of the exhaust mixer, as shown in Figure 1. The opening area of the orifice of the valve is adjusted by the signal given by the DCU after processing the pressure signal.

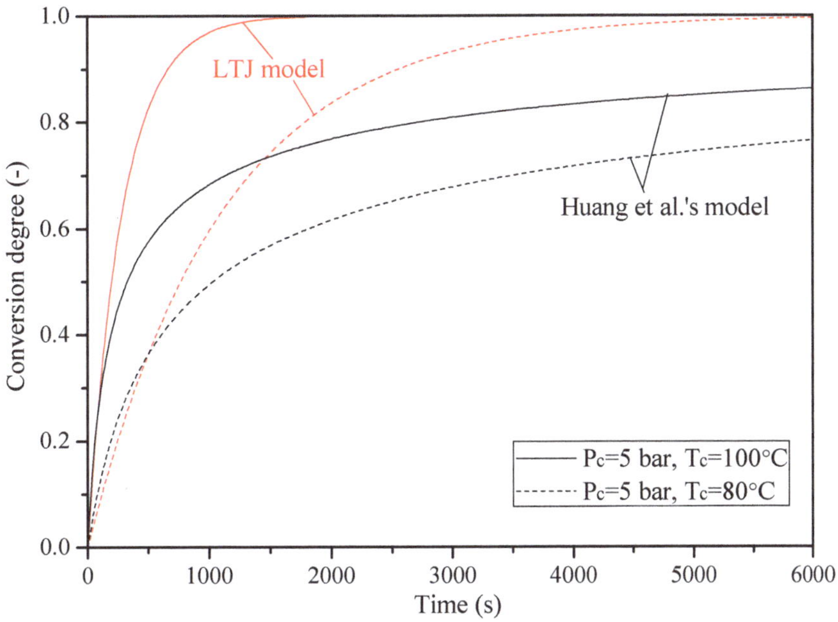

Figure 6. Kinetic curves of $SrCl_2$-ammonia desorption based on LTJ model and Huang et al.'s model.

The constraint pressure that should be imposed on the sorbent to generate the ammonia stream at a required flow rate is shown in Figure 7 when the engine operates with the highest NOx emissions (Mode 8). If using the LTJ model, i.e., the heat and mass transfer of the reactor are under ideal conditions, the unrestrained kinetics enable ammonia release for a longer time, and all the ammonia stored in the sorbent composite can be released, regardless of different reactor layouts; the backpressure of desorption can be almost constant, and there is no need for it to go down for the purpose of creating a higher pressure difference to propel ammonia release. The operation duration of the whole unit is almost the same, independent of the heating temperature and layout design. It is noticeable that there is a small point-down tail on the curves of the 10 L reactor; this small tail gradually disappears as the reactor size becomes larger, as shown in Figure 7. Since the LTJ kinetic model has no item directly representing the conversion degree, the calculated conversion degree is solely time-dependent. However, for a comparatively small reactor, the constraint pressure has to decrease to a relatively lower level to release the last bit of ammonia. For a relatively larger reactor, the larger total reactive ammonia ensures that the ammonia dosing can be persistently supplied at a required rate, even when the desorption rate is low as it approaches reaction completion. This is more evident in Equation (9).

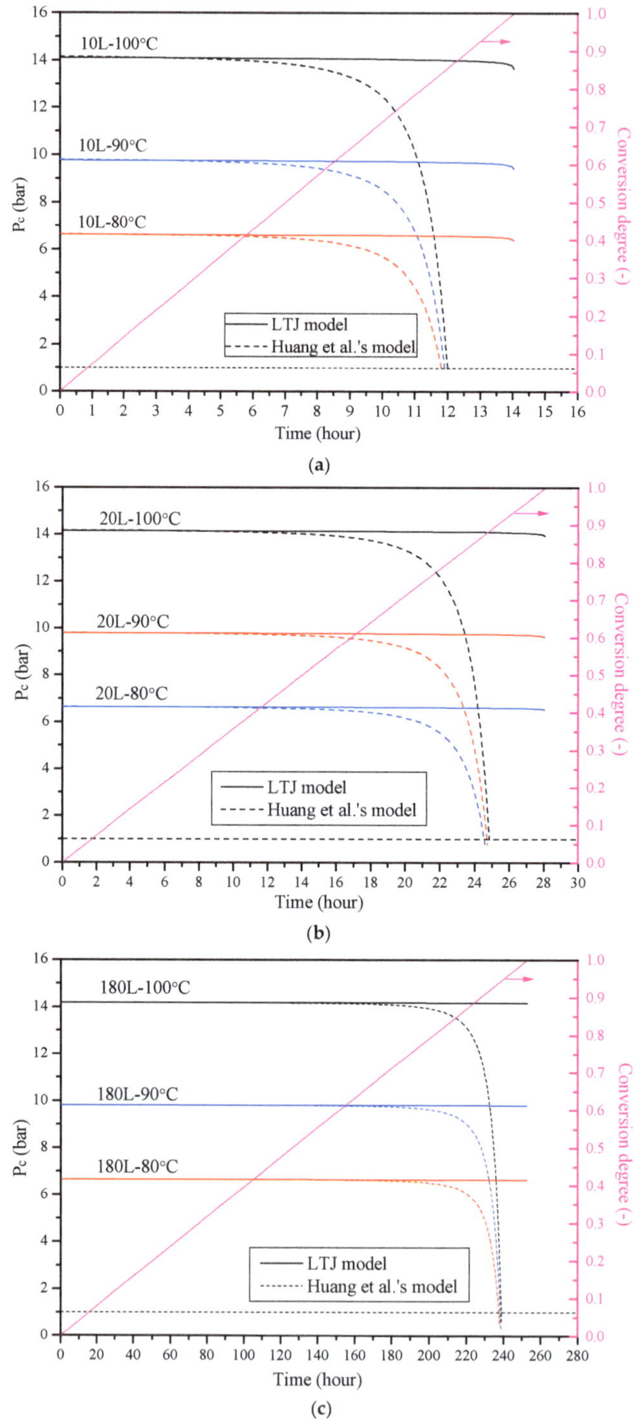

Figure 7. Working pressure and desorption conversion degree of sorption-SCR system in Mode 8, calculated with LTJ model and Huang et al.'s model: (**a**) 10 L reactor; (**b**) 20 L reactor; (**c**) 180 L reactor.

The conversion degree of the sorbent has an evident impact on the reaction rate in Huang et al.' model. In the case of a bulk reactor/sorbent with limited heat transfer, as the ammonia content within the sorbent decreases with the evolving desorption, the equilibrium pressure drop should increase to retain the continuously stable ammonia stream at the required flow rate with a given desorption temperature. For this instance, the constraint pressure P_c should decline to create the equilibrium drop required. Once the ammonia conversion degree reaches around 0.4–0.5, it can be observed in Figure 7 that P_c declines drastically; when P_c decreases to 1 bar (exhaust pressure), there is still some ammonia detained within the sorbent.

Figure 8 shows the final desorption conversion degree in each reactor with different volumes and heating temperatures. The value of the conversion degree in a reactor with 100 °C heating temperature for Mode 6 and Mode 8 is an average of 0.86% and 1.05% higher than that with 80 °C heating temperature, respectively, because higher heating temperature leads to higher equilibrium pressure, resulting in a higher equilibrium drop to maintain the required ammonia dosing rate for a longer time. Meanwhile, the reactor volume has a significant influence on the conversion degree. To meet the required ammonia flow rate, a relatively larger desorption rate (dx/dt) is required for a smaller reactor, according to Equation (9), leading to a quicker drop in ammonia content. In a mutual coupling of a vicious circle, the decrease in the value of x results in a declining reaction rate. In this instance, as the desorption is ongoing, a larger equilibrium drop, according to Equation (8), is the only way to achieve the desired desorption rate. However, since there is a floor limit of desorption backpressure, the equilibrium drop cannot infinitely be enlarged with a given desorption temperature. Consequently, when the conversion degree reaches a certain value, the smaller reactor becomes incapable of providing the required ammonia flow rate. For example, for the operation of Mode 6 at the same heating temperature of 80 °C, when the conversion degree reaches 69.1%, the ammonia output from 2.5 L reactor cannot meet the flow rate demand (107.25 mg/s). In contrast, a 10 L reactor can still provide the required ammonia until the conversion degree grows to 80.5%. The actual achievable ammonia storage capacity of a small reactor deviates more from its theoretical expectation than a large reactor. Another obvious indication is that, unlike the results of the LTJ model, the duration of the effective ammonia release in a 30 L reactor is not straightforwardly half that of a 60 L reactor but less; likewise, a 180 L reactor lasts longer than two 90 L reactors combined. Hence, using a layout consisting of small reactors requires a large total volume of the main unit for a certain storage capacity or requires more frequent reactor replacement or ammonia charging. For example, for the operation of Mode 8, a system with a 6 × 30 L reactor layout would need an extra 25 L sorbent composite to fulfill the total 180 L capacity, not to mention the collateral volume penalty of fitting and piping arrangements between reactors.

6.3. The Start-Up Unit

The start-up unit was adopted to deliver ammonia during the warm-up time of the main unit. As a matter of fact, a layout with a few small replaceable reactors composing the main unit is preferable over one large reactor if considering the warm-up time for the main unit. Since the storage tank and the whole bulk of sorbent should be heated up until the sorbent reaches the desorption point with proper working pressure in order to activate ammonia release, the less its thermal mass, the faster the unit can begin to function effectively. Alternatively, to offset the relatively retarded action at the outset, the size and capacity of the start-up unit can be enlarged so that it can last a longer time before the main reactor is ready.

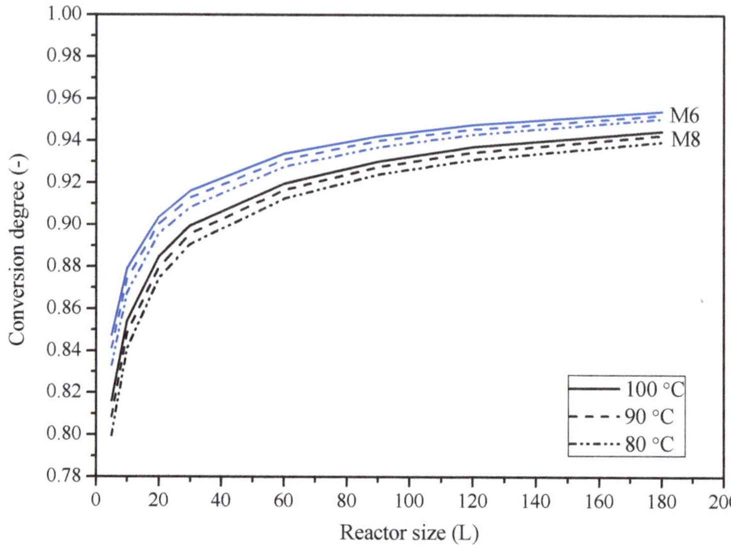

Figure 8. Desorption conversion degree using three different heating temperatures.

This warm-up time has been calculated using the common design of double-tube reactors for the main unit as an example. In a double-tube reactor, the heat-exchange fluid flows through the inner tube and the solid sorbent is compressed into the annular space between the inner and outer tubes. The geometric parameters of the reactors in different sizes are assumed and given in Table 4. The reactor is not necessarily straight; meandering in multiple U shapes or spiral shapes could be possible, especially for large volume design. According to the ratio of coolant water flow rate to brake power as a function of engine speed [38], the coolant water flow rate can be estimated for engine operation Mode 6 and Mode 8 at about 0.31 ± 0.01 kg/s. The temperature of the coolant water entering the reactor is assumed to be 90 °C. The start-up unit using NH_4Cl/ENG-TAS composite has an equilibrium pressure of around 5 bar at 25 °C and begins to adsorb ammonia when it is subject to a pressure equilibrium drop of 1 bar. This means that the main reactor should be heated and pressurized up to around 6 bar before it can release ammonia for both NOx conversion and for charging the start-up unit. After that, the required volume of the start-up unit can be estimated based on the ammonia storage capacity of the used start-up MHA and the NOx emissions during the warm-up period.

Table 4. Parameters of different reactors and the corresponding start-up units.

	Inner Tube Diameter (mm)	Outer Tube Diameter (mm)	Length (m)	Total System Volume (litre)	Volume of the Start-Up Unit Required (mL)	Warm-Up Time (min)
10 L	20	100	1.35	10.60	91	5.95
20 L	20	100	2.7	21.20	104	6.75
30 L	20	100	4	31.42	117	7.63
60 L	20	100	8	62.83	165	10.78
90 L	20	100	11.95	93.85	219	14.25
120 L	20	100	15.95	125.27	274	17.9
180 L	20	100	23.9	187.71	387	25.25

Table 4 shows the warm-up time required for different sizes of reactors and the required volume of the start-up unit corresponding to different layouts. The warm-up time of the main unit is from 5.95 min to 25.25 min, with the reactor size of the main unit ranging from 10 L to 180 L. Although the start-up volume increases from 0.104 L to 0.387 L, this volume is at least two orders of magnitude smaller than that of the main unit. As aforementioned, a smaller main unit requires less warm-up time, so the start-up unit can be smaller. However, the volume difference between start-up units for different layouts cannot offset the extra volume required for the layout with smaller reactors compared to that with a larger reactor. In fact, the capacity of the start-up unit onboard the vehicle should be rationally designed greater than that calculated in Table 4 for more reliable operation and resistance to unexpected driving conditions. On the other hand, considering the potential safety issue of ammonia leakage from high-pressure containers (>1 bar) at the ambient temperature in the scenario of eruption, the capacity of the start-up unit should be as small as possible.

Therefore, choosing a layout with a higher conversion degree and shorter warm-up time would be preferable. However, these two values are contradictory. Consequently, a trade-off is necessary for optimal performance, as one or the other is compromised. It is reasonable to compromise the desorption conversion slightly and choose the smaller reactor, i.e., a 60 L reactor, with a much shorter warm-up time and an acceptable conversion degree larger than 0.9.

7. Conclusion

This work studied a new design of sorption-SCR system for improved ammonia storage density and delivery for high NOx conversion efficiency at a relatively lower driven temperature (<100 °C) compared to urea-SCR systems. The system mainly consists of a main unit and a start-up unit that contain, respectively, the main metal halide ammine and a secondary metal halide ammine as sorbents. The selection criteria of metal halide ammine for each unit have been discussed. Taking the $SrCl_2$ and NH_4Cl as working pairs for an example, the dynamic performance of the sorption system in different layouts has been investigated. Ammonia demand for 100% NOx conversion efficiency was compared between two kinetic models.

Although the actual NOx emissions from engines in real driving conditions are more complex, the European stationary cycle test data of a diesel engine was used to explore the feasibility of this sorption-SCR system as a preliminary study. The modeling results indicated that this system is promising. Some conclusions are given as follows:

(a) Among the commonly used metal halide ammines, NH_4Cl and NaBr can be the ideal candidates for the start-up unit due to their relatively high equilibrium pressure at ambient temperature; and $SrCl_2$ and LiCl have the potential to be competent as the main metal halide ammine.

(b) A 20 °C growth in desorption temperature can lead to a 1% improvement of utilization efficiency of ammonia storage. Although the volume of the reactor had a significant influence on the conversion degree of the sorbent, larger reactors benefit from the final desorption conversion degree. The conversion degree of a system with a 10 L reactor is 16.6% higher than that with a 2.5 L reactor under a heating temperature of 80 °C for Mode 8. Namely, a larger reactor can release more ammonia than the combination of several small reactors with the same overall sorbent volume.

(c) When the volume of main reactor was increased from 10 L to 180 L, the warm-up time was extended by 19.3 min. Due to the thermal mass of the reactor and the solid sorbent, larger reactors require longer warm-up times before they can effectively deliver the ammonia at the desired flow rate. This necessitates a larger capacity of the start-up unit to cover up the warm-up period and increases the risk of unsatisfied ammonia delivery during non-ideal driving conditions. The optimal design should be a trade-off between the total desorption conversion degree and the warm-up time.

(d) For further work, a more suitable kinetic model based on a reactor design similar to a real system will be needed. A better kinetic model influenced by more practical factors and closer to true kinetic behavior can provide more insights for system design.

Author Contributions: Conceptualization, C.W. and S.C.; methodology, C.W.; software, C.W.; validation, C.W., S.C. and L.J.; formal analysis, C.W.; investigation, S.C.; resources, Y.W., H.B. and L.J.; data curation, C.W. and S.C.; writing—original draft preparation, C.W.; writing—review and editing, Y.W. and H.B.; visualization, C.W.; supervision, Y.W. and H.B.; project administration, H.B.; funding acquisition, H.B. All authors have read and agreed to the published version of the manuscript.

Funding: This research received no external funding.

Acknowledgments: The authors gratefully acknowledge the support from the Solar S&HP project (EP/T023090/1), the H + C Zero Network (EP/T022906/1), the GEMS project (EP/V042564/1), and Durham University EPSRC IAA (EP/R511523/1) funded by the Engineering and Physical Science Research Council of the UK.

Conflicts of Interest: The authors declare no conflict of interest.

Nomenclature

ΔN	Ammonia exchange amount [mol]
ASDS/AdAmmine	Ammonia storage and delivery system
AT	Ammonia transfer
AW	Ammonia warning
ΔH_r	Chemisorption enthalpy [J/mol (NH_3)]
x	Conversion degree [−]
DESs	Deep eutectic solvents
ρ	Density [kg/m^3]
DCU	Dosing control unit
ENG-TSA	Expanded graphite treated with sulfuric acid
R	Gas constant [J/(mol K)]
S	Heat exchange area (m^2)
α	Heat transfer coefficient [−]
ILs	Ionic liquids
LTJ	Large temperature jump
LNT	Lean NOx trap
$\dot{m}$	Mass flow rate [kg/s]
MHA	Metal halide ammine
M	Molar mass [kg/mol]
NOx	Nitrogen oxides
m	Parameter in kinetic model [−]
E	Parameter in kinetic model [J/mol]
c	Parameter in kinetic model [K^{-1} s^{-1}]
s	Parameter in kinetic model [s^{-1}]
P	Pressure [Pa]
k	Rate constant [s^{-1}]
f	Salt mass fraction in composite [−]
SCR	Selective catalytic reduction
γ	Stoichiometric coefficient [−]
T	Temperature [°C]
ΔT	Temperature difference [°C]
t	Time [s]
V	Volume [m^3]
Subscripts	
amb	Ambient
NH_3	Ammonia
c	Constrains
d	Desorption
eq	Equilibrium
exh	Exhaust
ms	Main salt
sou	Source
ss	Start-up salt

References

1. Johnson, T. Diesel engine emissions and their control: An overview. *Platin. Met. Rev.* **2008**, *52*, 23–37. [CrossRef]
2. Nova, I.; Tronconi, E. (Eds.) *Urea-SCR Technology for deNOx after Treatment of Diesel Exhausts*; Springer: New York, NY, USA, 2014. [CrossRef]
3. Fulks, G.; Fisher, G.B.; Rahmoeller, K.; Wu, M.-C.; D'Herde, E.; Tan, J. *A Review of Solid Materials as Alternative Ammonia Sources for Lean NOx Reduction with SCR*; SAE Technical Papers; SAE: Warrendale, PA, USA, 2009. [CrossRef]
4. Ma, Y.; Zhang, J.; Huang, K.; Jiang, L. Highly efficient and selective separation of ammonia by deep eutectic solvents through cooperative acid-base and strong hydrogen-bond interaction. *J. Mol. Liq.* **2021**, *337*, 116463. [CrossRef]
5. Lauer, T. Preparation of Ammonia from Liquid AdBlue—Modeling Approaches and Future Challenges. *Chem. Ing. Tech.* **2018**, *90*, 783–794. [CrossRef]
6. Deng, X.; Duan, X.; Gong, L.; Deng, D. Ammonia Solubility, Density, and Viscosity of Choline Chloride-Dihydric Alcohol Deep Eutectic Solvents. *J. Chem. Eng. Data* **2020**, *65*, 4845–4854. [CrossRef]
7. Santiago, R.; Mossin, S.; Bedia, J.; Fehrmann, R.; Palomar, J. Methanol-Promoted Oxidation of Nitrogen Oxide (NOx) by Encapsulated Ionic Liquids. *Environ. Sci. Technol.* **2019**, *53*, 11969–11978. [CrossRef]
8. Zolotareva, D.; Zazybin, A.; Rafikova, K.; Dembitsky, V.M.; Dauletbakov, A.; Yu, V. Ionic liquids assisted desulfurization and denitrogenation of fuels. *Vietnam J. Chem.* **2019**, *57*, 133–163. [CrossRef]
9. Wazeer, I.; Hadj-Kali, M.K.; Al-Nashef, I.M. Utilization of Deep Eutectic Solvents to Reduce the Release of Hazardous Gases to the Atmosphere: A Critical Review. *Molecules* **2021**, *26*, 75. [CrossRef]
10. Jiang, W.-J.; Zhang, J.-B.; Zou, Y.-T.; Peng, H.-L.; Huang, K. Manufacturing Acidities of Hydrogen-Bond Do-nors in Deep Eutectic Solvents for Effective and Reversible NH3Capture. *ACS Sustain. Chem. Eng.* **2020**, *8*, 13408–13417. [CrossRef]
11. Johannessen, T.; Schmidt, H.; Svagin, J.; Johansen, J.; Oechsle, J.; Bradley, R. *Ammonia Storage and Delivery Systems for Automotive NOx Aftertreatment*; SAE Technical Papers; SAE: Warrendale, PA, USA, 2008. [CrossRef]
12. Elmøe, T.D.; Sørensen, R.Z.; Quaade, U.; Christensen, C.H.; Nørskov, J.K.; Johannessen, T. A high-density ammonia storage/delivery system based on Mg(NH3)6Cl2 for SCR–DeNOx in vehicles. *Chem. Eng. Sci.* **2006**, *61*, 2618–2625. [CrossRef]
13. Koebel, M.; Elsener, M.; Kleemann, M. Urea-SCR: A promising technique to reduce NOx emissions from automotive diesel engines. *Catal. Today* **2000**, *59*, 335–345. [CrossRef]
14. Kruger, M.; Nisius, P.; Scholz, V.; Wiartalla, A. A compact solid SCR system for NOx reduction in passenger cars and light duty trucks. *MTZ Worldw.* **2003**, *64*, 14–17. [CrossRef]
15. Tatur, M.; Tomazic, D.; Lacin, F.; Sullivan, H.; Kotrba, A. Solid SCR demonstration truck application. In Proceedings of the Directions in Engine Efficiency and Emissions Research (DEER) Conference, Dearborn, MI, USA, 27–30 September 2009.
16. Lacin, F.; Kotrba, A.; Hayworth, G.; Sullivan, H.; Tatur, M.; Jacques, J.; Tomazic, D.; Cho, H. *SOLID SCR: Demonstrating an Improved Approach to NOx Reduction via a Solid Reductant*; SAE Technical Papers; SAE: Warrendale, PA, USA, 2011. [CrossRef]
17. Marko, A.; Wahl, T.; Alkemade, U.; Brenner, F.; Bareis, M.; Harndorf, H.; Method and Device for Selective Catalytic NOx Reduction. 14 May 2002. Available online: https://patents.google.com/patent/US6387336B2/en (accessed on 21 October 2022).
18. Liu, C.Y.; Aika, K.-I. Effect of the Cl/Br Molar Ratio of a CaCl2−CaBr2 Mixture Used as an Ammonia Storage Material. *Ind. Eng. Chem. Res.* **2004**, *43*, 6994–7000. [CrossRef]
19. Johannessen, T. Next generation SCR system for fuel-Efficient NOx reduction. In *IQPC Conference Selective Catalytic Reduction*; IQPC: Wiesbaden, Germany, 2011.
20. Johannessen, T.; Hviid, C.C.; Kehlet, N.J.; Zink, S.R.; Ulrich, Q.; High Density Storage of Ammonia. 14 May 2009. Available online: https://patents.google.com/patent/US7964163/en (accessed on 21 October 2022).
21. Johannessen, T. 3rd generation SCR system using solid ammonia storage and direct gas dosing:—Expanding the SCR win-dow for RDE. In Proceedings of the Directions in Engine Efficiency and Emissions Research (DEER) Conference, Dearborn, MI, USA, 27–30 September 2012.
22. Mauran, S.; Lebrun, M.; Prades, P.; Moreau, M.; Spinner, B.; Drapier, C. Active Composite and Its Use as Reaction Medium. U.S. Patent 5,283,219, 1 February 1994.
23. Han, J.H.; Lee, K.H.; Kim, D.H.; Kim, H. Transformation Analysis of Thermochemical Reactor Based on Thermophysical Properties of Graphite−$MnCl_2$ Complex. *Ind. Eng. Chem. Res.* **2000**, *39*, 4127–4139. [CrossRef]
24. Jiang, L.; Xie, X.L.; Wang, L.W.; Wang, R.Z.; Roskilly, A.P. Performance analysis on a novel self-adaptive sorption system to reduce nitrogen oxides emission of diesel engine. *Appl. Therm. Eng.* **2017**, *127*, 1077–1085. [CrossRef]
25. Wang, L.W.; Metcalf, S.J.; Critoph, R.E.; Thorpe, R.; Tamainot-Telto, Z. Thermal conductivity and permeability of consolidated expanded natural graphite treated with sulphuric acid. *Carbon* **2011**, *49*, 4812–4819. [CrossRef]
26. Zhang, P.; Xiao, X.; Ma, Z.W. A review of the composite phase change materials: Fabrication, characterization, mathematical modeling and application to performance enhancement. *Appl. Energy* **2016**, *165*, 472–510. [CrossRef]
27. Wang, Z.; Wang, L.; Gao, P.; Yu, Y.; Wang, R. Analysis of composite sorbents for ammonia storage to eliminate NOx emission at low temperatures. *Appl. Therm. Eng.* **2018**, *128*, 1382–1390. [CrossRef]
28. Zhang, C.; An, G.; Wang, L.; Wu, S. Multi-stage ammonia production for sorption selective catalytic reduction of NOx. *Front. Energy* **2022**. [CrossRef]
29. Jiang, L.; Wang, L.; Wang, R. Investigation on thermal conductive consolidated composite CaCl2 for adsorption refrigeration. *Int. J. Therm. Sci.* **2014**, *81*, 68–75. [CrossRef]

30. Bao, H.S.; Wang, R.Z. A review of reactant salts for resorption refrigeration systems. *Int. J. Air-Cond. Refrig.* **2010**, *18*, 165–180. [CrossRef]
31. Lu, H.-B.; Mazet, N.; Spinner, B. Modelling of gas-solid reaction—Coupling of heat and mass transfer with chemical reaction. *Chem. Eng. Sci.* **1996**, *51*, 3829–3845. [CrossRef]
32. Neveu, P.; Castaing, J. Solid-gas chemical heat pumps: Field of application and performance of the internal heat of reaction recovery process. *Heat Recover. Syst. CHP* **1993**, *13*, 233–251. [CrossRef]
33. Veselovskaya, J.V.; Tokarev, M.M. Novel ammonia sorbents 'porous matrix modified by active salt' for adsorptive heat transformation: 4. Dynamics of quasi-isobaric ammonia sorption and desorption on BaCl2/vermiculite. *Appl. Therm. Eng.* **2011**, *31*, 566–572. [CrossRef]
34. Huang, H.J.; Wu, G.B.; Yang, J.; Dai, Y.C.; Yuan, W.K.; Lu, H.B. Modeling of gas–solid chemisorption in chemical heat pumps. *Sep. Purif. Technol.* **2004**, *34*, 191–200. [CrossRef]
35. How Much AdBlue Do I Need? | Find AdBlue®Consumption | Greenchem. Available online: https://www.greenchem-adblue.com/informations/how-much-adblue-needed/ (accessed on 22 October 2022).
36. Wang, R.; Wang, L.; Wu, J. *Adsorption Refrigeration Technology: Theory and Application*; John Wiley & Sons: Singapore, 2014.
37. Lebrun, M.; Spinner, B. Models of heat and mass transfers in solid—Gas reactors used as chemical heat pumps. *Chem. Eng. Sci.* **1990**, *45*, 1743–1753. [CrossRef]
38. Heywood, J.B. *Internal Combustion Engine Fundamentals*; McGraw-Hill Education: New York, NY, USA, 2018.

energies

MDPI

Article

AC Loss Characteristics of HTS Novel Twisted Cables Composed of Soldered-Stacked-Square (3S) Wires

Zhuyong Li [1], Zhixuan Zhang [1], Mingshuo Wang [2], Yingying Lv [3] and Kyungwoo Ryu [4,*]

[1] Department of Electrical Engineering, Shanghai Jiao Tong University, Shanghai 200240, China
[2] Chaoyang Electric Power Company, State Grid Corporation of China, Beijing 122000, China
[3] Chengnan Power Supply Company, State Grid Tianjin Electric Power Company, Tianjin 300201, China
[4] Department of Electrical Engineering, Chonnam National University, Gwangju 500-757, Korea
* Correspondence: kwryu@chonnam.ac.kr

Abstract: Compared with traditional cables, superconducting multi-stage cables have the natural advantages of greater transmission power and less energy loss, which have gradually attracted attention. However, conventional multi-stage cables are based on low temperature superconducting (LTS) technology and there is considerable scope for improvement in their performance. In this paper, a novel structure of the multi-stage high temperature superconducting (HTS) twisted cable prepared by the soldered-stacked-square (3S) wire is proposed. The AC loss characteristics of the twisted cable are deeply studied by experiments and simulation. Through the experiment, the influence of the voltage-leads on the AC loss measurement accuracy is eliminated, and frequency dependent is shown in the AC loss of the twisted cable. Besides, the simulated value of AC loss is consistent with the experimental value, which verifies the accuracy of the simulation. The AC loss of twisted cable is only 20% of that of the thin strip model, which reveals its outstanding advantages in AC loss.

Keywords: AC loss; HTS twisted cable; 3S wire; experiment; simulation

Citation: Li, Z.; Zhang, Z.; Wang, M.; Lv, Y.; Ryu, K. AC Loss Characteristics of HTS Novel Twisted Cables Composed of Soldered-Stacked-Square (3S) Wires. *Energies* **2022**, *15*, 7454. https://doi.org/10.3390/en15207454

Academic Editors: Alessandro Cannavale and Ubaldo Ayr

Received: 14 September 2022
Accepted: 1 October 2022
Published: 11 October 2022

1. Introduction

Due to the significant features of high critical current density and high irreversible field, REBCO tapes are gradually applied in high temperature superconducting (HTS) cables [1,2]. Researchers have been working to further improve the current-carrying capacity of REBCO tapes by twisting or stacking, such as twisted stacked-tapes cable (TSTC) [3], conductor on round core cables (CORC) [4], Roebel coated conductor cables [5], HTS cross conductor (HTS-CroCo) [6], quasi-isotropic strand (QIS) [7,8], soldered-stacked-square (3S) wire [9], highly flexible REBCO cable (HFRC) [10], and so on. They also promote possible wide applications, such as electrical machine [11] and fault current-limiting superconducting cable [12].

Based on the design of multistage twisted CICC cable used in the superconducting fusion magnet of ITER and other projects, an HTS twisted cable composed of soldered-stacked-square (3S) wires is designed. Compared with the low temperature superconducting (LTS) wires utilized in traditional CICC design, HTS 3S wires effectively makes up for its deficiencies in current-carrying capacity, mechanical strength, AC loss, and cooling cost. The structure of the HTS twisted cable is shown in Figure 1. First, several 2-mm-wide HTS narrow tapes and copper tapes are stacked in a certain order and packaged into an HTS 3S wire. Then, seven 3S wires are twisted to form the first-stage cable, which is named 2 mm-7-cable. Six first-stage cables can be further twisted into the second-stage cable. As shown in Figure 2, the 3S wire indicated by the red rectangle is made up of two HTS tapes and ten copper tapes with 2 mm width, so we named it 3S (2s + 10c)-wire. The HTS tapes used in this paper are the YBCO materials without artificial pinning centers, manufactured by Shanghai Superconducting Technology Company (SSTC). Specifications of the used HTS tapes, 3S (2 + 10c)-wire and 2 mm-7-cable are listed in Table 1. The fabrication and critical current evaluation of the twisted cable have been studied specifically in our previous work [13]. In this study, copper tapes are used instead of the HTS tapes to reduce experimental costs while realizing the cable structures. Meanwhile, copper tapes can also enhance the mechanical strength and provide a certain overcurrent protection.

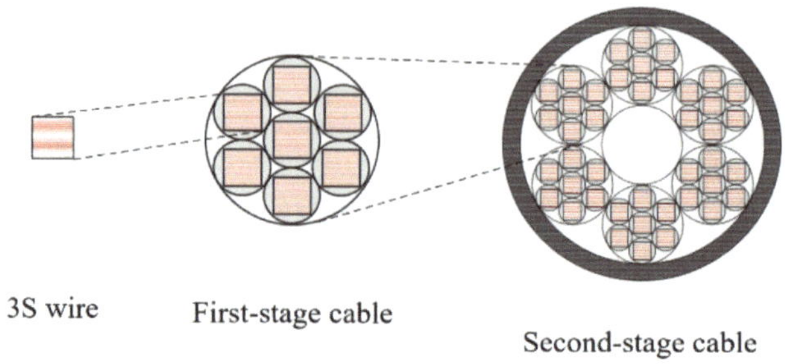

Figure 1. Conceptual illustration of multistage cable structure using 3S wires.

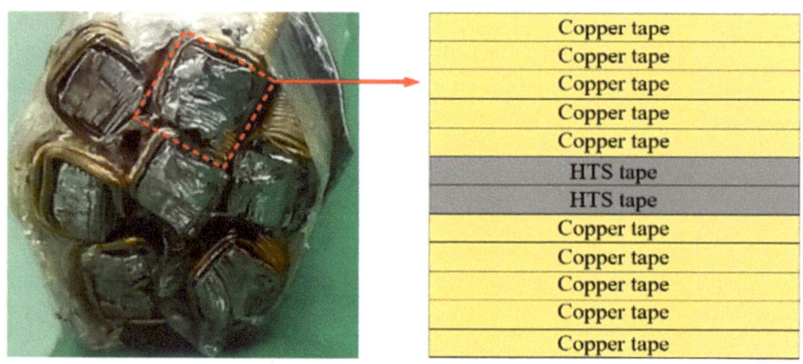

Figure 2. Cross-section view of 2 mm-7-cable and structure of 3S (2s + 10c)-wire with 2 mm width.

Table 1. Specifications of HTS tape, 3S (2s + 10c)-wire, and 2 mm-7-cable.

HTS tape	Width and thickness	2.0 mm × 70 μm
	Thickness of superconducting layer	~1.3 μm
	Thickness of silver layers	~1 μm
	Thickness of substrate layer (plus buffer)	~10 μm
	Thickness of copper plating layer, each side	90 A @ 77 K
	Critical current of HTS tapes	30 A ± 5 A
3S (2 + 10c)-wire	Width and thickness	2.09 mm × 1.85 mm
	Number of HTS tapes	2
	Number of copper tapes	10
	Width and thickness of copper tape	2.00 mm × 150 μm
2mm-7-cable	Diameter	9.7 mm
	Critical current, self-field, average	300 A @ 77 K
	Twist pitch	100 mm
	Number of 3S wires	7

For superconducting cables that carry large current, AC loss is another important characteristic to be investigated. AC loss will lead to heating and increase the burden on the cooling system, and in serious cases, it may also decrease the critical current of the cable, affecting the stability of the system. Therefore, we focus on the AC loss of the 2 mm-7-cable in this paper. The measuring method of the AC loss is firstly introduced, and the influences of different voltage leads arrangement and different voltage-lead heights on AC loss are discussed. Then, the frequency dependence of AC loss of the twisted cable is tested. Finally, the finite element method (FEM) is used to calculate AC loss of the 2 mm-7-cable and the simulation and experimental results are contrasted in detail.

2. AC Loss Measurement

The AC loss measurement system used in this paper is mainly composed of a waveform generator, a current source, a current transformer, an amplifier, an oscilloscope, and a compensation coil. The schematic diagram of the whole measurement system is shown in Figure 3a, and the actual picture is shown in Figure 3b. The experimental steps are as follows: Controlled by the waveform generator, the current source sends out AC current signals with specified cycle, frequency and amplitude. The transformed current signal is collected by the oscilloscope through the filter amplifier by adjusting the relative position of the compensation coil. The voltage signal in the same phase as the current signal is also collected by the oscilloscope after being filtered and amplified. The experimental value of AC loss can be obtained by integrating the compensated voltage and current data. In this study, all the AC loss measurements were carried out in 77 K liquid nitrogen.

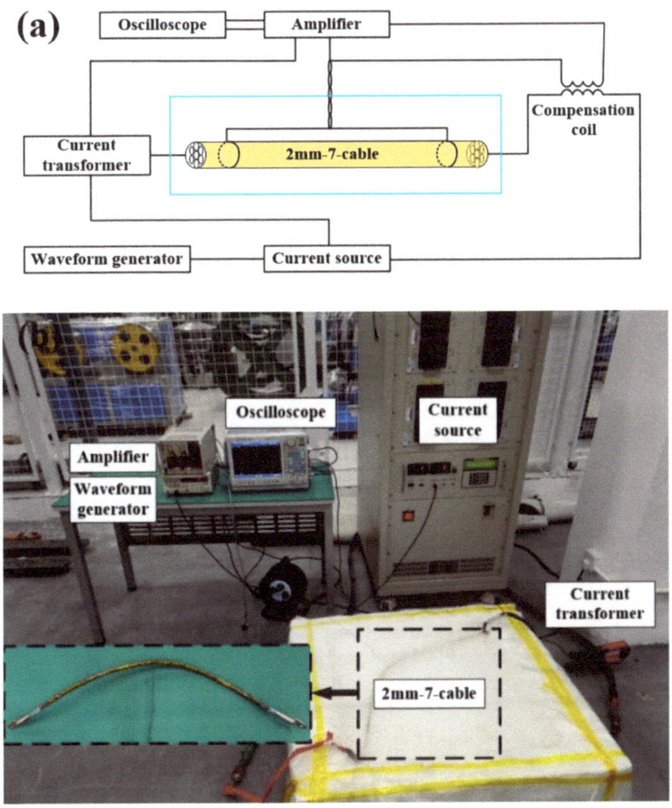

Figure 3. AC loss measurement system. (**a**) Schematic diagram; (**b**) Actual picture.

Four-point method is adopted for AC loss measurement. The two ends of 2 mm-7-cable are inserted into copper terminals which is slightly larger than the cable, and solder is poured to fill the gap between 2 mm-7-cable and copper terminals for fixation and electric conduction. The current-leads are connected to the copper terminals of 2 mm-7-cable. For voltage-leads, as there is no precedent to measure the AC loss of this twisted cable structure, two connection methods are designed. The first connection method is the same as the traditional voltage-lead connection. Keep the voltage-leads parallel to the surface of 2 mm-7-cable and wind them into one at the place where the voltage-leads meet, as shown in Figure 4a. A rectangular gap is formed between the voltage-leads and the surface of 2 mm-7-cable. This voltage-lead arrangement is named as square-lead arrangement. Another method is to keep the voltage-leads parallel to the surface of the 3S wires and let it spirally wrap on the 2 mm-7-cable, as shown in Figure 4b. The voltage-leads also meet in the middle. The second connection method is named spiral-lead arrangement. Only when the voltage-lead is high enough and the rectangular area is large enough can the complete loss voltage induced by saturated magnetic flux be collected on the voltage-lead.

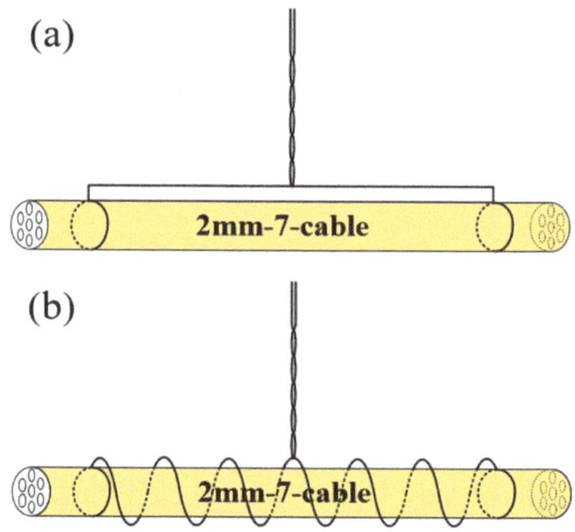

Figure 4. Schematic diagram of two voltage lead arrangements. (**a**) Square-lead arrangement; (**b**) Spiral-lead arrangement.

Comparative experiments are carried out on these two connection methods. The AC loss is measured and compared under the condition of keeping the same distance between the voltage-leads and surface of the 2 mm-7-cable and the same frequency. The distance between the voltage-leads and surface of the 2 mm-7-cable ranges from 0 mm to 5 mm, 10 mm, and 15 mm. The measured frequencies include 25 Hz, 50 Hz, 100 Hz, and 200 Hz. In sixteen groups of comparative tests, all the results show that the AC loss is almost the same under the two voltage-leads arrangements. Figure 5 shows the AC loss measurement results of two groups of comparative experiments. Therefore, it can be seen that these two arrangements of voltage-leads have no effect on AC loss measurement.

In addition, the influence of voltage-lead height on AC loss measurement is studied. For a single HTS tape, in order to measure the loss voltage correctly, the distance between the sample surface and the voltage lead should be kept at the distance about three times the tape half-width [14]. However, for the twisted cable structure in this study, the influence of voltage-lead height on AC loss is not clear, so we design an experiment to evaluate the influence of voltage-lead height to guide the application of high current HTS twisted cable in the future. As shown in Figure 6, four voltage-leads with different heights are welded on the same 2 mm-7-cable at the same time, and the two ends of the voltage lead are fixed on the same solder joint. The voltage leads are kept parallel to the surface of the 2 mm-7-cable. The distance between the voltage-taps is 770 mm. As we can see in Figure 7, when the height of voltage-leads increases from 0 mm to 5 mm, 10 mm, and 15 mm, the value of AC loss does not change significantly.

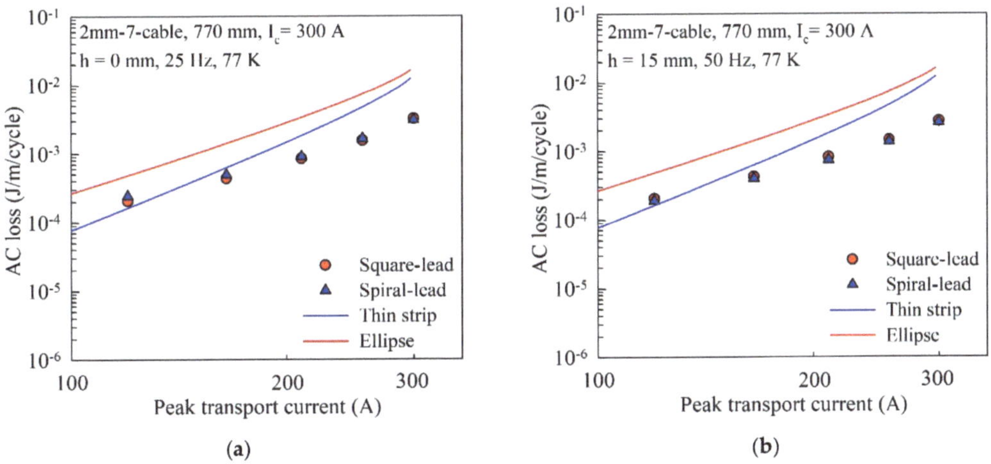

Figure 5. Comparison of AC loss of 2 mm cable under two voltage−lead arrangements. (**a**) AC loss values when the distance between the voltage−leads and the cable sample is 0 mm and the frequency is 25 Hz; (**b**) AC loss values when the distance between the voltage−leads and the cable sample is 15 mm and the frequency is 50 Hz.

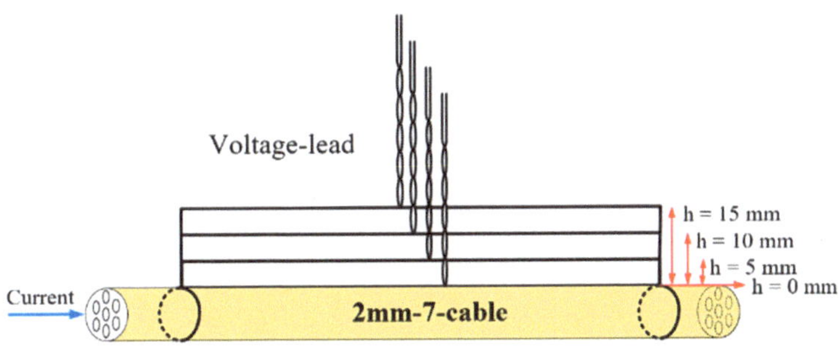

Figure 6. Schematic diagram of voltage-lead height of 2 mm-7-cable.

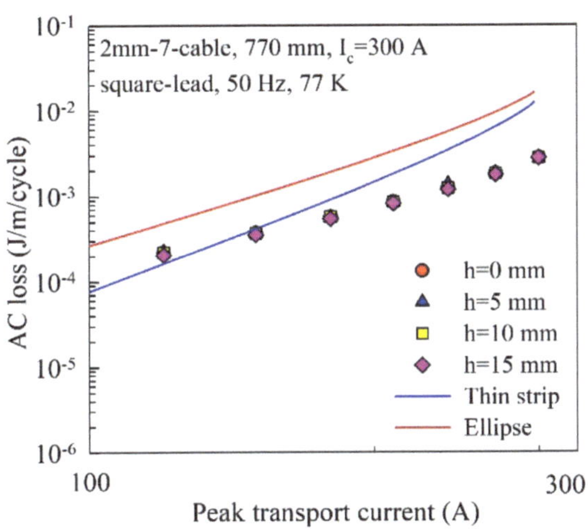

Figure 7. Dependence of height of voltage-lead configuration on AC loss for 2 mm−7−cable.

Next, the frequency dependence of the AC loss of the twisted cable was tested, and the measurement results are shown in Figure 8. It can be seen that the AC loss of the cable has obvious frequency dependence, and the value is approximately proportional to the current frequency, indicating that the eddy current loss accounts for a large proportion of the AC loss of the cable.

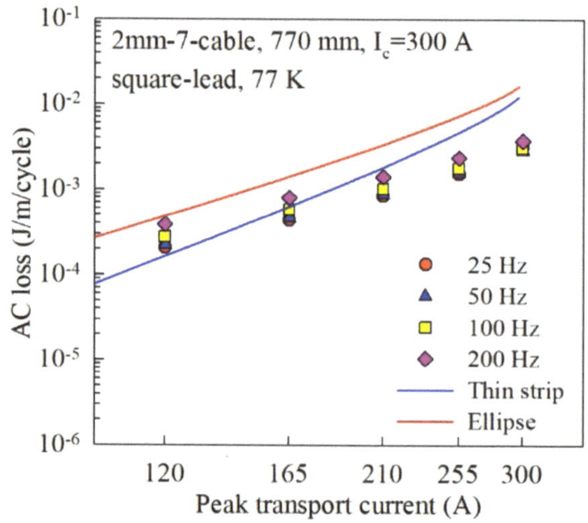

Figure 8. Effect of frequency on AC loss of 2 mm−7−cable.

3. AC Loss Simulation

The electromagnetic field distribution inside the superconductor under the complex geometric model is often solved by the finite element method (FEM) [15]. FEM modeling can separate the different components of AC loss in HTS cables, which helps to judge the hysteresis loss in superconducting layers and eddy current loss in copper layers and further analyze the AC loss characteristics. The FEM modeling normally has three numerical methods, all named after the variables used in the partial differential formulations: one is the A-V formulation based on the magnetic vector potential, the second is the T-Ω formulation based on the current vector potential, and the third is the H-formulation based on the magnetic field. Considering the transition process of superconductors from the superconducting state to the normal conducting state, the nonlinear E-J relationship is often used concurrently with these formulations [16]. For the simulation of superconducting cables, the H-formulation is selected as the mathematical basis of the three-dimensional model, and its outstanding advantages are as follows:

(1) For the A-V formulation and the T-Ω formulation, four variables are required for5three-dimensional problems, while the H-formulation has only three variables, namely H_x, H_y, and H_z. Fewer variables and no second derivative make the H-formulation less computationally intensive and faster than the other two formulations.

(2) In many scenarios, the model needs to consider both the transport current and the background field. The H-formulation is most used in integral boundary conditions. The background field can be achieved by setting the boundary conditions of the H-formulation, while the transport current can be injected into the model by Ampere's law. Furthermore, there is no need to distinguish vector potentials that could lead to computational errors.

(3) The A-V formulation uses the electric field E to calculate the current density J, a small change in E will cause a dramatic change in J, which makes the formulation very unstable. The H-formulation is just the opposite, using the current density J to

calculate the electric field E, successfully avoiding the non-convergence problem that may occur in the A-V formulation.

The H-formulation is essentially another representation of Maxwell's formulations, which is equivalent to a quasi-static process for HTS. The specific formula is shown as follow:

$$\begin{array}{c} \nabla \times E = -\mu_0\mu_r\frac{\partial H}{\partial t} \\ J = \nabla \times H \\ E_{norm} = \rho_\Omega \cdot J_{norm} \end{array} \quad (1)$$

In this model, the Cartesian three-dimensional coordinate system is used, and the magnetic field strength H is regarded as an independent variable, defined as $H = [H_x, H_y, H_z]^T$; the current density J is defined as $J = [J_x, J_y, J_z]^T$, current density modulus value $J_{norm} = \sqrt{J_x^2 + J_y^2 + J_z^2}^T$; electric field intensity E is defined as $E = [E_x, E_y, E_z]^T$, the electric field strength modulus $E_{norm} = \sqrt{E_x^2 + E_y^2 + E_z^2}$. According to Ampere's law, the relationship between J and H can be obtained as:

$$\begin{array}{c} J_x = \frac{\partial H_z}{\partial y} - \frac{\partial H_y}{\partial z} \\ J_y = \frac{\partial H_x}{\partial z} - \frac{\partial H_z}{\partial x} \\ J_z = \frac{\partial H_y}{\partial x} - \frac{\partial H_x}{\partial y} \end{array} \quad (2)$$

For the second-generation HTS tapes used in this study, the nonlinear relationship between current and voltage can be described by the E-J characteristic:

$$E = E_0\left(\frac{J}{J_c}\right)^n \quad (3)$$

According to Faraday's law, the relationship between E and H can be described as:

$$\begin{array}{c} \frac{\partial E_z}{\partial y} - \frac{\partial HE_y}{\partial z} = -\mu_0\mu_r\frac{\partial H_x}{\partial t} \\ \frac{\partial E_x}{\partial z} - \frac{\partial E_z}{\partial x} = -\mu_0\mu_r\frac{\partial H_y}{\partial t} \\ \frac{\partial E_y}{\partial x} - \frac{\partial E_x}{\partial y} = -\mu_0\mu_r\frac{\partial H_y}{\partial t} \end{array} \quad (4)$$

The entire model is divided into three subdomains: HTS domain, copper domain, and air domain. Different subdomains are assigned different resistivities:

$$\begin{array}{c} \rho_{HTS} = \frac{E_c}{J_c}\cdot\left(\frac{J_{norm}}{J_c}\right)^{n-1} \\ \rho_{Cu} = 1.97 \times 10^{-9}\ \Omega{\cdot}m \\ \rho_{Air} = 1\ \Omega{\cdot}m \end{array} \quad (5)$$

The influence of the external magnetic field and the pinning centers is important for the critical current [17,18]. In the complex HTS cables, each HTS tape is affected by the magnetic field produced by other tapes in the same cable. To simplify the calculation, the critical current in FEM model is set as the measured value of the cable instead of the initial value of the HTS tape. Therefore, the overall critical current of the HTS cable is considered here, and the measured value of the critical current has represented the self-field influence caused by the interaction of the HTS tapes.

The geometric dimensions and parameters of 2 mm-7-cable are set as shown in Table 1 so as to correctly compare the difference between the experimental value and the simulation value. In order to speed up the calculation speed, some parameters have been optimized: the width of the HTS layer in the REBCO tape is 2 mm, while the thickness is only 1 μm. In COMSOL, such a large width-to-thickness ratio will lead to a surge in the number of meshes, increasing the calculation convergence difficulty. Thus, in this model, the thickness of the superconducting layer is increased by 10 times to 10 μm, which will not affect the accuracy of the calculation results and greatly speed up the calculation speed [19]. In addition, the cable length is set as the half-twist pitch, which also greatly reduces the degree of freedom and speed up the calculation. The substrate layer, copper layer, and other parts of the HTS tape except the superconducting layer are regarded as the copper domain, and the distance between the two superconducting layers is 150 μm. The 3D view and side view of the geometric model are shown in Figure 9.

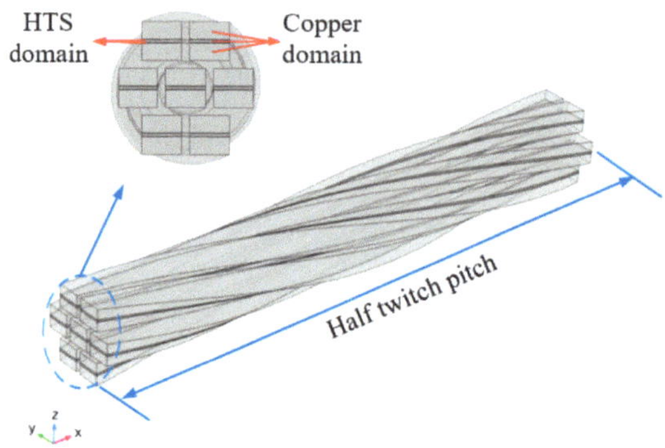

Figure 9. Three-dimensional geometric model of 2 mm-7-cable.

The solution adopts parametric sweep, and five groups of 50 Hz sinusoidal alternating currents pass through the cable, ranging from 0.4 times the critical current to one time the critical current, respectively 120 A, 165 A, 210 A, 255 A, and 300 A. The formula for calculating the AC loss is as follow:

$$Q_{AC\ loss} = 2\int_{T/2}^{T}\left(\int_{V} E \cdot J\, dV/L\right)dt \tag{6}$$

Next, the simulated and experimental values of AC loss are compared, and the results are shown in Figure 10. The experimental value is a set of data with the voltage leads arranged in parallel, with a height of 0 mm and a current of 300 A. The parameters of the simulation are kept consistent with the experimental values. From the figure, we can see that the overall simulation value of AC loss is very close to the experimental value, which verifies the accuracy of the simulation. Since there are many copper layers in this cable, although there is almost no current flow in the copper layers, the loss of the copper layers also accounts for about 20% to 30% of the total loss. In future designs, as the overall proportion of the superconducting layer in the cable increases, the AC loss of the copper layers can be ignored. In addition, when the transport current of the twisted cable is greater than 0.4 times the critical current, compared with the theoretical value of the thin strip of the Norris model, the twisted structure significantly reduces the AC loss. The larger the transport current value, the greater the decrease in AC loss. When the current increases to the critical current, the AC loss is only 20% of the thin tape, which indicates the unique advantages of twisted cables in reducing AC loss.

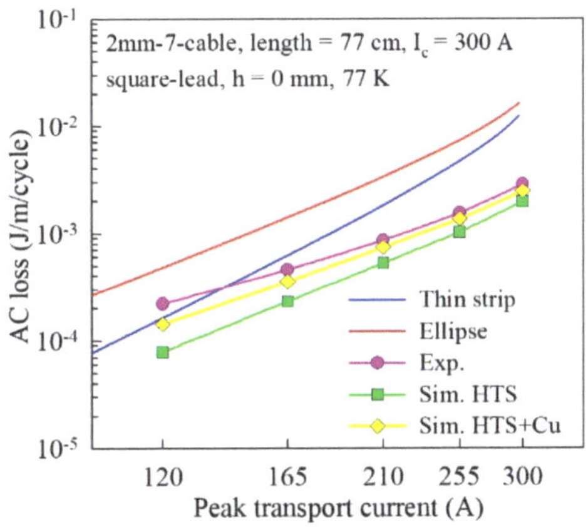

Figure 10. Comparison between experimental and simulation values of AC loss of 2 mm−7−cable.

4. Conclusions

In this paper, the AC loss characteristics of the HTS novel twisted cables are studied by both experimental and simulation methods. Firstly, the devices and steps of AC loss experimental measurement are introduced. Then, the experimental results are analyzed, and the following important conclusions are drawn: The parallel arrangement of the voltage leads and the spiral arrangement have little effect on the experimental measurement results of AC loss. The voltage-leads are close to the surface of the cable to measure the saturated magnetic flux of the sample is obtained. The AC loss of the cable has a certain frequency dependence, indicating that the eddy current loss accounts for a large proportion of the AC loss of the cable. The simulation value of AC loss is highly consistent with the experimental value, which proves the effectiveness and accuracy of the simulation. Compared with the thin strip model, the twisted structure can reduce the AC loss by up to 80%, highlighting the outstanding advantages of the twist structure in terms of AC loss. This paper successfully combined the CICC structure and 3S wire to form a second-stage HTS cable. The AC loss reduction is also confirmed as a promising result, which enhances the application potential of this cable design.

Author Contributions: Conceptualization, Z.L. and M.W.; methodology, Z.L. and M.W.; software, M.W.; validation, Z.L., M.W., and Z.Z.; formal analysis, Z.L. and M.W.; investigation, Z.L.; resources, Z.L.; data curation, Z.L.; writing—original draft preparation, M.W.; writing—review and editing, Y.L.; visualization, M.W.; supervision, Z.L. and K.R.; project administration, Z.L.; funding acquisition, Z.L. All authors have read and agreed to the published version of the manuscript.

Funding: This work was supported by the National Nature Science Foundation of China under grant No. 52077134.

Data Availability Statement: Not applicable.

Conflicts of Interest: The authors declare no conflict of interest.

References

1. Martucciello, N.; Giubileo, F.; Grimaldi, G.; Corato, V. Introduction to the focus on superconductivity for energy. *Supercond. Sci. Technol.* **2015**, *28*, 070201. [CrossRef]
2. Wu, Y.; Zhao, Y.; Han, X.; Jiang, G.; Shi, J.; Liu, P.; Khan, M.Z.; Huhtinen, H.; Zhu, J.; Jin, Z.; et al. Ultra-fast growth of cuprate superconducting films: Dual-phase liquid assisted epitaxy and strong flux pinning. *Mater. Today Phys.* **2015**, *18*, 100400. [CrossRef]
3. Takayasu, M.; Chiesa, L.; Bromberg, L.; Minervini, J.V. HTS twisted stacked-tape cable conductor. *Supercond. Sci. Technol.* **2012**, *25*, 014011. [CrossRef]
4. Van der Laan, D.C.; Lu, X.F.; Goodrich, L.F. Compact $GdBa_2Cu_3O_{7-\delta}$ coated conductor cables for electric power transmission and magnet applications. *Supercond. Sci. Technol.* **2011**, *24*, 042001. [CrossRef]

5. Goldacker, W.; Frank, A.; Heller, R.; Schlachter, S.I.; Ringsdorf, B.; Weiss, K.-P.; Schmidt, C.; Schuller, S. ROEBEL assembled coated conductors (RACC): Preparation, properties and progress. *IEEE Trans. Appl. Supercond.* **2007**, *17*, 3398–3401. [CrossRef]
6. Wolf, M.J.; Fietz, W.H.; Bayer, C.M.; Schlachter, S.I.; Heller, R.; Weiss, K.-P. HTS CroCo: A stacked HTS conductor optimized for high currents and long-length production. *IEEE Trans. Appl. Supecond.* **2016**, *26*, 6400106. [CrossRef]
7. Li, Y.; Wang, Y.; Miao, J.; Shi, C.; Ju, P.; Huang, P.; Xue, J.; Hasegawa, T. Investigation on critical current properties of quasi-isotropic strand made from coated conductor. *IEEE Trans. Appl. Supercond.* **2015**, *25*, 6600805. [CrossRef]
8. Pi, W.; Ma, S.; Kang, Q.; Liu, Z.; Meng, Y.; Wang, Y. Study on Mechanical Properties of Quasi-Isotropic Superconducting Strand Stacked by 2-mm-Wide REBCO and Copper Tapes. *IEEE Trans. Appl. Supecond.* **2020**, *30*, 6600105. [CrossRef]
9. Li, Z.; Hu, D.; Zhang, L.; Xie, Z.; Sun, L.; Liu, B.; Hong, Z.; Jin, Z.; Ryu, K. Development of a Novel Soldered-Stacked-Square (3S) HTS Wire Using 2G Narrow Tapes with 1 mm Width. *IEEE Trans. Appl. Supercond.* **2017**, *27*, 6600904. [CrossRef]
10. Guo, Z.; Qin, J.; Lubkemann, R.; Wang, K.; Jin, H.; Xiao, G.; Li, J.; Zhou, C.; Nijhuis, A. AC loss and contact resistance in highly flexible rebco cable for fusion applications. *Superconductivity* **2022**, *2*, 100013. [CrossRef]
11. Bruno, D.; Berger, K.; Ivanov, N. Characterization of High-Temperature Superconductor Bulks for Electrical Machine Application. *Materials* **2021**, *14*, 1636. [CrossRef]
12. Xia, Y.; Song, Y.; Ma, T.; Zheng, J.; Liu, H.; Liu, F.; Song, M. Design and Performance Tests of a Fault Current-Limiting-Type Tri-Axial HTS Cable Prototype. *Electronics* **2022**, *11*, 1242. [CrossRef]
13. Wang, M.; Qi, H.; She, M.; Zhang, L.; Ryu, K.; Li, Z.; Hong, Z.; Jin, Z. Fabrication and Critical Current Evaluation for HTS Twisted Cables Consisting of Soldered-Stacked-Square (3S) Wires. *IEEE Trans. Appl. Supercond.* **2021**, *31*, 4804004. [CrossRef]
14. Ciszek, M.; Ashworth, S.P.; James, M.P.; Glowacki, B.A.; Campbell, A.M.; Garre, R.; Conti, S. Self-field AC losses and critical currents in multi-tube Ag-Bi-2223 conductors. *Supercond. Sci. Technol.* **1996**, *9*, 379–384. [CrossRef]
15. Krzysztof, K.; Kampik, M.; Stępień, M. Characterization of high-temperature superconducting tapes. *IEEE Trans. Instrum. Meas.* **2019**, *69*, 2959–2965. [CrossRef]
16. Zhang, M.; Coombs, T.A. 3D modeling of high-Tc superconductors by finite element software. *Supercond. Sci. Technol.* **2011**, *25*, 015009. [CrossRef]
17. Patrick, P.; Sieger, M.; Ottolinger, R.; Lao, M.L.; Eisterer, M.; Meledin, A.; Van Tendeloo, G.; Haenisch, J.; Holzapfel, B.; Schultz, L.; et al. Influence of artificial pinning centers on structural and superconducting properties of thick YBCO films on ABAD-YSZ templates. *Supercond. Sci. Technol.* **2018**, *31*, 044007. [CrossRef]
18. Krzysztof, K.; Grilli, F.; Kario, A.; Godfrin, A.; Zermeno, V.M.R.; Stepien, M.; Kampik, M. Length Uniformity of the Angular Dependences of I c and n of Commercial REBCO Tapes with Artificial Pinning at 77 K. *IEEE Trans. Appl. Supercond.* **2018**, *29*, 8000309. [CrossRef]
19. Wang, M.; Li, Z.; Zhang, Y.; Sun, C.; Dong, F.; Pan, Y.; Yao, L.; Hong, Z.; Jin, Z. Three-Dimensional Numerical Study on Transport AC Loss of Soldered-Stacked-Square-Twisted (3S-T) Wire. *IEEE Trans. Appl. Supercond.* **2019**, *29*, 5900105. [CrossRef]

energies

Article

Photovoltaic Performance of Dye-Sensitized Solar Cells with a Solid-State Redox Mediator Based on an Ionic Liquid and Hole-Transporting Triphenylamine Compound

Minseon Kong †, Da Hyeon Oh †, Baekseo Choi and Yoon Soo Han *

Department of Advanced Materials and Chemical Engineering, Daegu Catholic University, Gyeongsan-si 38430, Korea; qewr1666@naver.com (M.K.); ws5963@naver.com (D.H.O.); skdiwlsdnr@naver.com (B.C.)
* Correspondence: yshancu@cu.ac.kr
† These authors contributed equally to this work.

Abstract: An ionic liquid, 1-methyl-3-propylimidazolium iodide (MPII), was solidified with an organic hole-transporting material, 4,4′,4′′-tris[(3-methylphenyl)phenylamino]triphenylamine (m-MTDATA), and the resulting solid-state redox mediator (RM) (m-MTDATA-solidified MPII) was employed in solar devices to realize solid-state dye-sensitized solar cells (sDSSCs). Solar devices with only MPII or m-MTDATA as an RM showed almost 0 mA/cm^2 of short-circuit current (J_{sc}) and thus 0% power conversion efficiency (PCE). However, an sDSSC with the m-MTDATA-solidified MPII exhibited 4.61 mA/cm^2 of J_{sc} and 1.80% PCE. It was found that the increased J_{sc} and PCE were due to the formation of I_3^-, which resulted from a reaction between the iodie (I^-) of MPII and m-MTDATA cation. Further enhancement in both J_{sc} (9.43 mA/cm^2) and PCE (4.20%) was observed in an sDSSC with 4-tert butylpyridine (TBP) as well as with m-MTDATA-solidified MPII. We attributed the significant increase (about 230%) in PCE to the lowered diffusion resistance of I^-/I_3^- ions in the solid-state RM composed of the m-MTDATA-solidified MPII and TBP, arising from TBP's role as a plasticizer.

Keywords: dye-sensitized solar cell; solid-state redox mediator; ionic liquid; 4-tert butylpyridine; m-MTDATA

Citation: Kong, M.; Oh, D.H.; Choi, B.; Han, Y.S. Photovoltaic Performance of Dye-Sensitized Solar Cells with a Solid-State Redox Mediator Based on an Ionic Liquid and Hole-Transporting Triphenylamine Compound. *Energies* **2022**, *15*, 2765. https://doi.org/10.3390/en15082765

Academic Editors: Alessandro Cannavale and Ubaldo Ayr

Received: 19 February 2022
Accepted: 6 April 2022
Published: 9 April 2022

1. Introduction

A conventional dye-sensitized solar cell (DSSC) typically comprises an organic-solvent-based liquid redox mediator (RM) and two electrodes (i.e., dye-adsorbed TiO_2 photoanode and Pt-coated counter electrode). It has a three-layered structure in which the liquid RM is sandwiched between the two electrodes [1–4]. For the last three decades, numerous studies have been performed, and a power conversion efficiency (PCE) as high as 14.3% has been achieved [5]. However, the potential problems caused by organic-solvent-based liquid RMs, such as their leakage, the volatilization of solvents at high temperatures, and the precipitation of salts at low temperature, are considered critical reasons for limiting the commercialization and practicable use of DSSCs [6]. By replacing the liquid RM with a solid-state (s) RM or hole-transporting material (HTM) in DSSCs, the above problems can be solved, and dye desorption can also be prevented [7]. There are two types of sDSSCs: one is a sandwich-structured cell with a basic configuration of glass/F-doped tin oxide (FTO)/mesoporous TiO_2:dye/sRM/platinized FTO/glass, and the other is a metal-back-contact-structured device with a configuration of glass/FTO/mesoporous TiO_2:dye/sHTM/metal electrodes (Au or Ag). The sandwich-structured sDSSCs include several-micrometer-thick sRM layers. Therefore, they are disadvantageous in improving PCE compared to the metal-back-contact-structured sDSSCs, in which the thickness of sHTM layers is typically under 1 μm.

Recently, photovoltaic properties of sandwich-structured sDSSCs have been reported [8–12]. When inorganic sRMs, such as $CsSnI_{2.95}F_{0.05}$ doped with SnF_2 [8] and Cs_2SnI_6 with additives [9], were applied, sandwich-structured sDSSCs showed PCEs of up to 10.2% and 7.8%, respectively. It has also been reported that PCEs of sandwich-structured sDSSCs ranged from 5.68% to up to 11%, when metal complexes ($Cu^{2+/+}$ or $Co^{3+/2+}$) in combination with lithium bis(trifluoromethanesulfonyl)imide (LiTFSI) and 4-tert-butylpyridine (TBP) [or 4-(trifluoromethyl)pyridine] were employed as sRMs [10–12]. As another approach to creating sRMs for sandwich-structured sDSSCs, the solidification of ionic liquids has been attempted [13–19]. Ionic liquid, a salt in liquid state at room temperature, features high ionic conductivity, negligible vapor pressure, high thermal stability, and a wide electrochemical window [6,15]. Dialkylimidazolium iodides, such as 1-methyl-3-propylimidazolium iodide (MPII), 1-butyl-3-methylimidazolium iodide (BMII), and 1-ethyl-3-methylimidazolium iodide (EMII), have been used for the liquid and/or sRMs of DSSCs as an iodide (I^-) source. These ionic liquids were solidified by mixing them with conductors, semiconductors, or insulators to apply them to sRMs for sDSSCs. A 5.4% PCE was achieved in a sandwich-structured sDSSC when MPII in combination with LiTFSI and TBP was cast on a solid-state polymerized poly(3,4-ethylenedioxythiophene) (P3HT) as a solidifying agent [13]. In addition, several ionic liquid/solidifier systems, such as MPII (without I_2)/polyaniline-loaded carbon black (5.81%) [6], BMII + $BMISO_3CF_3$(with I_2)/SiO_2 (4.83%) [14], BMII(with I_2)/TiC (1.68%) [15], EMITFSI + EMII(with I_2)/SiO_2 (3.7%) [16], $BMIBF_4$(with I_2)/silica (4.98%) [17], EMITFSI + EMII(with I_2)/TiO_2 (5.0%) [18], EMITFSI + EMII(with I_2)/carbon fiber (4.97%) [18], and MPII(with I_2)/silica (7.0%) [19], were also reported. In most of these reports, iodine (I_2) was added to solidified ionic liquid systems to produce triiodide (I_3^-) via a reaction with the iodide (I^-) [14–19]. However, the corrosive nature of iodine can limit the use of metallic-grid electrodes in the manufacturing of multi-cell modules [6,13].

This study aims at providing an iodine-free and volatile-organic-solvent-free sRM, which applies to even several-micrometer-thick hole transporting layers for applications to sandwich-structured sDSSCs. We prepared an iodine-free sRM based on an ionic liquid (MPII) as an iodide source and a hole-transporting triphenylamine compound, 4,4′,4″-tris[phenyl(*m*-tolyl)amino]triphenylamine (m-MTDATA), as a solidifier. m-MTDATA is one of the most well-known arylamine compounds applied in organic light-emitting diodes and perovskite solar cells as a hole injection or transporting material, and devices with m-MTDATA have shown good performance [20–22]. Compared with polymeric solidifiers such as P3HT [13] and polyaniline [6], low-molecular-weight m-MTDATA is highly advantageous in terms of pore-filling property because polymers are poorly penetrated into the TiO_2 mesopores, which arises from mismatches between the polymer and mesopore sizes. Uniform pore-filling property is essential to ensuring higher PCEs of DSSCs. The viscous liquid of MPII was solidified with m-MTDATA by simply mixing them and then applying them to sDSSCs as an sRM. We fabricated sandwich-structured sDSSCs with the solidified MPII, and their photovoltaic performance was investigated. To the best of our knowledge, sandwich-structured sDSSCs with m-MTDATA-solidified MPII as an sRM have never been reported. The reported photovoltaic parameters [i.e., short-circuit current density (J_{sc}), open-circuit voltage (V_{oc}), fill factor (*FF*) and PCE] of sandwich-structured sDSSCs with solidified ionic liquid are summarized in Table 1, including those of our device with the m-MTDATA-solidified MPII.

Table 1. Reported photovoltaic performance of sandwich-structured sDSSCs with an sRM based on ionic liquids and solidifying agents.

Ionic Liquids	Solidifying Agents	J_{sc} (mA/cm^2)	V_{oc} (V)	FF (%)	PCE (%)	Ref.
MPII (without I_2)	P3HT	14.2	0.64	60	5.4	[13]

Table 1. *Cont.*

Ionic Liquids	Solidifying Agents	J_{sc} (mA/cm^2)	V_{oc} (V)	FF (%)	PCE (%)	Ref.
MPII (without I_2)	polyaniline-loaded carbon black	12.20	0.737	65	5.81	[6]
MPII (without I_2)	m-MTDATA	9.43	0.610	73.0	4.20	This study
BMII + $BMISO_3CF_3$ (with I_2)	SiO_2	11.3	-	-	4.83	[14]
BMII (with I_2)	TiC	3.40	0.6686	74	1.68	[15]
EMITFSI + EMII (with I_2)	SiO_2	10.4	0.592	62	3.7	[16]
$BMIBF_4$ (with I_2)	Silica	8.60	0.621	69.9	4.98	[17]
EMITFSI + EMII (with I_2)	TiO_2	11.45	0.675	65	5.00	[18]
EMITFSI + EMII (with I_2)	Carbon fiber	11.11	0.688	65	4.97	[18]
MPII (with I_2)	Silica	13.67	0.700	73.1	7.0	[19]

2. Experimental Details

2.1. Materials

FTO glass with a sheet resistance of ~7 Ω/square (TCO22-7), TiO_2 paste for the mesoporous layer (Ti-nanoxide T/SP), TiO_2 paste for the scattering layer (Ti-nanoxide R/SP), N719 dye (Ruthenizer 535-bisTBA), and hot-melt adhesive (Metlonix 1170-25, DuPont Surlyn) were purchased from Solaronix (Aubonne, Switzerland). Titanium diisopropoxide bis(acetylacetonate) (TPA), $TiCl_4$, MPII, m-MTDATA, TBP, and LiTFSI were procured from Sigma-Aldrich (St. Louis, MO, USA). Platinum paste (PT-1) from Dyesol-Timo JV (Seoul, Korea) was selected as the source for the Pt-coated counter electrode. All of the chemicals for DSSC fabrications were used without further purification.

2.2. Fabrication of Sandwich-Structured sDSSCs

Except for the preparation of the sRM, the same procedures mentioned in our previous reports were employed to fabricate working (glass/FTO/TiO_2:dye) and counter (glass/platinized FTO) electrodes for sandwich-structured sDSSCs [23,24]. A 25-μm-thick hot-melt adhesive was sandwiched between the working and counter electrodes and then annealed for 10 min at 120 °C to seal the two electrodes.

The preparation processes of the sRM are as follows: m-MTDATA (50 mg) as a solidifying agent was dissolved in 1 mL of chloroform. An ionic liquid solution was separately prepared by dissolving MPII (1 M, 252 mg) in 1 mL of chloroform. For comparison, TBP- and TBP/LiTFSI-containing ionic liquid solutions were also prepared by adding TBP (0.2 M, 27 mg) and TBP (0.2 M, 27 mg)/LiTFSI (0.066 M, 19 mg) into the MPII solution, respectively. The m-MTDATA and ionic liquid solutions were mixed with a volume ratio of 3:1 to apply to the sRM. The mixed solutions were then injected into the sealed cells through a pre-drilled hole formed on the counter electrodes, and the cells were dried in a vacuum oven for 60 min at 50 °C. The injection and drying process was repeated four times to fully fill a space between the mesoporous TiO_2 layer and the platinized FTO layer, thereby fabricating sandwich-structured sDSSCs with a 25 mm^2 active area. The detailed fabrication conditions of sandwich-structured sDSSCs are provided in the electronic supplementary information (ESI).

2.3. Fabrication of Hole-Only Devices

Hole-only devices (HODs) with a layer configuration of glass/platinized FTO/TiO_2: N719 dye/sRM with or without additives/platinized FTO/glass were fabricated using the same procedures presented in our earlier work [24]. The detailed fabrication conditions are provided in the ESI.

2.4. Measurements

The cross-sectional morphology of the sandwich-structured sDSSCs was visualized via field-emission scanning electron microscopy (FE–SEM; S-4800, Hitachi High-Technology; Tokyo, Japan) equipped with energy-dispersive X-ray spectroscopy (EDS) (Horiba EX-250, Horiba; Kyoto, Japan). The photocurrent–voltage measurements were performed using a CompactStat potentiostat (Ivium Technologies B.V.; Eindhoven, The Netherlands) and a PEC-L01 solar simulator system equipped with a 150-W xenon arc lamp (Peccell Technologies, Inc.; Yokohama, Japan). The light intensity was adjusted to 1 sun (100 mW/cm^2) using a silicon photodiode (PEC-SI01, Peccell Technologies, Inc.; Yokohama, Japan). The monochromatic incident-photon-to-current conversion efficiencies (IPCEs) were plotted as a function of light wavelength using an IPCE measurement instrument (PEC-S20, Peccell Technologies, Inc.; Yokohama, Japan). The UV–vis absorption spectra were obtained using a SINCO NEOSYS-2000 spectrophotometer (Seoul, Korea). Electrochemical impedance spectroscopic (EIS) analyses were performed using an electrochemical analyzer (CompactStat, Ivium Technologies B.V.; Eindhoven, The Netherlands). The active areas of the dye-adsorbed TiO_2 films were estimated using a digital microscope camera (SZ61, OLYMPUS Corporation; Tokyo, Japan) equipped with image analysis software.

3. Results and Discussion

3.1. Photovoltaic Performance of sDSSCs with m-MTDATA-Solidified MPII

We could prepare the sRM composed of MPII (iodide source) and m-MTDATA (solidifier) by simply mixing and drying the corresponding solutions, i.e., a viscous liquid phase of MPII was solidified with a solid powder of m-MTDATA. As shown in Figure S1 of the ESI, the viscous flow behavior of MPII disappeared by mixing it with m-MTDATA. In addition, a solid-state was maintained when additives (TBP and LiTFSI) were added to the m-MTDATA-solidified MPII. We first investigated the pore-filling property of the solidifier (m-MTDATA) into TiO_2 layers in DSSCs. As shown in Figure 1, carbon atoms from the low-molecular-weight m-MTDATA are uniformly distributed throughout the TiO_2 layer. This is beneficial to obtain higher PCE in DSSCs due to effective hole collection at the counter electrode and dye regeneration near the TiO_2:dye layer. In our previous report, the pore-filling into the TiO_2 layers was not completed when polymeric hole-transporting materials were used [24].

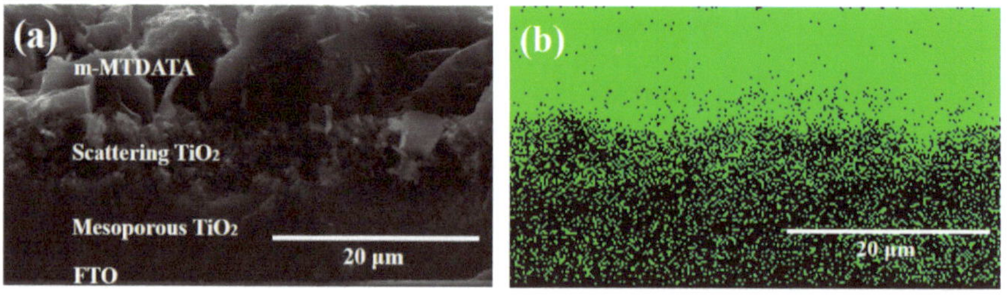

Figure 1. Cross-sectional SEM (**a**) and EDS mapping images showing the distribution of carbon (**b**) in a decapped sDSSC with m-MTDATA. N719 dyes containing carbon atoms were not adsorbed on TiO_2 surface.

Using the MPII, m-MTDATA, or m-MTDATA-solidified MPII, sandwich-structured DSSCs with or without additives were fabricated, as shown in Table 2, and their photovoltaic properties were investigated. The average photovoltaic properties measured using four cells are compared in Figure 2 and Table 3, and their raw data are presented in Table S1 of the ESI. Solar devices with only MPII (DSSC-a) or m-MTDATA (sDSSC-b) showed 0.22 ± 0.05 and 0% PCE, respectively; on the other hand, a significant increase in average PCE (1.42 ± 0.42%) was observed in the sDSSC-c with m-MTDATA-solidified MPII. When TBP was added to m-MTDATA-solidified MPII, the average PCE value (3.75 ± 0.43%) of the sDSSC-d was further increased compared to that of the sDSSC-c without TBP. Unexpectedly, by additional incorporation of LiTFSI as well as TBP into the m-MTDATA-solidified MPII, the photovoltaic performance (1.50 ± 0.12%) of the sDSSC-e was again decreased from that of the sDSSC-d.

Table 2. Fabricated DSSCs with five different types of RM for comparison.

Solar Cells	Components of RM	Phase of RM
DSSC-a	MPII	Viscous liquid
sDSSC-b	m-MTDATA	Solid powder
sDSSC-c	MPII, m-MTDATA	Solid-state (nonfluidic composite)
sDSSC-d	MPII, m-MTDATA, TBP	Solid-state (nonfluidic composite)
sDSSC-e	MPII, m-MTDATA, TBP, LiTFSI	Solid-state (nonfluidic composite)

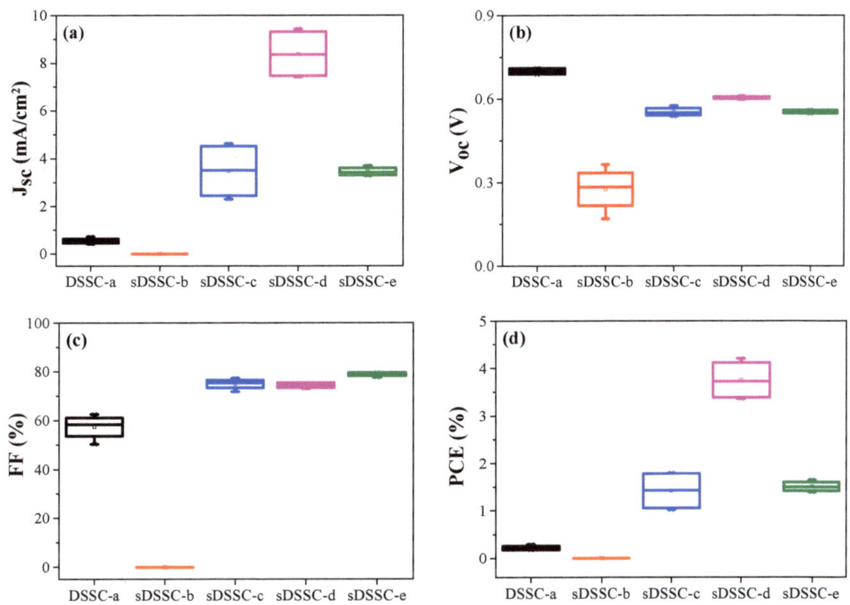

Figure 2. Box plots for photovoltaic performance of the DSSCs with five different types of RM; (**a**) J_{SC}, (**b**) V_{OC}, (**c**) *FF*, and (**d**) PCE measured under the AM 1.5 condition.

Among the four cells prepared in each condition, we selected the champion cells to compare their photovoltaic properties. Here, we denote the champion cell with the m-MTDATA-solidified MPII by sDSSC-c′. The same denotation is applied for the champion sDSSC-d and e. Figure 3a,b shows the current density (*J*)-voltage (*V*) and dark-current curves of the champion devices, respectively, and their photovoltaic parameters are listed in Table 4. Photovoltaic performance largely varied with the types of RM, i.e., PCEs ranged from 0% to 4.20%. We investigated the origins of the performance variations, and the examination results are described in detail in the following sections.

Table 3. Averages and standard deviations of cell performance measured using four cells with five different types of RM.

Solar Cells	J_{sc} (mA/cm^2)	V_{oc} (V)	*FF* (%)	PCE (%)
DSSC-a	0.54 ± 0.13	0.722 ± 0.052	57.38 ± 5.25	0.22 ± 0.05
sDSSC-b	0.001 ± 0.002	0.275 ± 0.082	0	0
sDSSC-c	3.48 ± 1.20	0.553 ± 0.017	74.90 ± 2.30	1.42 ± 0.42
sDSSC-d	8.38 ± 1.08	0.603 ± 0.006	74.34 ± 1.19	3.75 ± 0.43
sDSSC-e	3.45 ± 0.19	0.553 ± 0.007	78.82 ± 0.91	1.50 ± 0.12

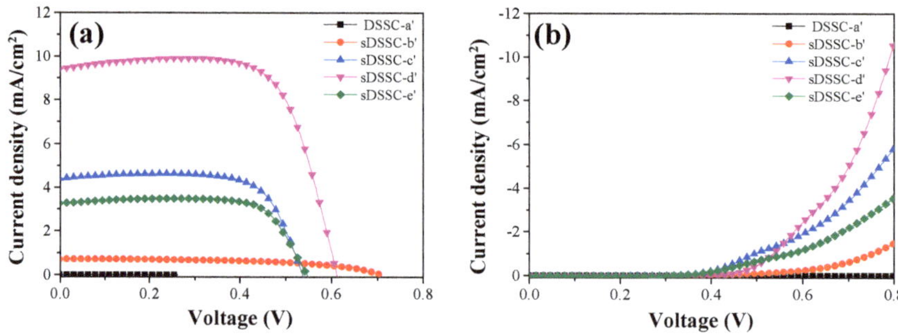

Figure 3. *J–V* characteristics (**a**) and dark-current (**b**) curves for the champion devices with MPII, m-MTDATA, or solidified MPII layer with or without additives.

Table 4. Photovoltaic performance of champion cells with MPII, m-MTDATA, or solidified MPII layer with or without additives.

Champion Cells	J_{sc} (mA/cm^2)	V_{oc} (V)	FF (%)	PCE (%)
DSSC-a′	0.72	0.708	56.86	0.29
sDSSC-b′	0.004	0.304	0	0
sDSSC-c′	4.61	0.544	71.81	1.80
sDSSC-d′	9.43	0.610	73.03	4.20
sDSSC-e′	3.28	0.546	77.53	1.39

3.2. Hole Conduction Mechanism in sDSSCs with m-MTDATA-Solidified MPII

As mentioned earlier, DSSCs with only MPII or m-MTDATA showed almost 0 mA/cm^2 of J_{sc} and thus 0% PCE. In the DSSC-a′ with MPII, triiodide (I_3^-) was barely present in the RM because iodine (I_2) was not added [25]. This fact indicates that hole transportation from the oxidized dye to the platinized counter electrode cannot be completed, leading to very low J_{sc} and PCE. In the case of the sDSSC-b′ with m-MTDATA, hole conduction cannot occur because the carrier diffusion length in a solid organic semiconductor is very short, typically around 100 nm [24,26,27]. When considering that the layer thickness of m-MTDATA in the sDSSC-b′ is about 10.6 μm (Figure S2 of the ESI), effective hole transportation is impossible, inducing approximately 0 mA/cm^2 of J_{sc}. However, the sDSSC-c′ with the m-MTDATA-solidified MPII as an RM exhibited 4.61 mA/cm^2 of J_{sc} and 1.80% PCE (Table 4). By adopting the m-MTDATA-solidified MPII, the PCE of the sDSSC-c′ was significantly enhanced compared to that of the DSSC-a′ (only MPII) or the sDSSC-b′ (only m-MTDATA). This tendency for variations in the J_{sc} value is quite consistent with that in the dark currents, as shown in Figure 3b.

It is important to elucidate the origins of the considerable improvement in the photovoltaic performance of the sDSSC-c′. Figure 4 depicts the UV-visible absorption spectra of

MPII and m-MTDATA solutions in chloroform. The MPII solution exhibited an absorption peak at 245 nm, attributable to iodide (I^-) [28,29], and two absorption peaks at 316 nm and 345 nm were observed from the m-MTDATA solution [20,30]. Surprisingly, the absorbance of the two peaks derived from m-MTDATA was increased by adding pure MPII to the m-MTDATA solution, even though the concentration of m-MTDATA was maintained. It has been reported that the absorption peaks of triiodide (I_3^-) are around 290 nm and 360 nm [28,29]. The triiodide's absorption regions are substantially overlapped with the two absorptions peaks of m-MTDATA. We thus noticed that the increased absorbance at both 316 nm and 345 nm was due to the generation of triiodide by any reaction between MPII and m-MTDATA. Meanwhile, although the concentration of MPII in the MPII/m-MTDATA mixture was increased, the absorbance at 316 nm and 345 nm was not further increased. In contrast, the absorbance of iodide (MPII) at 245 nm was increased, as shown in Figure S3 of the ESI. This fact also indicates that the increase in absorbance (at 316 nm and 345 nm) for the MPII/m-MTDATA mixture, compared to that of the pristine m-MTDATA, is not attributable to MPII but instead to the triiodide produced by a chemical reaction between MPII and m-MTDATA. To obtain other evidence of the triiodide formation, IPCEs of the sDSSC-c′ were measured as a function of wavelength. Relatively low IPCE at around 360 nm is generally observed in I^-/I_3^--based DSSCs, because triiodides (I_3^-) absorb the 360 nm light, and therefore N719 dye cannot effectively absorb the light [28]. As displayed in Figure S4 of the ESI, a valley ranging from 350 to 400 nm was observed in the IPCE spectrum of the sDSSC-c′, indicating the presence of triiodides in the sRM (i.e., m-MTDATA-solidified MPII) layer.

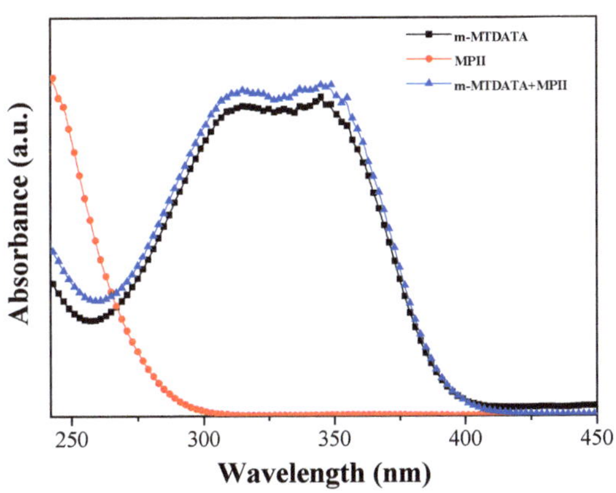

Figure 4. UV–visible absorption spectra of MPII, m-MTDATA, and their mixture in chloroform.

A possible mechanism for triiodide formation can be explained by chemical reactions (1)–(4): m-MTDATA can absorb near UV and blue regions of visible light and can then be excited, as expressed in reaction (1). The excited m-MTDATA reacts with the oxygen molecules infiltrated during the fabrication of cells, and therefore m-MTDATA cations and oxygen anions are formed. It is known that the reduction potential of oxygen molecules is −0.35 V versus a normal hydrogen electrode (NHE) [31], and the lowest-unoccupied-molecular-orbital (LUMO) energy level of m-MTDATA is −2.0 eV of the absolute vacuum scale (AVS) [32], corresponding to −2.5 V versus NHE ($E_{AVS} = -E_{NHE} - 4.50$ eV) [33]. Thus, reaction (2) can occur spontaneously because of the low-lying energy level of oxygen molecules compared to the LUMO level of m-MTDATA. Similar results, i.e., reactions between excited semiconductors and oxygen gases, have been reported [34–36]. Finally, triiodides can be formed via reaction (3) between m-MTDATA cations and some parts of iodides (MPII), and thereby they exist throughout the RM (m-MTDATA-solidified MPII) layer. This occurs because the highest-occupied-molecular-orbital energy level of m-MTDATA

(0.60 V versus NHE or 5.1 eV versus AVS) is approximately 0.25 V higher than the reduction potential of the I^-/I_3^- redox couple (0.35 V versus NHE). Reaction (4) means that the generated oxygen anions can be stabilized with the imidazolium cations (MPI^+) to a certain degree. Overall, by mixing MPII and m-MTDATA, triiodides were formed in the m-MTDATA-solidified MPII. The residual (unreacted) I^- and the produced I_3^- led to the completion of the hole conduction, i.e., the dye regeneration ($3I^- + 2D^+ \rightarrow I_3^- + 2D$, where D indicates N719 dye), near the TiO_2:dye layer and the hole collection ($I_3^- + 2e^- \rightarrow 3I^-$) at the platinized counter electrode. As another contribution to hole conduction, m-MTDATA can transport holes from oxidized dyes to platinized counter electrodes. As presented in Equation (2), oxidized species (m-MTDATA$^+$) was formed by a reaction between m-MTDATA and oxygen molecule, resulting in p-doping of m-MTDATA. This indicated that holes were presented throughout the sRM layer under illumination, enhancing conductivity and faster hole mobility [34].

$$\text{2m-MTDATA} + hv \rightarrow \text{2m-MTDATA}^* \quad (1)$$

$$\text{2m-MTDATA}^* + 2O_2 \rightarrow \text{2m-MTDATA}^+ + 2O_2^- \quad (2)$$

$$\text{2m-MTDATA}^+ + 3I^- \rightarrow \text{2m-MTDATA} + I_3^- \quad (3)$$

$$2MPI^+ + 2O_2^- \rightarrow 2MPI^+ \cdots O_2^- \quad (4)$$

Meanwhile, oxygen molecules permeated into cells during their fabrication can cause the degradation of organic semiconductors [37,38]. It has been reported that the quantum yield for degradation of N719 dye bound on TiO_2 surfaces increases with increasing amounts of oxygen molecules, probably due to the generation of active oxygen species. Thus, oxygen gases in the cells can function as beneficial p-dopant and a harmful medium for dye degradation.

3.3. Effects of Additives on Performance of sDSSCs with m-MTDATA-Solidified MPII

In both metal-back-contact-structured and sandwich-structured sDSSCs, additives such as TBP and LiTFSI were widely utilized for improving photovoltaic performance [7,9–13]. To examine the effects of TBP and LiTFSI incorporated into the m-MTDATA-solidified MPII, we fabricated the sDSSC-d and the sDSSC-e (Table 2), and photovoltaic performance was compared, as shown in Figure 3 and Table 4. By incorporating TBP into the m-MTDATA-solidified MPII, a PCE of 4.20% (J_{sc} = 9.43 mA/cm^2, V_{oc} = 0.610 V, and FF = 73.03%) was achieved in the sDSSC-d′, which corresponded to a 230% enhancement in efficiency compared to that (1.80%) of the sDSSC-c′ without TBP. It has been reported that TBP can be adsorbed onto the free area of the TiO_2 surface, resulting in the shifting of the TiO_2′s conduction band edge (CBE) to a negative direction [39]. This negative shift of the CBE can lead to an increase in V_{oc} due to a broadened potential gap between the CBE (TiO_2) and the redox potential (electrolyte) and a decrease in J_{sc} due to a reduced electron injection efficiency [23,39]. However, in this study, by the addition of TBP, the J_{sc} value was dramatically increased, from 4.61 mA/cm^2 for the sDSSC-c to 9.43 mA/cm^2 for the sDSSC-d. As a reference, weight ratios of TBP, m-MTDATA, and MPII in the sDSSC-d′ were calculated to be 6.0, 37.9, and 56.1 wt%, respectively. We attributed the large increase (around 205%) in J_{sc} in the sDSSC-d′ to the plasticizer effect of TBP. In other words, it was considered that the liquid phase of TBP (T_m = −41.0 °C) at room temperature played the role of a plasticizer in the m-MTDATA-solidified MPII. When a small amount of bulky TBP (6.0 wt% or 8.7 vol%) was added to the m-MTDATA-solidified MPII, it could be inserted between the m-MTDATA molecules, thereby broadening the distance between the solidifying agents. This could induce the faster diffusion of both I^- and I_3^- through the m-MTDATA-modified MPII layer.

To confirm this, we conducted EIS analysis for the sDSSC-c′, the sDSSC-d′, and the sDSSC-e′. The Nyquist plots of the EIS spectra for the sDSSCs measured under AM 1.5 one-sun illumination are shown in Figure 5, providing the sheet resistance (R_s) and interface resistances [40]. Three distinct semicircles were observed. The first and second semicircles corresponded to the carrier transport resistances at the Pt/RM (R_1) and TiO_2/N719/RM (R_2) interfaces, respectively. The last semicircle was ascribed to the Warburg diffusion resistance (w_1) for the ionic transport within the RM. The fitted resistances using Z-view software are compared in Table 5. The w_1 value of the sDSSC-d′ with TBP was lowered to 22.02 Ω from 46.98 Ω of the sDSSC-c′ without TBP. By adding TBP, the ionic diffusion resistance was largely decreased, indicating that more efficient ion (I^- and I_3^-) diffusions between the Pt counter electrode and the dye-adsorbed TiO_2 layer were accomplished in the sDSSC-d′. This fact suggests that TBP acts as a plasticizer in the m-MTDATA-solidified MPII. As a result, redox reactions, such as the dye regeneration ($3I^- + 2D^+ \rightarrow I_3^- + 2D$) near the TiO_2:dye layer and the hole collection [$I_3^- + 2e^-$ (Pt) $\rightarrow 3I^-$] at the platinized counter electrode, can occur more effectively. This can lower both R_1 and R_2 values in the sDSSC-d, as shown in Table 5. Eventually, the sharp reduction in the ionic diffusion resistance in the DSSC-d′, compared to that in the sDSSC-c′, induced about a 205% increase in J_{sc} (Table 4). As a reference, when the EIS measurement is conducted under illumination (open-circuit condition), the R_2 value is affected by the entire resistances, i.e., the electron injection from excited dyes to the conduction band of TiO_2 (R_{inj}), transportation through TiO_2 layer (R_{tra}), recombination between injected electron and electrolyte (R_{rec}), and dye regeneration (R_{reg}). The R_{inj}, R_{rec}, and R_{reg} values are relatively low because electrons move to lower energy levels. However, in the case of R_{tra}, electrons move through the same energy level of TiO_2. Thus, when the EIS measurement is conducted at open-circuit voltage (under illumination), the R_2 value almost depends on the electron transfer resistance [41–43]. As tabulated in Table 5, a lower R_2 value (15 Ω) in the sDSSC-d′ indicates a more efficient charge transfer through the TiO_2 layer, and, thereby, a lower recombination rate between photoinjected electrons and sRM, compared to those of sDSSC-c′ and e′.

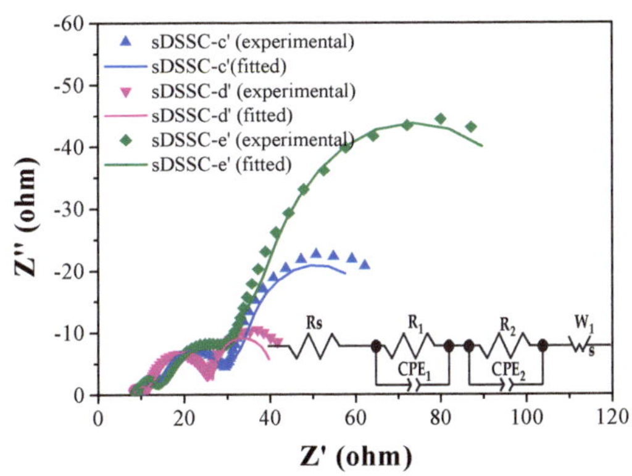

Figure 5. Nyquist plots of EIS spectra for the sDSSC-c′, sDSSC-d′, and sDSSC-e′, measured in an open-circuit condition under the illumination of simulated AM 1.5 solar light. Inset shows the equivalent circuit for sDSSCs; where CPE indicates constant phase element.

Table 5. Fitted resistances for the sDSSC-c′, sDSSC-d′, and sDSSC-e′ based on the m-MTDATA-solidified MPII with or without additives.

Champion Cells	Additive	R_s (Ω)	R_1 (Ω)	R_2 (Ω)	w_1 (Ω)
sDSSC-c′	None	8.36	7.43	19.78	46.98
sDSSC-d′	TBP	8.39	4.41	15.00	22.02
sDSSC-e′	TBP and LiTFSI	8.89	6.53	29.39	103.81

Meanwhile, it has been reported that LiTFSI as a p-dopant can increase conductivity and hole mobility when it is doped in various organic small-molecular and polymeric

HTMs [34,35,44]. We thus fabricated the sDSSC-e with an sRM composed of m-MTDATA-solidified MPII, TBP, and LiTFSI, and compared its photovoltaic performance to that of the sDSSC-d without LiTFSI. The PCE of the sDSSC-e′ was decreased from that of the sDSSC-d′, mainly due to a decrement in J_{sc} (Table 4). In addition, the ion diffusion resistance surged with the addition of LiTFSI (Table 5). We first considered that the LiTFSI addition would lead to an improvement in cell performance by its doping effect, but a decrement in PCE was observed in the sDSSC-e′ with LiTFSI. Xiong and Meng et al. reported that a wax-like complex was formed by mixing TBP with LiTFSI via an interaction between Li^+ (LiTFSI) and a nonbonding electron pair of nitrogen (TBP). The TBP-LiTFSI complex could alleviate the hygroscopicity of LiTFSI and the corrosive effect of TBP, achieving enhanced device efficiency and stability [44]. From this point of view, it is believed that TBP and LiTFSI form a wax-like complex in the sDSSC-e′, and thus TBP can no longer act as a plasticizer. As a result, we can attribute the decreased PCE and increased diffusion resistance to the formation of a wax-like TBP-LiTFSI complex in the sDSSC-e′.

To further confirm the influence of additives (TBP and LiTFSI), we fabricated HOD-c (with m-MTDATA-solidified MPII), HOD-d (with m-MTDATA-solidified MPII and TBP), and HOD-e (with m-MTDATA-solidified MPII, TBP, and LiTFSI), and their *J-V* characteristics were measured. *J-V* curves of HODs fabricated using three different sRMs are shown in Figure S5 of the ESI. We selected the best-performing HODs among 3 devices in each condition to compare their *J-V* properties. As compared in Figure 6, by incorporating TBP into m-MTDATA-solidified MPII, dark currents of the HOD-d′ were sharply increased from those of the HOD-c′ without TBP, indicating that TBP acted as a plasticizer in m-MTDATA-solidified MPII. Furthermore, dark currents of the HOD-e′ were again decreased by adding LiTFSI, probably due to the formation of wax-like TBP-LiTFSI complexes, causing a loss of TBP's function as a plasticizer. These tendencies in the dark currents of the HODs were quite consistent with the J_{sc} values of their counterparts, i.e., sDSSC-c′, sDSSC-d′, and sDSSC-e′.

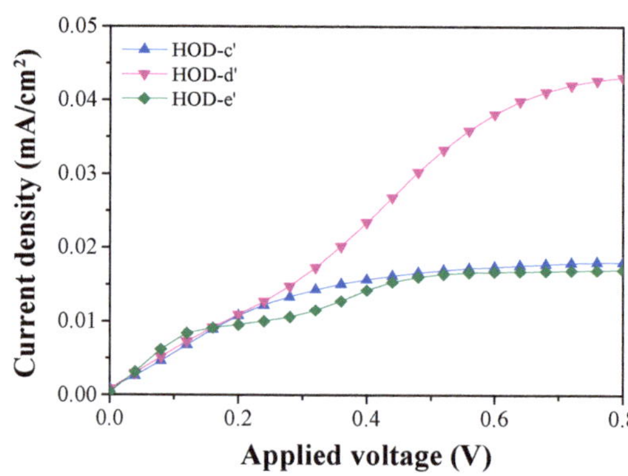

Figure 6. *J-V* characteristics of the HOD-c′, HOD-d′, and HOD-e′.

Figure 7 compares IPCE spectra for the champion cells such as sDSSC-c′, d′, and e′. As can be expected, higher IPCEs were recorded in the sDSSC-d′ by incorporation of TBP into the sDSSC-c′ with m-MTDATA-solidified MPII. IPCE values of the sDSSC-e′ were again reduced by adding LiTFSI. The tendencies of IPCE variations with additives were well consistent with the measurement results of the J_{sc} values (Figure 3a), dark currents (Figure 3b), interface resistances (Figure 5), and HOD's dark currents (Figure 6).

As a result, the highest PCE of 4.20% in an sDSSC was achieved by adopting an sRM composed of m-MTDATA-solidified MPII and TBP. Although this efficiency was lower than those of iodine-free P3HT/MPII (5.40%) [13] and polyaniline-loaded carbon black/MPII (5.81%) [6], this study's result has great importance. It is because the efficiency was accomplished in sandwich-structured sDSSCs, in which the thickness of the hole-transporting layer was approximately 10 μm as shown in Figure S2. In P3HT/MPII-based sDSSC, adhesives for sealing working and counter electrodes were not used, and thus TiO_2 photoanode was directly contacted with a counter electrode, allowing very short ion diffusion length [13]. In sDSSC with polyaniline-loaded carbon black/MPII, carbon blacks coated on counter electrode functioned electrical conductive pathway from counter electrode to TiO_2 photoanode, and therefore ion diffusion length could be minimized [6]. Overall, it is believed that the PCE of 4.20% is an impressive achievement because it is extracted from the sDSSC with a 10 μm-thick hole transporting layer.

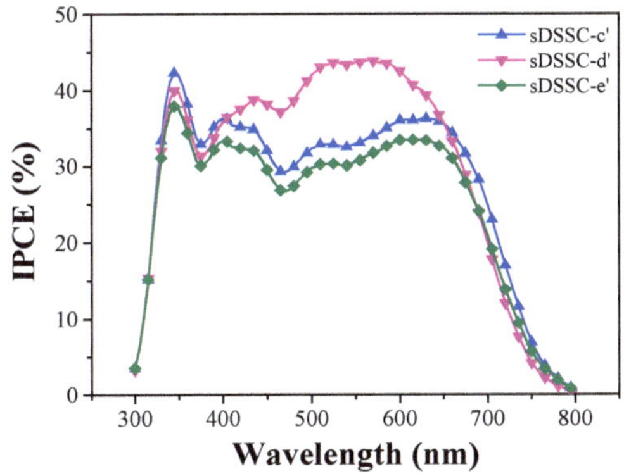

Figure 7. IPCE spectra for sDSSC-c′, d′ and e′ as a function of wavelength.

4. Conclusions

An sRM based on an ionic liquid (MPII) and a hole-transporting triphenylamine compound (m-MTDATA) was successfully prepared and applied for applications to sDSSCs. The sDSSC-c′ with the m-MTDATA-solidified MPII showed a PCE of 1.80% with 4.61 mA/cm^2 of J_{sc}, 0.544 V of V_{oc}, and 71.81% of *FF*. The UV-visible absorption and IPCE studies disclosed that a chemical reaction between MPII and m-MTDATA cation resulted in the formation of triiodides (I_3^-), allowing hole conduction from the oxidized dye to the counter electrode. In addition, by incorporating TBP as an additive into the m-MTDATA-solidified MPII, the PCE of the sDSSC-d′ was sharply increased by 4.20% (J_{sc} = 9.43 mA/cm^2, V_{oc} = 0.610 V, and *FF* = 73.03%), due to improvements in all photovoltaic parameters, and the ion diffusion resistance was considerably decreased. These results were ascribed to the plasticizer effect of the bulky-structured TBP. As a result, the sRM composed of m-MTDATA-solidified MPII and TBP appears to be a promising material for replacing conventional organic-solvent-based liquid RMs, and thereby for realizing the sDSSCs.

Supplementary Materials: The following supporting information can be downloaded at https://www.mdpi.com/article/10.3390/en15082765/s1. Figure S1: Photographs of redox mediators; (a) MPII, (b) m-MTDATA-solidified MPII (MPII + m-MTDATA), (c) m-MTDATA-solidified MPII + TBP, and (d) m-MTDATA-solidified MPII + TBP + LiTFSI. Table S1: Photovoltaic parameters of DSSCs with five different types of redox mediator. Figure S2: Cross-sectional SEM image of the sDSSC-b′ with only m-MTDATA. Figure S3: Absorbance variations of the m-MTDATA/MPII mixtures with increasing concentrations of MPII; the concentration of m-MTDATA in the mixture was constant. For comparison, the absorption spectra of m-MTDATA and MPII were included. Figure S4: IPCE spectrum of the sDSSC-c′ with m-MTDATA-solidified MPII. The arrow-marked valley ranging from 350 to 400 nm was attributed to absorption by I_3^- ions in sRM (i.e., m-MTDATA-solidified MPII) layer. Figure S5: Dark current density-voltage characteristics of HODs with three different types of RM; (a) HOD-c with m-MTDATA-solidified MPII, (b) HOD-d with m-MTDATA-solidified MPII/TBP, and (c) HOD-e with m-MTDATA-solidified MPII/TBP/LiTFSI.

Author Contributions: Conceptualization, Y.S.H.; methodology, M.K. and D.H.O.; formal analysis, M.K., D.H.O. and B.C.; writing—original draft preparation, Y.S.H.; writing—review and editing, Y.S.H.; funding acquisition, Y.S.H. All authors have read and agreed to the published version of the manuscript.

Funding: This work was supported by research grants from Deagu Catholic University in 2021 (20211058).

Data Availability Statement: Data are contained within the article or the Supplementary Material.

Conflicts of Interest: The authors declare no conflict of interest. The funders had no role in the design of the study; in the collection, analyses, or interpretation of data; in the writing of the manuscript; or in the decision to publish the results.

References

1. O'Regan, B.; Grätzel, M. A low-cost, high-efficiency solar cell based on dye-sensitized colloidal TiO_2 films. *Nature* **1991**, *353*, 737–740. [CrossRef]
2. Gong, J.; Sumathy, K.; Qiao, Q.; Zhou, Z. Review on dye-sensitized solar cells (DSSCs): Advanced techniques and research trends. *Renew. Sustain. Energ. Rev.* **2017**, *68*, 234–246. [CrossRef]
3. Sharma, K.; Sharma, V.; Sharma, S.S. Dye-sensitized solar cells: Fundamentals and current status. *Nanoscale Res. Lett.* **2018**, *13*, 381. [CrossRef] [PubMed]
4. Devadiga, D.; Selvakumar, M.; Shetty, P.; Santosh, M.S. Recent progress in dye sensitized solar cell materials and photo-supercapacitors: A review. *J. Power Sources* **2021**, *493*, 229698. [CrossRef]
5. Kakiage, K.; Aoyama, Y.; Yano, T.; Oya, K.; Fujisawa, J.I.; Hanaya, M. Highly-efficient dye-sensitized solar cells with collaborative sensitization by silyl-anchor and carboxy-anchor dyes. *Chem. Commun.* **2015**, *51*, 15894–15897. [CrossRef]
6. Lee, C.P.; Chen, P.Y.; Vittal, R.; Ho, K.C. Iodine-free high efficient quasi solid-state dye-sensitized solar cell containing ionic liquid and polyaniline-loaded carbon black. *J. Mater. Chem.* **2010**, *20*, 2356–2361. [CrossRef]
7. Benesperi, I.; Michaels, H.; Freitag, M. The researcher's guide to solid-state dye-sensitized solar cells. *J. Mater. Chem.* **2018**, *6*, 11903–11942. [CrossRef]
8. Chung, I.; Lee, B.; He, J.; Chang, R.P.H.; Kanatzidis, M.G. All-solid-state dye-sensitized solar cells with high efficiency. *Nature* **2012**, *485*, 486–489. [CrossRef]
9. Lee, B.; Stoumpos, C.C.; Zhou, N.; Hao, F.; Malliakas, C.; Yeh, C.Y.; Marks, T.J.; Kanatzidis, M.G.; Chang, R.P.H. Air-stable molecular semiconducting iodosalts for solar cell applications: Cs_2SnI_6 as a hole conductor. *J. Am. Chem. Soc.* **2014**, *136*, 15379–15385. [CrossRef]
10. Cao, Y.; Saygili, Y.; Ummadisingu, A.; Teuscher, J.; Luo, J.; Pellet, N.; Giordano, F.; Zakeeruddin, S.M.; Moser, J.E.; Freitag, M.; et al. 11% efficiency solid-state dye-sensitized solar cells with copper(II/I) hole transport materials. *Nat. Commun.* **2016**, *8*, 15390. [CrossRef]
11. Freitag, M.; Daniel, Q.; Pazoki, M.; Sveinbjornsson, K.; Zhang, J.; Sun, L.; Hagfeldt, A.; Boschloo, G. High-efficiency dye-sensitized solar cells with molecular copper phenanthroline as solid hole conductor. *Energy Environ. Sci.* **2015**, *8*, 2634–2637. [CrossRef]
12. Kashif, M.K.; Milhuisen, R.A.; Nippe, M.; Hellerstedt, J.; Zee, D.Z.; Duffy, N.W.; Halstead, B.; Angelis, F.D.; Fantacci, S.; Fuhrer, M.S.; et al. Cobalt polypyridyl complexes as transparent solution-processable solid-state charge transport materials. *Adv. Energy Mater.* **2016**, *6*, 1600874. [CrossRef]
13. Koh, J.K.; Kim, J.; Kim, B.; Kim, J.H.; Kim, E. Highly efficient, iodine-free dye-sensitized solar cells with solid-state synthesis of conducting polymers. *Adv. Mater.* **2011**, *23*, 1641–1646. [CrossRef] [PubMed]
14. Lee, K.M.; Chen, P.Y.; Lee, C.P.; Ho, K.C. Binary room-temperature ionic liquids based electrolytes solidified with SiO_2 nanoparticles for dye-sensitized solar cells. *J. Power Sources* **2009**, *190*, 573–577. [CrossRef]
15. Lee, C.P.; Lee, K.M.; Chen, P.Y.; Ho, K.C. On the addition of conducting ceramic nanoparticles in solvent-free ionic liquid electrolyte for dye-sensitized solar cells. *Sol. Energy Mater. Sol. Cells* **2009**, *93*, 1411–1416. [CrossRef]
16. Katakabe, T.; Kawano, R.; Watanabe, M. Acceleration of redox diffusion and charge-transfer rates in an ionic liquid with nanoparticle addition. *Electrochem. Solid-State Lett.* **2007**, *10*, F23–F25. [CrossRef]
17. Chen, Z.; Yang, H.; Li, X.; Li, F.; Yi, T.; Huang, C. Thermostable succinonitrile-based gel electrolyte for efficient, long-life dye-sensitized solar cells. *J. Mater. Chem.* **2007**, *17*, 1602–1607. [CrossRef]
18. Usui, H.; Matsui, H.; Tanabe, N.; Yanagida, S. Improved dye-sensitized solar cells using ionic nanocomposite gel electrolytes. *J. Photochem. Photobiol. A Chem.* **2004**, *164*, 97–101. [CrossRef]
19. Wang, P.; Zakeeruddin, S.M.; Comte, P.; Exnar, I.; Grätzel, M. Gelation of ionic liquid-based electrolytes with silica nanoparticles for quasi-solid-state dye-sensitized solar cells. *J. Am. Chem. Soc.* **2003**, *125*, 1166–1167. [CrossRef]
20. Shan, M.; Jiang, H.; Guan, Y.; Sun, D.; Wang, Y.; Hua, J.; Wang, J. Enhanced hole injection in organic light-emitting diodes utilizing a copper iodide-doped hole injection layer. *RSC Adv.* **2017**, *7*, 13584–13589. [CrossRef]
21. Kim, J.Y.; Kwak, G.; Choi, Y.C.; Kim, D.H.; Han, Y.S. Enhanced performance of perovskite solar cells by incorporation of a triphenylamine derivative into hole-transporting poly(3-hexylthiophene) layers. *J. Ind. Eng. Chem.* **2019**, *73*, 175–181. [CrossRef]

22. Liu, C.; Zhang, D.; Li, Z.; Zhang, X.; Shen, L.; Guo, W. Efficient 4,4′,4″-tris(3-methylphenylphenylamino)triphenylamine (m-MTDATA) hole transport layer in perovskite solar cells enabled by using the nonstoichiometric precursors. *Adv. Funct. Mater.* **2018**, *28*, 1803126. [CrossRef]
23. Baek, G.W.; Kim, Y.J.; Jung, K.H.; Han, Y.S. Enhancement of solar cell performance through the formation of a surface dipole on polyacrylonitrile-treated TiO_2 photoelectrodes. *J. Ind. Eng. Chem.* **2019**, *73*, 260–267. [CrossRef]
24. Kong, M.; Kim, K.S.; Nga, N.V.; Lee, Y.; Jeon, Y.S.; Cho, Y.; Kwon, Y.; Han, Y.S. Molecular weight effects of biscarbazole-based hole transport polymers on the performance of solid-state dye-sensitized solar cells. *Nanomaterials* **2020**, *10*, 2516. [CrossRef]
25. Boschloo, G.; Hagfeldt, A. Characteristics of the iodide/triiodide redox mediator in dye-sensitized solar cells. *Acc. Chem. Res.* **2009**, *42*, 1819–1826. [CrossRef] [PubMed]
26. Longeaud, C.; Allah, A.F.; Schmidt, J.; Yaakoubi, M.E.; Berson, S.; Lemaitre, N. Determination of diffusion lengths in organic semiconductors: Correlation with solar cell performances. *Org. Electron.* **2016**, *31*, 253–257. [CrossRef]
27. Mikhnenko, O.V.; Blom, P.W.M.; Nguyen, T.Q. Exciton diffusion in organic semiconductors. *Energy Environ. Sci.* **2015**, *8*, 1867–1888. [CrossRef]
28. Kim, M.J.; Lee, C.R.; Jeong, W.S.; Im, J.H.; Ryu, T.I.; Park, N.G. Unusual enhancement of photocurrent by incorporation of bronsted base thiourea into electrolyte of dye-sensitized solar cell. *J. Phys. Chem.* **2010**, *114*, 19849–19852. [CrossRef]
29. Bonomo, M.; Carlo, A.D.; Dini, D. Study of the influence of the I-based electrolyte composition on the photoconversion properties of p-type dye-sensitized solar cell. *J. Electrochem. Soc.* **2018**, *165*, H889–H896. [CrossRef]
30. Sakurai, M.; Kabe, R.; Fuki, M.; Lin, Z.; Jinnal, K.; Kobori, Y.; Adachi, C.; Tachikawa, T. Organic photostimulated luminescence associated with persistent spin-correlated radical pairs. *Commun. Mater.* **2021**, *2*, 74. [CrossRef]
31. Koppenol, W.H.; Stanbury, D.M.; Bounds, P.L. Electrode potentials partially reduced oxygen species, from dioxygen to water. *Free Radic. Biol. Med.* **2010**, *49*, 317–322. [CrossRef] [PubMed]
32. Chulkin, P.; Lapkowski, M.; Bryce, M.R.; Santos, J.; Data, P. Determination of standard redox rate constants of OLED active compounds by electrochemical impedance spectroscopy. *Electrochim. Acta* **2017**, *258*, 1160–1172. [CrossRef]
33. Sung, H.K.; Lee, Y.; Kim, W.H.; Lee, S.J.; Sung, S.J.; Kim, D.H.; Han, Y.S. Enhanced power conversion efficiency dye-sensitized solar cells by band edge shift of TiO_2 photoanode. *Molecules* **2020**, *25*, 1502. [CrossRef] [PubMed]
34. Abate, A.; Leijtens, T.; Pathak, S.; Teusher, J.; Avolio, R.; Errico, M.E.; Kirkpatrik, J.; Ball, J.M.; Docampo, P.; McPherson, I.; et al. Lithium salts as "redox active" p-type dopants for organic semiconductors and their impact in solid-state dye-sensitized solar cells. *Phys. Chem. Chem. Phys.* **2013**, *15*, 2575–2579. [CrossRef]
35. Cappel, U.B.; Daeneke, T.; Bach, U. Oxygen-induced doping of spiro-MeOTAD in solid-state dye-sensitized solar cells and its impact on device performance. *Nano Lett.* **2012**, *12*, 4925–4931. [CrossRef]
36. Xu, B.; Gabrielsson, E.; Safdari, M.; Cheng, M.; Hua, Y.; Tian, H.; Gardner, J.M.; Kloo, L.; Sun, L. 1,1,2,2-Tetrachloroethane (TeCA) as a solvent additive for organic hole transport materials and its application in highly efficient solid-state dye-sensitized solar cells. *Adv. Energy Mater.* **2015**, *5*, 1402340. [CrossRef]
37. Quan, V.A. Degradation of the Solar Cell Dye Sensitizer N719 Preliminary Building of Dye-Sensitized Solar Cell. Master's Thesis, Roskide University, Roskide, Denmark, June 2006.
38. Schafferhans, J.; Baumann, A.; Wagernpfahl, A.; Deibel, C.; Dyakonov, V. Oxygen doping of P3HT:PCBM blends: Influence on trap states, charge carrier mobility and solar cell performance. *Org. Electron.* **2010**, *11*, 1693–1700. [CrossRef]
39. Kusama, H.; Arakawa, H. Influence of aminotriazole additives in electrolytic solution on dye-sensitized solar cell performance. *J. Photochem. Photobio. A Chem.* **2004**, *164*, 103–110. [CrossRef]
40. Shaikh, S.F.; Mane, R.S.; Min, B.K.; Hwang, Y.J.; Joo, O.S. D-sorbitol-induced phase control of TiO_2 nanoparticles and its application for dye-sensitized solar cells. *Sci. Rep.* **2016**, *6*, 20103. [CrossRef]
41. Xu, J.; Fan, K.; Shi, W.; Li, K.; Peng, T. Application of ZnO micro-flowers as scattering layer for ZnO-based dye-sensitized solar cells with enhanced conversion efficiency. *Sol. Energy* **2014**, *101*, 150–159. [CrossRef]
42. Park, J.H.; Kim, J.Y.; Kim, J.H.; Choi, C.J.; Kim, H.; Sung, Y.E.; Ahn, K.S. Enhanced efficiency of dye-sensitized solar cells through $TiCl_4$-treated, nanoporous-layer-covered TiO_2 nanotube arrays. *J. Power Sources* **2011**, *196*, 8904–8908. [CrossRef]
43. Ambade, S.B.; Ambade, R.B.; Mane, R.S.; Lee, G.W.; Shaikh, S.F.; Patil, S.A.; Joo, O.S.; Han, S.H.; Lee, S.H. Low temperature chemically synthesized rutile TiO_2 photoanodes with high electron lifetime for organic dye-sensitized solar cells. *Chem. Commun.* **2013**, *49*, 2921–2923. [CrossRef] [PubMed]
44. Wang, S.; Huang, Z.; Wang, X.; Li, Y.; Gunther, M.; Valenzuela, S.; Parikh, P.; Cabreros, A.; Xiong, W.; Meng, Y.S. Unveiling the role of tBP−LiTFSI complexes in perovskite solar cells. *J. Am. Chem. Soc.* **2018**, *140*, 16720–16730. [CrossRef] [PubMed]

energies

Article

Modeling of Dielectric Electroactive Polymer Actuators with Elliptical Shapes

Jakub Bernat * and Jakub Kołota

Faculty of Control, Robotics and Electrical Engineering, Poznan University of Technology, 60-965 Poznan, Poland; jakub.kolota@put.poznan.pl
* Correspondence: jakub.bernat@put.poznan.pl

Abstract: Dielectric electroactive polymers have been widely used in recent applications based on smart materials. The many advantages of dielectric membranes, such as softness and responsiveness to electric stimuli, have lead to their application in actuators. Recently, researchers have aimed to improve the design of dielectric electroactive polymer actuators. The modifications of DEAP actuators are designed to change the bias mechanism, such as spring, pneumatic, and additional mass, or to provide a double cone configuration. In this work, the modification of the shape of the actuator was analyzed. In the standard approach, a circular shape is often used, while this research uses an elliptical shape for the actuator. In this study, it was shown that this construction allows a wider range of movement. The paper describes a new design of the device and its model. Further, the device is verified by the measurements.

Keywords: DEAP actuator; dielectric electroactive polymers; dielectric elastomer actuator; smart materials

Citation: Bernat, J.; Kołota, J. Modeling of Dielectric Electroactive Polymer Actuators with Elliptical Shapes. *Energies* **2021**, *14*, 5633. https://doi.org/10.3390/en14185633

Academic Editor: Alessandro Cannavale

Received: 21 July 2021
Accepted: 2 September 2021
Published: 8 September 2021

1. Introduction

Due to the dynamic development of automation and robotics, alternative actuator solutions have increasingly been used in recent times. This can be seen in particular in soft robotics, where actuators or sensors are used to convert energy in smart materials. One group of such smart materials are Dielectric Electroactive Polymers (DEAPs). DEAPs offer excellent performance, are flexible, lightweight and inexpensive. These materials have developed very dynamically in recent times and are increasingly used in innovative constructions, such as pumps, robots, valves or micro-positioning systems [1–3].

Accurate modeling of the DEAP actuator phenomena is important for designers working with these materials. It is an important problem which is still extensively analyzed in the literature [2,4–6]. One of the interesting issues is the analysis of actuator shapes. The most common actuators used in the solutions presented so far have the shape of a circle. This paper presents an elliptical actuator model described with a circular approximation. The paper presents the experiments carried out for three different actuators with different geometries. Finite element model (FEM) modeling of the behavior of DEAPs is useful to understand such systems better and help in the optimal design of prototypes [7,8]. An FEM-based simulation was performed, which demonstrated its applicability to compare stress distribution in the three configurations of DEAP actuators.

DEAP actuators can operate with a variety of input and output signals, but in the most common configuration the voltage is the input and displacement is the output [4,9,10]. In order to increase the operating range of DEAP actuators, various biasing mechanism are used, for example, a mechanical spring [3,4,9]. There are also other solutions to achieve greater tension, such as magnetic coupling [11]. In this research, the actuator membranes were loaded with additional mass, similar to the works [6,12,13].

Measurements of the displacement of elliptical and circular actuators were analyzed, and then the process of identifying model parameters was discussed. The process of identifying the dynamic parameters of the model with the use of the actuators' displacement responses to the step input voltage is presented and illustrated.

2. Motivation

In this section, the motivation of this study is presented. In recent times, the circular shape of DEAP actuator has commonly been applied in different configurations [14–16]. In the presented work, the idea is to exploit the anisotropy of many devices (such as, for instance, a pipe with an installed pump), which allows one direction to be extended while the second must be limited. The motivation of our modification is to provide more energy in the device. Let us consider two cases: a circle with radius r_1 and an ellipse with semi-minor axis r_1 and semi-major axis r_2, satisfying $r_2 > r_1$. The area of circle $S_c = \pi r_1^2$ and area of ellipse is $S_e = \pi r_1 r_2$; hence, $S_e > S_c$. If the circle or ellipse is covered by the electrodes like in the DEAP actuators, the capacity of both devises approximated by parallel plate capacitor will be:

$$C_c = \varepsilon \frac{S_c}{d} \quad C_e = \varepsilon \frac{S_e}{d}. \tag{1}$$

where d is the distance between the electrodes and ε is the permittivity of the material. It is clear that $C_c < C_e$; therefore, the energy stored in the elliptical device under a constant voltage ($E = \frac{1}{2}Cu^2$, [17]) is larger than in the circular device. Therefore, it is expected that the elliptical shape will cause different responses compared to the circular shape.

3. DEAP Elliptical Actuator Model

The DEAP actuator models have been widely studied in the literature [1,6,18,19], showing the most significant phenomena in these devices. This work applies the knowledge of these works to analyze the proposed elliptical actuator. The mass biased actuator was chosen for simplicity of biasing mechanics [6]. However, it is also possible to apply the proposed work with different bias mechanics [15,16,19]. The aim is to model the elliptical actuator as an extension of the circular one. The elliptical shape actuator model was defined using two circular model actuators ($N = 2$), as presented in Figure 1. The approximation was based on the use of a model consisting of two circular actuators with radii corresponding to the minor (r_1) and major (r_2) semi-axis of the elliptical actuator.

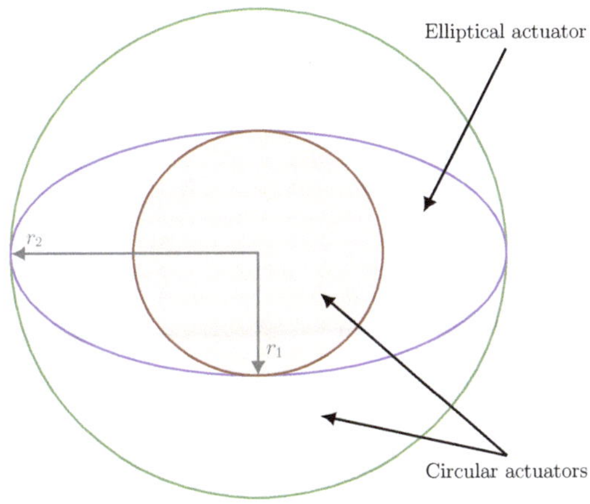

Figure 1. Elliptical DEAP actuator approximated by two circular actuators.

In the first step, the basic relationships for two circular models (indexed by $j = 1, 2$ for simplicity) are redefined. In general, the DEAP membrane has the following property:

$$\lambda_{rj} \lambda_{zj} \lambda_c = 1 \tag{2}$$

which comes from the assumption on the constant volume of material during deformation. The variables λ_{rj}, λ_c and λ_{zj} are radial, circumferential, and vertical stretches (taking into account also the initial prestretch of the DEAP membrane applied in the production process [20]).

The basic formulas of the DEAP actuator are defined for two circular actuators:

$$\lambda_{rj} = \frac{l_j}{l_{0j}} = \frac{\sqrt{l_{0j}^2 + y^2}}{l_{0j}} = \sqrt{s_j(y)},\ \lambda_{zj} = \frac{z_j}{z_0},\ \lambda_c = 1 \quad (3)$$

The variables for the undeflected state (without applied mass and voltage) are given by z_0 and l_{0j}, which are the membrane initial thickness and the electrode width. The membrane is assumed to be the same for all elliptical shapes (with circular prestretch); therefore, the initial thickness is common for both models. The electrode width depends on the length of minor/major semi-axis; hence, it must include the index of the actuator. In the case of the deflected state (with applied mass or voltage), the variables are z_j and l_j. In this case, both variables are different due to different stretch for small and large circular actuators. The function $s_j(y)$ is applied to clarify the notation. To simplify the analysis of relationship, the cross-sections of both actuators (for short and long axis) are presented in Figure 2. It is assumed that both actuators move over the same distance y because they express the single elliptical actuator. However, it is worth pointing out that angles θ_j will be different for both axes:

$$sin(\theta_j) = \frac{y}{\sqrt{l_{0j}^2 + y^2}}. \quad (4)$$

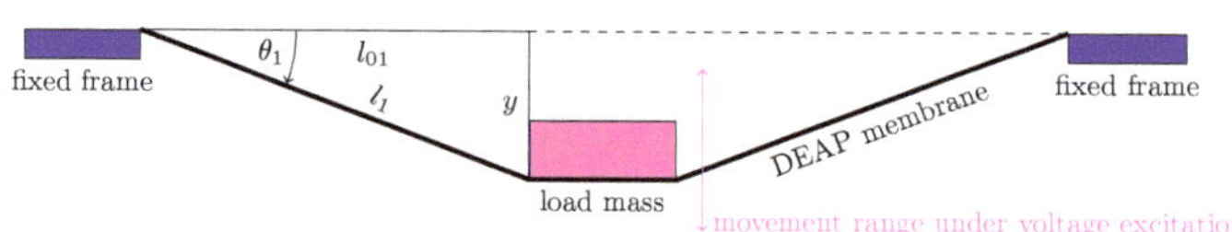

(**a**) short axis

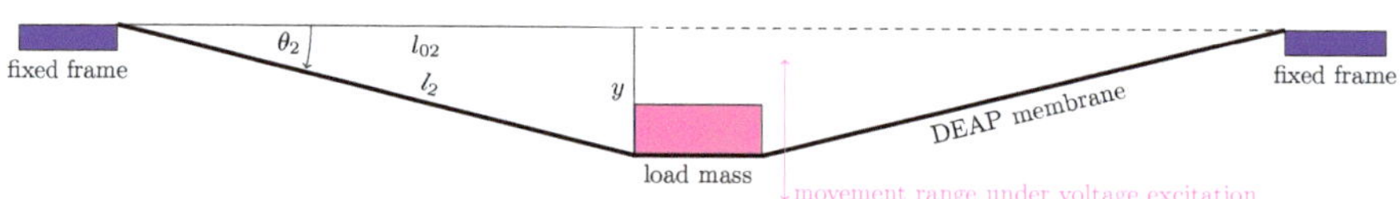

(**b**) long axis

Figure 2. The variables of DEAP actuator for cross-section through short/long axes of the elliptical actuator.

In the second step, the vertical force equilibrium is considered taking into account the biasing mass:

$$\ddot{y} = g - \sum_{j=1}^{N} \alpha_j \left[\frac{2\pi r z_j}{m} sin(\theta_j) \left(-\frac{\varepsilon_0 \varepsilon_r}{z_0^2} s_j(y) u^2 + \sigma_j \right) \right] \quad (5)$$

where g denotes the standard gravity and r is the radius of the internal mass m, ε_0 is vacuum permittivity, ε_r is the relative permittivity of the actuator membrane, u is the applied voltage and σ_j is the mechanical stress of the individual circular actuators.

The influence of small and large circular actuators on the elliptical one is weighted by the choice of the appropriate coefficient $\alpha_j \geq 0$. It was assumed that the behaviour of the elliptical actuator will be between the small and large actuator; hence, the sum of these factors is 1:

$$\sum_{j=1}^{N} \alpha_j = 1. \tag{6}$$

It is worth emphasizing that the presented model of the elliptical model extends the circular model. If it was considered that an ellipse is a circle ($r_1 = r_2$), then the model is reduced to single circle, as presented in [6].

Circular actuators approximating the description of the elliptical actuator have different mechanical stresses σ_j:

$$\sigma_j = \sigma_{hj} + \sigma_{v_1j} + \sigma_{v_2j} \tag{7}$$

where: σ_{hj} defines the Ogden model representing the hyperelastic properties of the DEAP membrane, $\sigma_{v_{1j}}$ is the viscoelastic stress and $\sigma_{v_{2j}}$ specifies the viscous damper.

The applied third level ($i = 1,2,3$) Ogden model, with two parameters, β_i and γ_i, integrates the prestretch into the strain energy function. The identification of the β_i and γ_i parameters was carried out taking into account prestretch; thus, the obtained values of the Ogden model take into account the phenomenon of initial stretching of the actuator membrane.

$$\sigma_{hj} = \sum_{i=1}^{3} \left(\beta_i \lambda_{rj}^{2i} - \gamma_i \lambda_{rj}^{-2i} \right) \tag{8}$$

The phenomenon of viscoelasticity of the individual circular cylinders was modeled by a series connection of a viscous damper and an elastic spring (σ_{v_1j}) and the viscous damper (σ_{v_2j}) connected parallel to them, as presented in Figure 3.

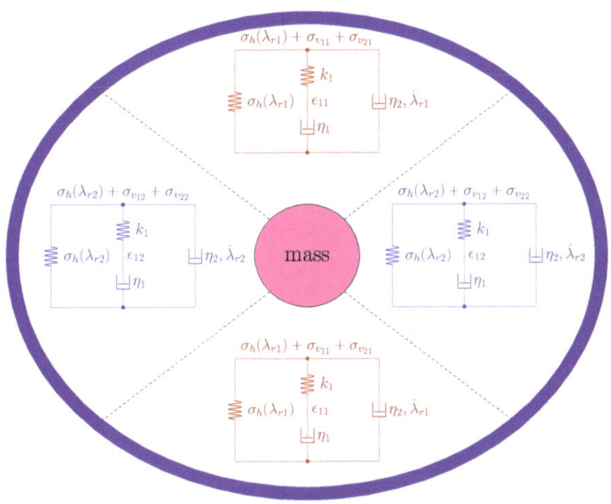

Figure 3. The composition of two material models and their placement in the elliptical actuator.

$$\begin{aligned} \sigma_{v_1j} &= -k_1\epsilon_{1j} + k_1(\lambda_{rj} - 1) \\ \dot{\epsilon}_{1j} &= -\frac{k_1}{\eta_1}\epsilon_{1j} + \frac{k_1}{\eta_1}(\lambda_{rj} - 1) \end{aligned} \tag{9}$$

$$\begin{aligned} \sigma_{v_2j} &= \eta_2\dot{\lambda}_{rj} = \eta_2\varphi_j(y)\dot{y} \\ \varphi_j(y) &= \frac{y}{l_{0j}\sqrt{l_{0j}^2 + y^2}} \end{aligned} \tag{10}$$

Each of the circular actuators has different λ_{rj}, so they have different strains of the damper ϵ_{1j}, while maintaining the same material properties (k_1, η_1 and η_2), whose values and other information are listed in Table 1 (the identification process is described in the following section).

The Stress Analysis of the Actuator's New Geometries

In this section, FEM analysis of the proposed actuators is performed. The FEM simulation was included to show the stress distribution in circular and elliptical actuators for the same material parameters. The goal of analysis is to see the influence of geometry change on the DEAP actuator. The main advantage of circular geometry is its radial symmetry. However, as will be shown in the experimental section, the elliptical shape allows a wider range of movement. As an example, three geometries are considered: circular, small elliptical and large elliptical. The internal plate has a constant shape of the cylinder. The outer side of the membrane is fixed and a pressure is applied to the internal plate. The Young modulus and Poisson ratio are the same for all configurations. The simulations were performed by the FEM module in FreeCAD software. In all cases, the membrane deformed to counteract pressure. The deformation was largest for the large elliptical shape and smallest for the circular shape. Using these simulations, it was shown that this distribution depends on the shape of the actuator. In Figure 4, the stress in all cases is shown. It is visible that for the circular shape the stress has a radial symmetry. In the case of the elliptical shape, the stress is larger for the shorter axis than for longer axis. It is especially visible for the large elliptical shape. However, the ultimate goal is to control the object-oriented model of elliptical actuators. The simulations included in this section are only a visualization of the distribution of these stresses in the three DEAP actuator configurations.

(**a**) circular (**b**) elliptical-small (**c**) elliptical-large

Figure 4. Comparison of stress in the three configurations (circular and small/large elliptical).

4. Experiments

The model presented above was experimentally verified with the use of three DEAP actuators with different geometries. The actuators were made of acrylic membrane (3M VHB tape), which was stretched on a plexiglass frame. The VHB tape was also prestretched, similarly to [20]. To obtain the same prestretch for all actuators, the prestretch was carried out for a circular frame larger than the largest elliptical actuator. The prestretch factor was 5, providing a change in thickness from 1 mm to 200 µm. The membrane is elastic (with a Poisson ratio slightly below 0.5), and hence the volume of the membrane during stretching process was constant. The electrodes were made of carbon grease and covered the entire surface of the actuators on both sides. The electrodes created a capacitor whose electrostatic field caused membrane compression. One actuator was circular, while the other two had elliptical shapes. The dimensions of the actuators are presented in Table 1. In the center of each actuator, there was a plexiglass circle on which additional mass was placed during the experiments. Two copper connections were glued to the carbon electrodes to apply the supply voltage (Figure 5).

The experiments were performed on the following hardware: a high voltage amplifier TREK MODEL 10/10B-HS, a laser distance sensor Micro-Epsilon optoNCDT ILD1320-10 with 1 μm accuracy, and an Inteco RT-DAC/USB data acquisition card. The laboratory set-up is presented in Figure 6. The system was measuring data with a probe time of 1 ms.

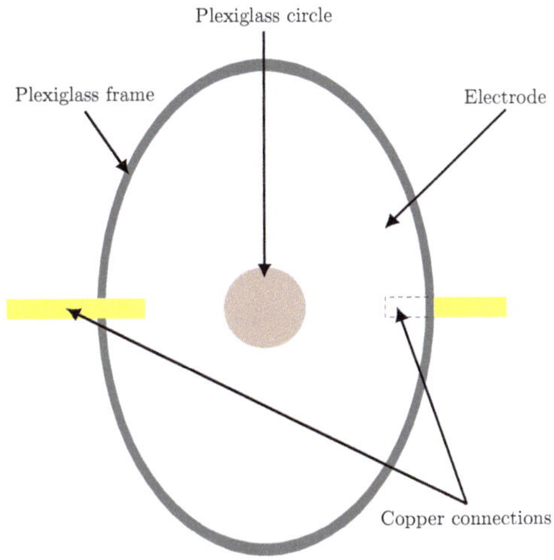

(**a**) schema

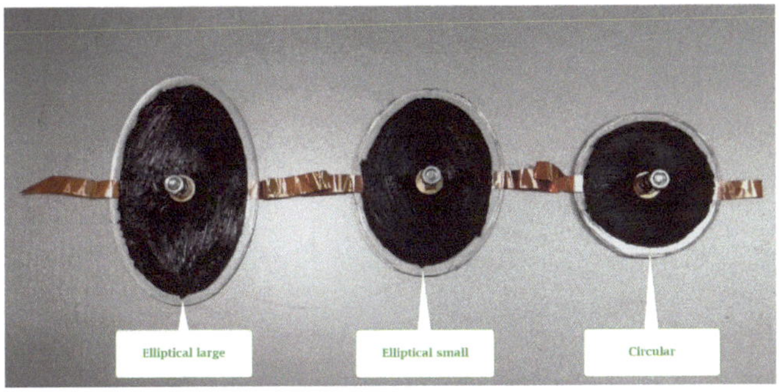

(**b**) samples

Figure 5. The schema of DEAP actuators and performed samples.

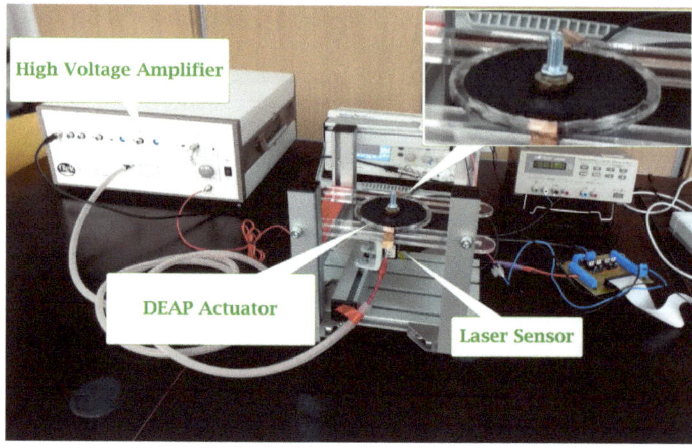

Figure 6. The laboratory setup of DEAP actuator measurement.

Table 1. Parameters of DEAP actuator models.

Parameter	Symbol	Value			Units
Actuator name		Circular	Elliptical small	Elliptical large	
Inner radius	r		1		cm
Outer radius (short)	r_1		4.5		cm
Outer radius (wide)	r_2	4.5	6	7.5	cm
Electrode surface	S	60.5	81.7	102.9	cm^2
Membrane initial thickness	z_{ini}		1		mm
Membrane final thickness	z_0		200		µm
Standard gravity	g		9.81		$m\,s^{-2}$
Coefficient	α_1		0.73		-
Coefficient	α_2		0.27		-
Vacuum permittivity	ε_0		$8.85 \cdot 10^{-12}$		$N\,m^{-1}$
Relative permittivity	ε_r		6.87		-
Coefficient of viscoelastic model	k_1		23.5		MPa
Coefficient of viscoelastic model	η_1		177.2		MPa s
Damping coefficients	η_2		62.4		kPa s
Hyperelastic model coefficient	β_1		13.4		kPa
Hyperelastic model coefficient	β_2		34.8		kPa
Hyperelastic model coefficient	β_3		35.8		kPa
Hyperelastic model coefficient	γ_1		−0.41		kPa
Hyperelastic model coefficient	γ_2		−113		kPa
Hyperelastic model coefficient	γ_3		−34.2		kPa

4.1. Identification-Static Parameters

The identification procedure of the circular actuator was studied in the previous works [6,19]. The presented approach extends the previous procedure to identify three models with the same set of mechanical and electromechanical parameters. Firstly, the measured static characteristics were exploited to search the values of the Ogden model coefficients, the weight coefficients α_1, α_2 and relative permittivity. To find the model parameters, the following optimization problem is solved:

$$J(\epsilon_r, \beta_1, \beta_2, \beta_3, \gamma_1, \gamma_2, \gamma_3, \alpha_1) = \sum_{p=1}^{P}\sum_{k=1}^{M}\frac{1}{2}\left[m_p g - \sum_{j=1}^{N}\alpha_j\left(c_{1j}c_2 y_k u_k^2 - \frac{c_{1j}y_k}{s_j(y_k)}\sigma_{hj}(y_k)\right)\right]^2. \quad (11)$$

where $c_{1j} = \frac{2\pi r z_0}{l_{0j}}$ is the geometry coefficient, M is the number of the steady state responses and P is the number of different masses applied in the identification process. The optimization process required only the determination of the α_1 value, because according to Formula (6), the α_2 parameter was the difference $1 - \alpha_1$.

The actuator displacement measurements were carried out for the range 0 w to 7 with loads of 17.9 and 22.45 g. After changing the voltage, there was a 30 s wait before the actuator displacement was stabilized. The steady state responses of distance obtained for the following voltages were used to optimize the parameters of the Ogden model, the weight coefficients α_1 and relative permittivity. The SciPy python package [21] was used to run a Nelder–Mead simplex algorithm which efficiently found the optimal parameters. The values of the obtained parameters are presented in Table 1. The figures illustrating the comparison of models and measurements for three different actuators and two values of additional masses prove that the values obtained in the process of identifying static parameters are correct (Figures 7 and 8). The experiments take into account the pure mechanics at the points of zero supply voltage. This applies to both analyzed weights 17.9 and 22.45 g. The process of identifying static parameters (hyperelastic model coefficients and relative permittivity) took place for all parameters simultaneously. This approach allowed us to perform only one optimization. This approach is in line with the literature [19]. It is worth pointing out that the placement of the distance versus voltage characteristics is correlated with the shape of the actuators. It can be seen that the large elliptical actuator is more sensitive for the same mass than the circular or small elliptical actuator.

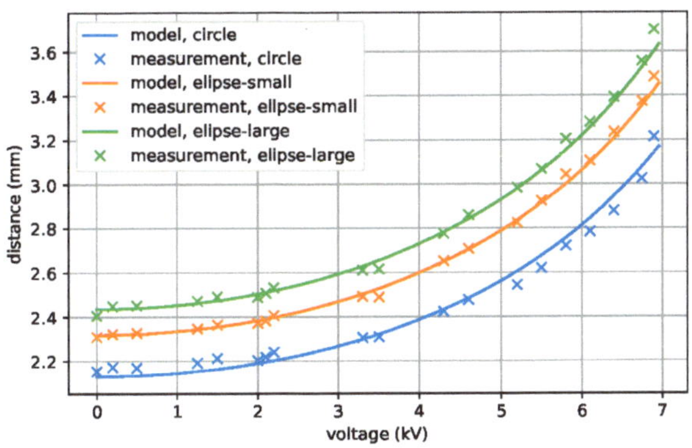

Figure 7. The static response of DEAP actuator for a mass of 17.9 g.

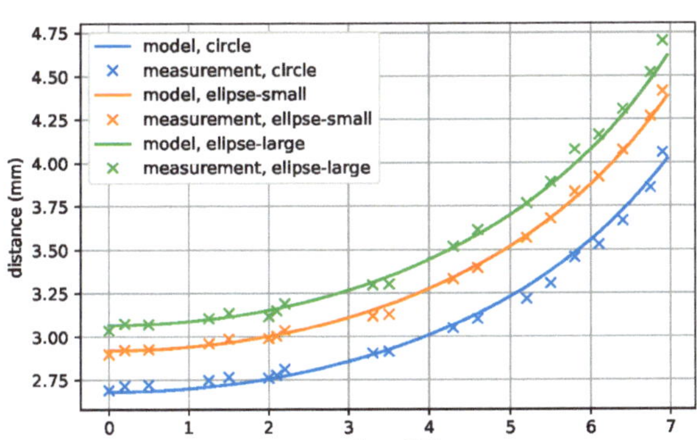

Figure 8. The static response of DEAP actuator for a mass of 22.45 g.

4.2. Identification of Viscoelastic Parameters

The viscoelastic parameters k_1, η_1 and η_2 can be only obtained by measuring the dynamic behavior of the DEAP models. Determining their values is a key issue in the process of identifying dynamic parameters. To achieve this, additional experiments were carried out in which the step responses of the DEAP actuators in the long duration horizon were analyzed. The responses of the actuators to the excitation by a step change in voltage from 0 kV to 3.5 kV were tested in two ranges of masses loading the actuators membranes (similar to the static parameter identification—17.9 and 22.45 g). To find the unknown parameters, the following optimization problem was tackled:

$$\min_{k_1,\eta_1,\eta_2} \sum_{j=1}^{N} \left[\frac{1}{N_s} \sum_{n=0}^{N_f} e_{j,y}^2(n) \right] \tag{12}$$

where N_s is the number of points in the step response and $e_{j,y}$ is the distance error between the j model and corresponding experiment.

The characteristic of the response for load 17.90 g is presented in Figure 9. Figure 10 shows the same response in the zoom version for a shorter time. The responses for the 22.45 g load are presented in an same way. Figure 11 shows the complete answer and Figure 12 shows its zoom version.

It can be seen that for all three actuators that the step responses are in good agreement with the experimental data for both loads. Further, the dynamic parameters k_1, η_1 and η_2 are also the same for all models.

The DEAP actuator has a relatively long settling time due to the relaxation process which exists in the VHB tape. The DEAP actuators are widely described by a few time constants [18]. In this work, the settling times between different shapes are similar and the differences are not correlated with the shape. For instance, the settling time of oscillations is 2.12 s, 2.38 s and 2.34 s for the circle, elliptical small and elliptical large cases, with a mass equal to 17.9 g. For the case with a mass of 22.45 g, the times are as follows: 2.28 s, 2.36 s and 2.27 s.

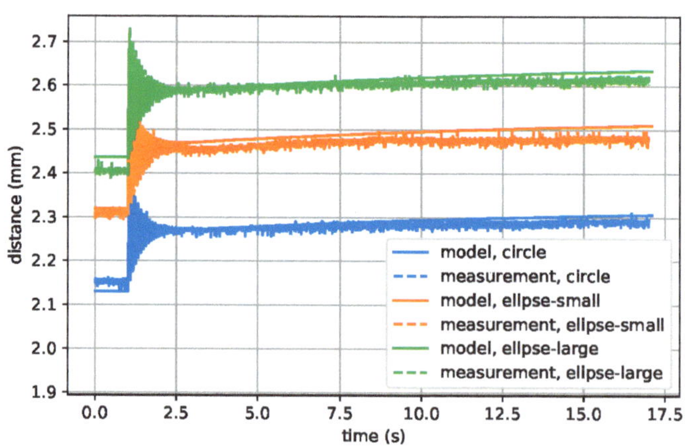

Figure 9. Response of the DEAP actuator for a step voltage signal from 0 kV to 3.5 kV with load of 17.9 g.

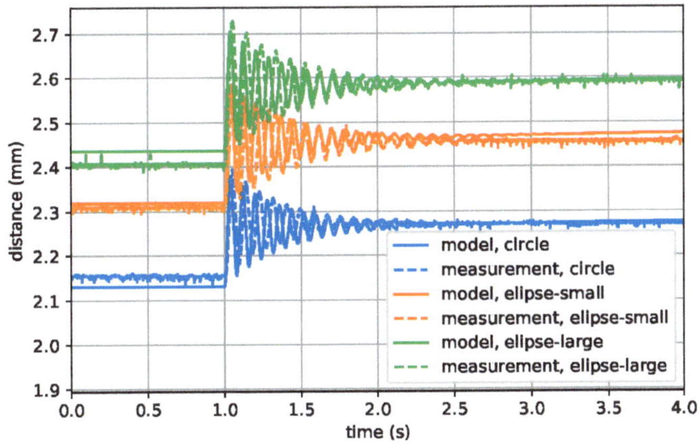

Figure 10. Response of the DEAP actuator for a step voltage signal from 0 kV to 3.5 kV with a load of 17.9 g (zoom version).

In the DEAP actuators exist a hysteresis between voltage and distance. The case of the circular geometry was reported in previous works [6]. The shape of hysteresis is different at varying frequencies and levels of signal. The identified models allow the calculation of the hysteresis between distance and voltage. All three models were tested with sinusoidal voltages $u(t) = u_0 + u_1 sin(\frac{2\pi}{T_s}t)$. In the first case, the constant signal was set $u_0 = 0$ kV and amplitude was equal to $u_1 = 1$ kV. In the second case, the constant signal was set $u_0 = 3.5$ kV and amplitude was equal to $u_1 = 3.5$ kV. The results for both cases are presented in Figure 13, showing that the influence of shape on hysteresis is minimal .

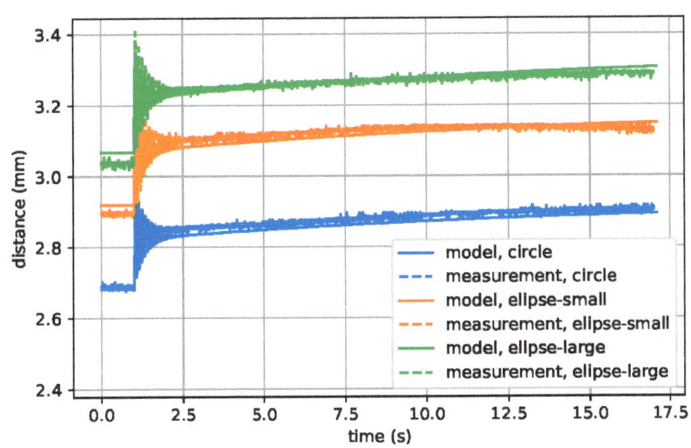

Figure 11. Response of the DEAP actuator for a step voltage signal from 0 kV to 3.5 kV with a load of 22.45 g.

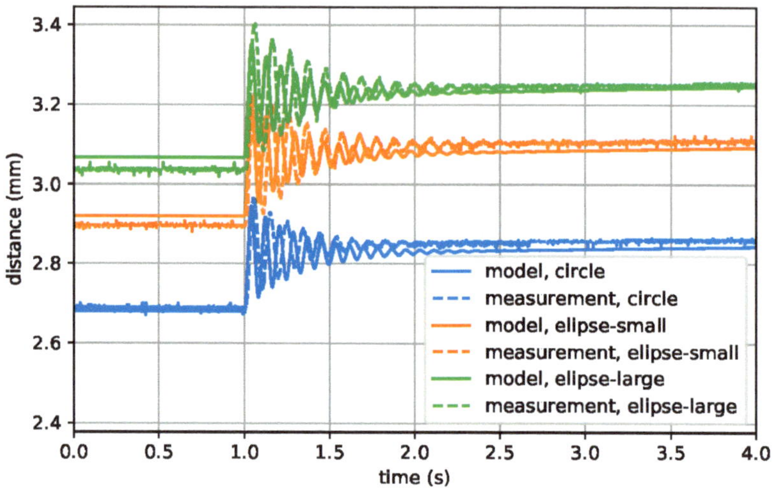

Figure 12. Response of the DEAP actuator for a step voltage signal from 0 kV to 3.5 kV with a load of 22.45 g (zoom version).

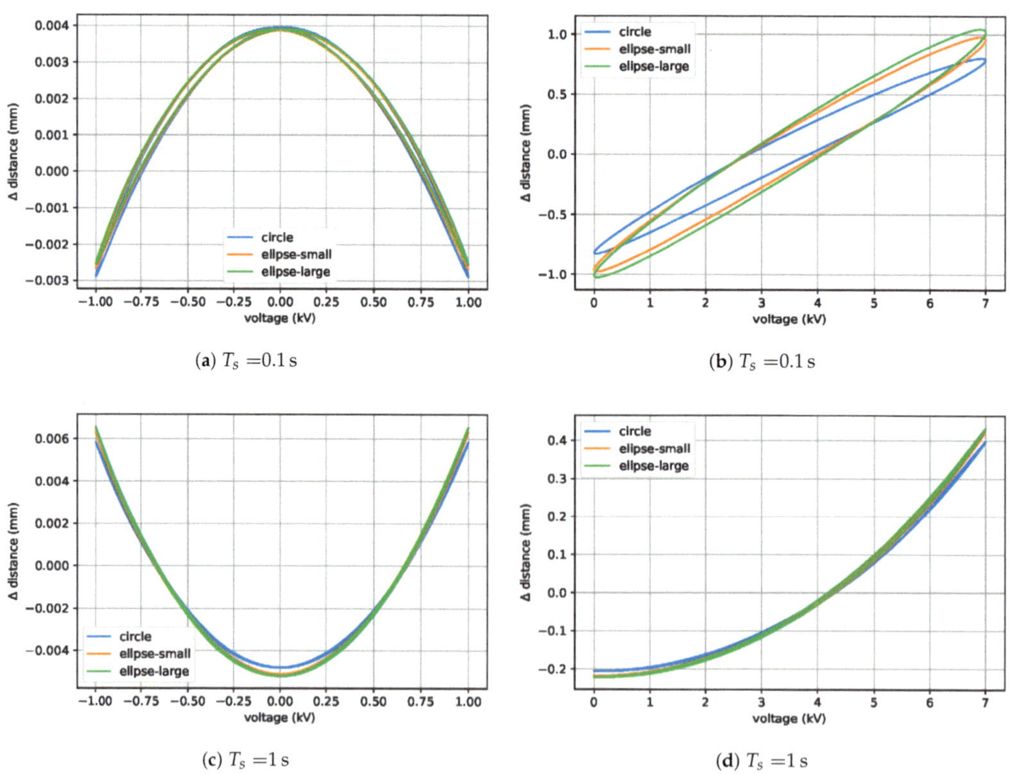

Figure 13. *Cont.*

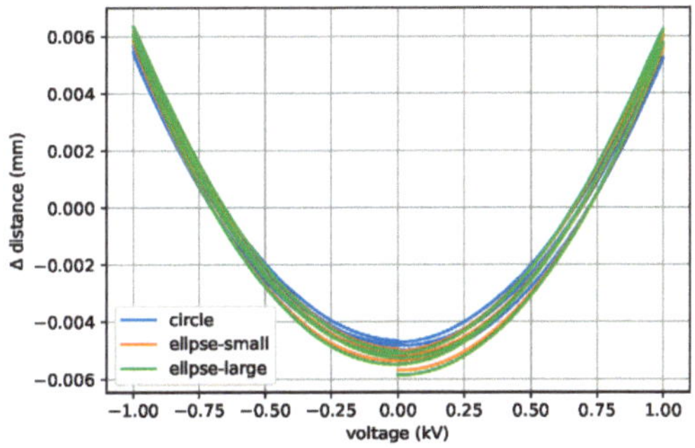

(e) T_s =10 s

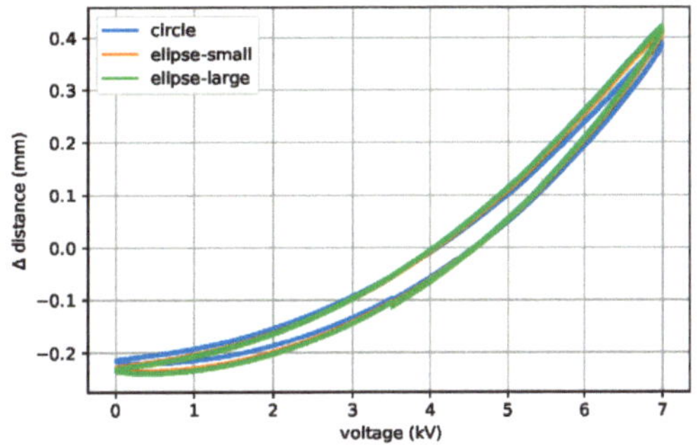

(f) T_s =10 s

Figure 13. The hysteresis between voltage and distance for different types of actuator models. Zero level input signal $u_0 = 0\,\text{kV}$, $u_1 = 1\,\text{kV}$ (**a**,**c**,**e**); nonzero level input signal $u_0 = 3.5\,\text{kV}$, $u_1 = 3.5\,\text{kV}$ (**b**,**d**,**f**).

4.3. Quantitative Comparison

In this section, a quantitative comparison between different shapes of actuators is described. To make said comparison, the random step responses with a voltage level between 0 kV to 7 kV with switch every 30 s were applied to actuators. The excitation and responses are presented in Figure 14. Further, for all responses, the difference between the minimal and maximal distances were measured. The results are presented in Table 2. The actuator large elliptical actuator has the larger movement range for both masses. It can be seen that the elliptical shape increases the moving range compared to the circular shape.

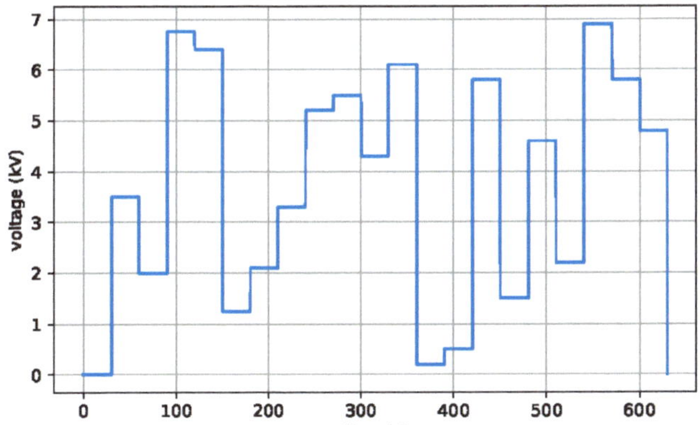

(**a**) transient of applied voltage

Figure 14. *Cont.*

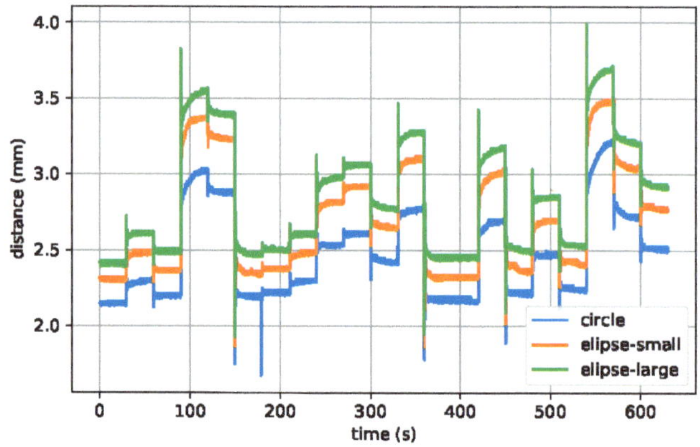

(**b**) distance response for mass 17.9 g

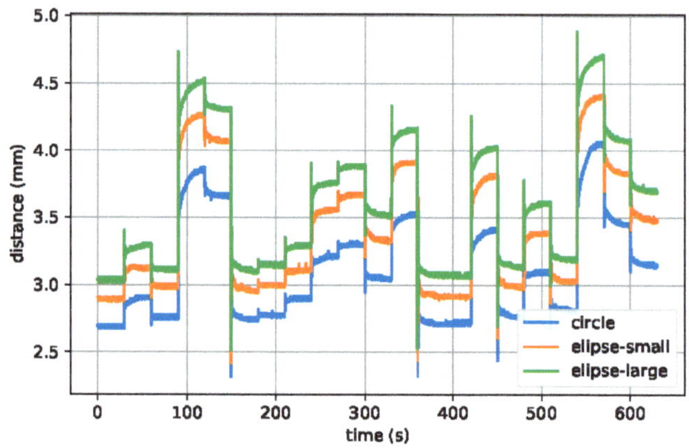

(**c**) distance response for mass 22.45 g

Figure 14. The response of actuators on voltage pseudo-random steps.

Table 2. The increase in moving range for different configurations.

	Mass of 17.9 g		Mass of 22.45 g	
Actuator Name	**Distance Range**	**Percentage**	**Distance Range**	**Percentage**
Circular	1.64 mm	100%	1.75 mm	100%
Elliptical small	1.86 mm	114%	2.16 mm	123%
Elliptical large	2.06 mm	126%	2.37 mm	135%

5. Conclusions

This paper has developed an elliptical DEAP actuator. An approximation of the elliptical actuator model was made with the use of circular models and the coefficients of individual components for the elliptical model were determined. The research was supplemented with the finite element models of the actuators used in the experiments. Finally, the elliptical model parameters were identified for three different geometries in two load modes. The experimental comparison of step responses for all actuators proves the correctness of the assumptions made in the model. Our research has shown that the elliptical shape of the actuator allows for a wider range of movement. This paper leaves many interesting research directions open for further investigations. In particular, it would be interesting to model other shapes of the actuator by analyzing the active surface of the membrane covered with electrodes. In further research, the authors would also like to verify various methods of control of elliptical DEAP actuators. The accurate modeling, as well as the analysis of performances carried out in this paper, are certainly useful steps in the direction of this challenging goal.

Author Contributions: Conceptualization, J.B. and J.K.; methodology, J.B. and J.K.; software, J.B.; validation, J.B. and J.K.; formal analysis, J.B. and J.K.; investigation, J.B. and J.K.; resources, J.B. and J.K.; data curation, J.B. and J.K.; writing—original draft preparation, J.B. and J.K.; writing—review and editing, J.B. and J.K.; visualization, J.B. and J.K.; supervision, J.B.; project administration, J.B.; funding acquisition, J.B. Both authors have read and agreed to the published version of the manuscript.

Funding: This work was realized with the frame of project SONATA 13 No. 2017/26/D/ST7/00092 from National Science Centre (Poland).

Institutional Review Board Statement: Not applicable.

Informed Consent Statement: Not applicable.

Conflicts of Interest: The authors declare no conflict of interest.

References

1. Kim, K.; Tadokoro, S. *Electroactive Polymers for Robotic Applications: Artificial Muscles and Sensors*; Springer: Berlin/Heidelberg, Germany, 2007; pp. 1–281. [CrossRef]
2. Plante, J.S.; Dubowsky, S. Large-scale failure modes of dielectric elastomer actuators. *Int. J. Solids Struct.* **2006**, *43*, 7727–7751. [CrossRef]
3. Hodgins, M.; York, A.; Seelecke, S. Experimental comparison of bias elements for out-of-plane DEAP actuator system. *Smart Mater. Struct.* **2013**, *22*, 094016. [CrossRef]
4. Rizzello, G.; Naso, D.; Turchiano, B.; York, A.; Seelecke, S. LMI-based design of PI controllers for micropositioning dielectric electro-active polymer membranes. In Proceedings of the 2015 American Control Conference (ACC), Chicago, IL, USA, 1–3 July 2015; pp. 5509–5514. [CrossRef]
5. Wang, S.; Decker, M.; Henann, D.L.; Chester, S.A. Modeling of dielectric viscoelastomers with application to electromechanical instabilities. *J. Mech. Phys. Solids* **2016**, *95*, 213–229. [CrossRef]
6. Bernat, J.; Kolota, J.; Rosset, S. Identification of a Nonlinear Dielectric Elastomer Actuator Based on the Harmonic Balance Method. *IEEE/ASME Trans. Mechatron.* **2020**. [CrossRef]
7. O'Brien, B.; McKay, T.; Calius, E.; Xie, S.; Anderson, I. Finite element modelling of dielectric elastomer minimum energy structures. *Appl. Phys. A: Mater. Sci. Process.* **2009**, *94*, 507–514. [CrossRef]
8. Loew, P.; Rizzello, G.; Simone, F.; Seelecke, S. Finite element simulation of plane strain dielectric elastomer membranes actuated by discretized electrodes. In *Electroactive Polymer Actuators and Devices (EAPAD) XXI;* International Society for Optics and Photonics: Bellingham, WA, USA, 2019; Volume 10966. [CrossRef]
9. Cao, C.; Conn, A. Performance optimization of a conical dielectric elastomer actuator. *Actuators* **2018**, *7*, 32. [CrossRef]
10. Bernat, J.; Kolota, J. Active Disturbance Rejection Control for Dielectric Electroactive Polymer Actuator. *IEEE Access* **2021**, *9*, 95218–95227. [CrossRef]
11. Kolota, J. The FEM model of the pump made of dielectric electroactive polymer membrane. *Appl. Sci.* **2020**, *10*, 2283. [CrossRef]
12. Kaal, W.; Herold, S. Electroactive polymer actuators in dynamic applications. *IEEE/ASME Trans. Mechatron.* **2011**, *16*, 24–32. [CrossRef]
13. Bernat, J.; Kolota, J. Adaptive Observer for State and Load Force Estimation for Dielectric Electro-Active Polymer Actuator. In Proceedings of the 11th IFAC Symposium on Nonlinear Control Systems, Vienna, Austria, 4–6 September2019.
14. Hau, S.; Rizzello, G.; Seelecke, S. A novel dielectric elastomer membrane actuator concept for high-force applications. *Extrem. Mech. Lett.* **2018**, *23*, 24–28. [CrossRef]
15. Cao, C.; Gao, X.; Conn, A. A Magnetically Coupled Dielectric Elastomer Pump for Soft Robotics. *Adv. Mater. Technol.* **2019**, *4*, 1900128. [CrossRef]
16. Bernat, J.; Kołota, J. DEAP Actuator Composed of a Soft Pneumatic Spring Bias with Pressure Signal Sensing. *Energies* **2021**, *14*, 1189. [CrossRef]
17. Ida, N. *Engineering Electromagnetics*; Springer International Publishing Switzerland: Cham, Switzerland; Heidelberg, Germany; New York, NY, USA; Dordrecht, The Netherlands; London, UK, 2015; pp. 1–1062. [CrossRef]
18. Sarban, R.; Lassen, B.; Willatzen, M. Dynamic Electromechanical Modeling of Dielectric Elastomer Actuators With Metallic Electrodes. *IEEE/ASME Trans. Mechatron.* **2012**, *17*, 960–967. [CrossRef]

19. Rizzello, G.; Naso, D.; York, A.; Seelecke, S. Modeling, Identification, and Control of a Dielectric Electro-Active Polymer Positioning System. *IEEE Trans. Control. Syst. Technol.* **2015**, *23*, 632–643. [CrossRef]
20. Rosset, S.; Araromi, O.A.; Schlatter, S.; Shea, H. Fabrication Process of Silicone-based Dielectric Elastomer Actuators. *J. Vis. Exp.* **2016**, *108*, 1–13. [CrossRef] [PubMed]
21. Virtanen, P.; Gommers, R.; Oliphant, T.E.; Haberland, M.; Reddy, T.; Cournapeau, D.; Burovski, E.; Peterson, P.; Weckesser, W.; Bright, J.; et al. SciPy 1.0: Fundamental Algorithms for Scientific Computing in Python. *Nat. Methods* **2020**, *17*, 261–272. [CrossRef] [PubMed]

energies

MDPI

Review

Effectiveness of Daytime Radiative Sky Cooling in Constructions

Alessandro Cannavale [1,2,*], Marco Pugliese [2], Roberto Stasi [1], Stefania Liuzzi [1], Francesco Martellotta [1], Vincenzo Maiorano [2] and Ubaldo Ayr [1]

[1] Dipartimento di Architettura, Costruzione e Design, Politecnico di Bari, Via Orabona 4, 70125 Bari, Italy; roberto.stasi@poliba.it (R.S.); stefania.liuzzi@poliba.it (S.L.); francesco.martellotta@poliba.it (F.M.); ubaldo.ayr@poliba.it (U.A.)

[2] National Research Council, Institute of Nanotechnology (CNR-NANOTEC), Via Monteroni, 73100 Lecce, Italy; marco.pugliese@nanotec.cnr.it (M.P.); vincenzo.maiorano@nanotec.cnr.it (V.M.)

* Correspondence: alessandro.cannavale@poliba.it; Tel.: +39-080-5963718

Abstract: In this paper, we have carefully studied the scientific literature dealing with the use of passive radiative surfaces within the construction industry. The aim of this paper is to highlight technologies and materials for daylight radiative cooling under study today—or already on the market—and to report their main characteristics, performance and, where possible, costs. Following a review of the available scientific literature, the advantages and limitations of such an option were highlighted, seeking to capture opportunities and future lines of research development. This review also provides the physical laws that evaluate the energy balance of passive radiative surfaces as well as the criteria to quantify all the terms of these equations.

Keywords: passive radiative cooling; buildings; sub-ambient temperature; urban heat island

Citation: Cannavale, A.; Pugliese, M.; Stasi, R.; Liuzzi, S.; Martellotta, F.; Maiorano, V.; Ayr, U. Effectiveness of Daytime Radiative Sky Cooling in Constructions. *Energies* **2024**, *17*, 3210. https://doi.org/10.3390/en17133210

Academic Editor: Anastassios M. Stamatelos

Received: 18 April 2024
Revised: 21 June 2024
Accepted: 27 June 2024
Published: 29 June 2024

1. Introduction

1.1. Background of the Review

The planetary emergence of climate change requires the search for innovative strategies and technologies, with reduced anthropogenic impact, to limit the effects of global warming. The European Directive 2018/844 states that buildings contribute to 36% of greenhouse gas emissions and that 50% of final energy consumption in the European Union is used for heating and cooling [1]. The latter is widely considered a challenging issue, simply due to the constraints of the second law of thermodynamics [2]. Furthermore, in hot climates, cooling is becoming more and more a driver of the electric energy demand. Much attention is being regarded worldwide to passive radiative materials, which provide a path to dissipate thermal energy from surfaces toward the ultracold extra-terrestrial space [3]. The idea of passive cooling of surfaces to sub-ambient temperatures, using the night sky as an effective heat sink dates to the earliest studies by Granqvist et al. [4]. Such cooling capacity has been exploited for climatization, since ancient times, in tropical areas [5]. A similar behaviour has been observed not only in artificial structures, but also in butterflies that regulate the wing temperatures using radiative cooling [6], and in nano-structured wild moth cocoon fibres [7], providing effective radiative cooling for the moth pupae by controlling optical reflection as well as radiative heat transfer. Finally, Shi et al. have demonstrated that Saharan silver ants [8] (*Cataglyphis bombycine*) efficiently dissipate heat back to the surroundings due to the high emissivity for wavelengths higher than 2.5 μm, and high reflectivity in the wavelength range of solar radiation.

If, on the one hand, night-time radiative cooling has been extensively studied so far [9], daytime cooling would indeed represent a challenging option to face peak cooling demand (mainly occurring during the day) [10] as well as heat island effects [11] and other similar challenging issues regarding the anthropogenic footprint due to energy consumption. Figure 1 summarizes the many interactions between radiation emitted by the Sun with the atmosphere and the Earth's surface [12–16]. Incident solar radiation (at low wavelengths) undergoes absorption, reflection and scattering phenomena as it interacts with the atmosphere and the Earth's surface. At the same time, the Earth's surface emits electromagnetic radiation at higher wavelengths (in the infrared range) towards the atmosphere. This radiation can be again scattered, absorbed or reflected, or can even be transmitted through the atmosphere to free space. All these phenomena contribute to the overall thermal balance of the Earth's surface and regulate its average temperature. Passive cooling may thus represent an interesting opportunity for sustainable development, mitigating both the causes and the effects of global warming, offering a source of off-grid energy and even a water source by night cooling of surfaces below the dew point temperature [17].

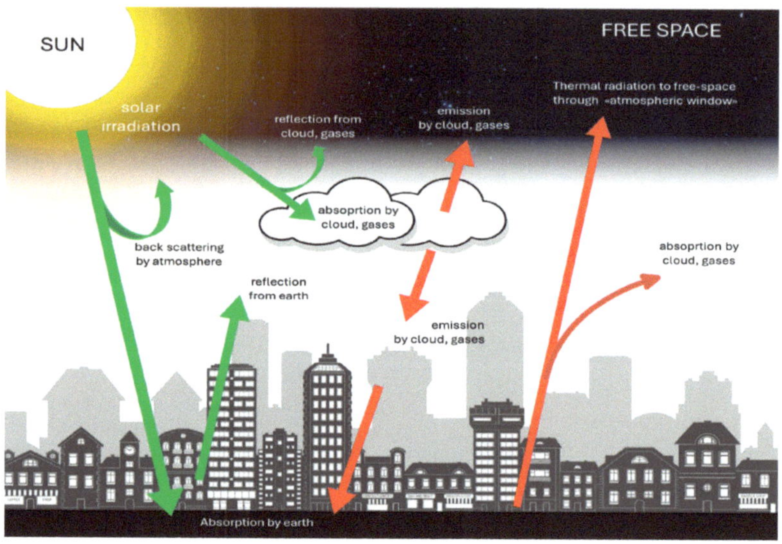

Figure 1. Energy exchange at the Earth's surface. The interactions of solar radiation and terrestrial radiation with the earth's atmosphere and surface are finely balanced to regulate the temperature of our planet.

1.2. Problem Statement and Motivation

A fundamental watershed in the research field regarding passive radiative coolers is undoubtedly represented by Aaswath P. Raman's seminal paper published in Nature in 2014 [18], which paved the way for research on radiative heat dissipation systems operating during daytime hours, exposed to solar radiation. This was a fundamental step, opening up new opportunities for the exploitation of heat dissipation. This paper aims to review the recent investigations in the field, with special reference to the construction sector, especially from that year to the present. This review may be useful to a wide audience of students, researchers and professionals who are interested in the effectiveness of new strategies to increase energy efficiency in buildings. With this aim, we have tried to identify the most suitable strategies to introduce radiative surfaces in buildings and to achieve significant energy savings in summer air conditioning operation. To this end, we took special consideration of publications that appeared after Aaswath P. Raman demonstrated, for the first time—supporting the assumption with an experimental activity—that it is possible to achieve sub-ambient temperature surfaces, even in the presence of solar radiation during daylight hours. This means that passive strategies may be exploited to achieve effective heat rejection without using energy. This is an important issue, especially in urban contexts, where urban heat islands represent more and more of an open issue [19].

1.3. Objective and Structure

This review is mainly devoted to gather the most relevant investigations about daytime passive coolers' technologies for the construction sector, within the time interval that separates us from the publication of Aaswath P. Raman's seminal paper published in Nature in 2014 [20], which fostered research relating to radiative heat dissipation systems operating during daytime hours. Hereafter, the design principles, the most investigated materials, and main applications of radiative sky cooling will be reviewed, with reference to the construction sector, with the intention of studying the effectiveness of such technologies, to ensure the reader's awareness of the benefits and limits of these technologies, with a highly quantitative approach. The structure of this review paper consists of an introduction reporting a background of the review, the problem statement and motivation faced in this activity, and the objective and structure of the manuscript; a short methodology section; principles of radiative cooling; a wide literature review section for building integration of radiative cooling surfaces; a part regarding switchable materials and systems applied to radiative sky cooling in buildings; and the concluding remarks, containing limitations and future directions for this investigation.

2. Methodology

For the reasons stated above, most of the publications reviewed in this investigation date from 2014 to 2024, beyond those strictly necessary to explain the principles underlying the design and physics of (passive and active) radiative systems. In the text, the main works that regarding building integration of radiative systems are reported, with special reference to materials, modes of operation, and cost, when the literature in the field also makes such elements available. We found the articles reported here by querying the Scopus database, using the most recurrent keywords in articles published in this field of investigation. Moreover, a table has been created, reporting the main technologies investigated in the current state of the art of building integrated radiative surfaces and the relevant figures of merit that express their performance.

3. Principles of Radiative Cooling

Though it was anticipated already in the 1980's, only in the last decade has passive radiative cooling been experimentally demonstrated by the seminal studies reported by Raman and co-workers [18]. This means that a passive cooling strategy may allow daytime cooling of structures to temperatures below ambient air, even in the presence of the Sun radiation. Such surfaces would be able to operate without any external energy, in a "passive" mode. In fact, the atmosphere of our planet shows a relatively wide transparency window for electromagnetic waves, in the infrared (IR) wavelength range between 8 μm and 13 μm (Figure 2b), due to the very low atmospheric emission in that spectral region [21]. This fact leads us to find suitable ways to exploit such a special feature of the atmosphere composition for the purpose of heat dissipation either from buildings or cities and industrial plants [22]. The existence of such "open window" paves the way towards the possibility to cool objects at an ambient temperature (≈300 K), by means of thermal radiation taking place between bodies on the Earth's surface and the extra-terrestrial space, roughly at ≈3 K. Coincidentally, according to Planck's law, the thermal radiation of a blackbody at 300 K shows its peak of radiation, precisely, at about 10 μm. In this way, the emitted radiation might escape to the extra-terrestrial space, which can be considered an outstanding heat sink. For daytime applications, the surface of a cooler must show selective features; at the same time, it should absorb as little solar radiation as possible but also emit the maximum radiation within the atmospheric window, using the extraterrestrial space as a thermal sink. This can be achieved by maximizing the reflectance of the cooler in the wavelength range from 250 nm to 2.5 μm (Figure 2a) and maximizing the emissivity in the wavelength range from 8 μm to 13 μm (Figure 2b). Structures capable of simultaneously working as broadband mirrors for the solar radiation (solar reflectors) and strong thermal emitters in the atmospheric window (IR radiators) may effectively act as daytime radiative coolers: achieving sub-ambient temperatures (for the whole day and night) may represent a milestone for engineering and materials science. Moreover, daytime cooling would match with the profile of peak energy uses for cooling in buildings and other applications.

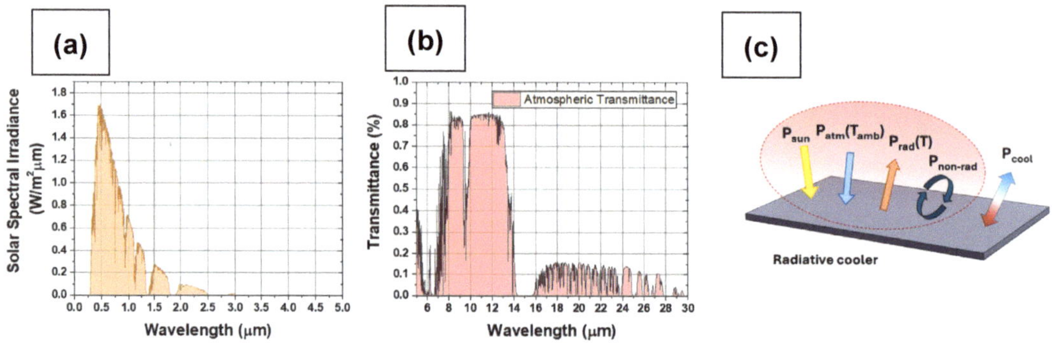

Figure 2. (**a**) Example of solar spectral irradiance at mid-latitude under a cloudless sky at sea level. (**b**) Atmospheric transmittance: the atmosphere shows high transmittance in the wavelength range between 8 and 13 μm. (**c**) Energy balance of a radiative cooler's surface: P_{sun} and P_{atm} represent the absorbed solar irradiation power and the absorbed atmospheric radiation power, respectively. Prad is the thermal radiation power from the surface, and $P_{non\text{-}rad}$ is the non-radiative heat transfer between the surface and the environment; all these terms contribute to the net cooling power P_{cool}, as expressed by Equation (1).

In essence, radiative coolers should maximize the radiative term of their full energy balance (Figure 2c), which is expressed as follows [20]:

$$P_{cool}(T) = P_{rad}(T) - P_{atm}(T_{amb}) - P_{sun} - P_{non-rad} \tag{1}$$

which not only takes into account the radiative term (P_{rad}) but also the one concerning non-radiative (convective/conductive) thermal exchange ($P_{non-rad}$), the absorbed solar irradiation (P_{sun}) and the absorbed power due to atmospheric thermal radiation (P_{atm}). Consequently, the term P_{cool} represents the net cooling power of a given structure that can be considered a radiator providing practical cooling only when the radiated power exceeds the sum of all the negative (absorbed) terms in Equation (1), which relates the net output power of a surface to its temperature. If $P_{cool} > 0$ when the ambient temperature equals the surface temperature, the surface may act as a daytime passive cooler. On the other hand, if Equation (1) is solved for $P_{cool} = 0$, it is possible to find the surface steady-state temperature, which must be lower than T_{amb}, to obtain effective radiative cooling [20]. The term $P_{rad}(T)$ is calculated as follows:

$$P_{rad}(T) = A \cdot \int_0^{\frac{\pi}{2}} (2\pi \cdot sin\theta \cdot cos\theta \cdot d\theta) \cdot \int_0^{\infty} \frac{2hc^2}{\lambda^5} \cdot \frac{1}{\left[e^{\left(\frac{hc}{\lambda k_B T}\right)} - 1\right]} \cdot \epsilon(\lambda, \theta) \cdot d\lambda \tag{2}$$

which integrates the spectral radiance of a blackbody at absolute temperature T, according to Planck's distribution law, embodying the angular integral over a hemisphere. For radiative sky cooling, it is generally acceptable to consider horizontal surfaces [23] and assume azimuthal symmetry. The term h represents Planck's constant, k_B is Boltzmann's constant, c is the speed of light in the vacuum and λ is the wavelength. On the other hand, $\varepsilon(\lambda, \theta)$ is the surface spectral angular emissivity, which is considered to vary only with the zenith angle θ and, according to the assumptions of Kirchhoff's law [24], replaces the surface absorptivity.

The term P_{atm} (T_{amb}) represents the absorbed power due to incident atmospheric thermal radiation, which can be calculated using the spectral angular emissivity of the atmosphere and the spectral radiance of a blackbody at temperature T_{amb} as follows:

$$P_{atm}(T_{amb}) = A\cdot\int_0^{\frac{\pi}{2}}(2\pi\cdot sin\theta\cdot cos\theta\cdot d\theta)\cdot\int_0^{\infty}\frac{2hc^2}{\lambda^5}\cdot\frac{1}{\left[e^{\left(\frac{hc}{\lambda k_B T_{amb}}\right)}-1\right]}\cdot\varepsilon_{atm}(\lambda,\theta)\cdot\varepsilon(\lambda,\theta)\cdot d\lambda \tag{3}$$

where again the surface absorptivity is replaced by the emissivity. To quantify the spectral atmospheric radiation, the spectral directional emissivity, $\varepsilon_{atm}(\lambda,\theta)$, of the atmosphere is used, which is defined as the ratio between the atmospheric spectral radiance and the blackbody spectral radiance calculated at the air temperature (T_{amb}) near the ground. This is because the downward atmospheric thermal radiation comes mostly from the lowest few hundred meters of the atmosphere for those wavelengths where water vapor and carbon dioxide are strongly absorbing [25–27]. The atmospheric emissivity is angle-dependent and is generally given by the following expression:

$$\epsilon_{atm}(\lambda,\theta) = 1 - t(\lambda)^{\frac{1}{cos\theta}} \tag{4}$$

where $t(\lambda)$ is the atmospheric transmittance in the zenith direction. The values of the $t(\lambda)$ depend mainly on the concentration of some gases (H_2O, CO_2, O_3, CH_4, N_2O) and aerosols in the stratigraphy of the atmosphere and on the related temperature profile. To model the atmospheric transmittance, commercial software can be used (e.g., MODTRAN version 6) which draws the radiative properties of individual gases from specific databases (e.g., HITRAN). It has been observed that the cooling effect is boosted in dry climates with clear sky, as clouds and humidity may drastically reduce the value of P_{cool} [11,28].

The P_{sun} term integrates the incident solar irradiance, considering the spectrum corresponding to a value of 1.5 for Air Mass ($I_{AM1.5}$). and again, the spectral absorptivity equal to the spectral emissivity.

$$P_{sun} = A\cdot\int_0^{\infty} I_{AM1.5}\cdot\varepsilon_{atm}(\lambda,\theta_{sun})\cdot d\lambda \tag{5}$$

Furthermore, the term reporting thermal power lost by non-radiative heat transfer is lumped as follows:

$$P_{non-rad}(T,T_{amb}) = h_c\cdot A\cdot(T_{amb} - T) \tag{6}$$

where h_c is a suitable heat transfer coefficient that considers both the convective heat exchange with air and the conductive heat exchange with external surfaces, in contact with the radiative cooler.

To achieve high values of $P_{cool}(T)$, the thermal emissivity of the cooler structure must be as high as possible within the atmospheric window, and, at the same time, its absorption should be very low throughout the wavelength range of the solar spectrum. Consequently, a correct design of passive cooling structures aiming at dissipating thermal energy during daytime hours minimizes the negative terms of the equation.

4. Literature Review for Building Integration of Radiative Cooling Surfaces

Several approaches have been explored towards the design of daytime radiative coolers, as reviewed in several works [29]. An effective approach is represented by one-dimensional photonic films obtained by depositing alternating layers of materials with different refractive index, deposited on top of a highly reflective material substrate [30]. With this approach, Raman et al. [20] cooled their structures about 5 °C below the ambient temperature, providing experimental evidence of the feasibility of daylight cooling. Chen et al. [31] theoretically showed that ultra-large temperature reductions to 60 °C below ambient can be achieved minimizing the parasitic heat losses (conduction and convection) by means of a vacuum chamber evacuated to a pressure of about 1.333×10^{-4} Pa. Their structure consisted of layers of silicon nitride, amorphous silicon, and aluminium, with thicknesses of 70 nm, 700 nm and 150 nm, respectively. A current trend of research for daytime passive cooling involves polymers. Zhai et al. [32] demonstrated the passive radiative cooling power of 93 W/m^2 of glass–polymer hybrid polymethylpenthene (TPX) membranes, embodying silicon dioxide microspheres, directly involved in the thermal emission. The highly transparent polymer reflects solar irradiation when backed with a silver thin film, deposited by E-Beam evaporation. Wang et al. [33] adopted the electrospinning technique to fabricate polyvinylidene fluoride/tetraethoxysilane (PVDF/TEOS) fibres with nanopores inside and silicon dioxide microspheres distributed on their surfaces. An average radiative cooling power of 61 W/m^2 was obtained, in May, in Shanghai. A radiative cooling structural material was developed by Li et al. [2], who demonstrated a highly resistant cellulose-based material embodying cellulose nanofibers capable of backscattering solar radiation and can emit in infrared wavelengths, providing continuous daytime and nocturnal cooling. More recently, Chen et al. reported the fabrication of a cellulose/SiO_2 bulk material, with significant passive cooling properties (sub-ambient temperature decline of about 7 °C) and mechanical properties. Current research trends also reported the use of nanophotonic structures, reviewed by Ko et al. [29]. This review reports an exhaustive list of best practices and research activities dealing with daytime radiative cooling applied to the construction sector.

The main limit to practical applications of cooling materials is represented by the low energy density provided, compared with design requirements. Apart from the intrinsic spectroscopic properties, cooling performance is dramatically affected by location, weather conditions, cooling profiles, cost of materials and maintenance, etc. Nevertheless, this field of investigation is widely recognized as an intriguing technological challenge [34,35].

Energy consumption in buildings is a relevant topic with regards to energy transition and reduction of environmental impacts of the construction sector responsible of 40% of total energy use in developed countries, where most of this consumption is related to HVAC (heating, ventilation, and air conditioning) systems. The impact of completely passive technologies for heat dissipation would be welcome both in terms of energy saving and in terms of CO_2 emissions reduction. In fact, in Zhao et al. [23], the adjective "passive" refers to technologies with zero energy input (i.e., with no fans or pumps to enhance cooling power).

Generally, the use of passive cooling in buildings can be achieved by exploiting passive and active approaches. The main difference is represented by the possibility of controlling performance according to seasonal or operation requirements, which is only possible in active systems. On the other hand, lower cost and complexity are the main advantages of passive systems.

4.1. *Passive Cooling Surfaces from Traditional "Cool Roofs" to Daytime Radiative Coolers*

Large rooftop surfaces are ideal candidates to host passive cooling surfaces. Cooling loads are compatible with values in the order of 100 W/m^2, as reported by Pacheco et al. [36]. With regards to passive technologies for buildings, cool roofs can be considered indeed an energy efficient solution [37,38]. Roof materials can be considered "cool" if they show—at the same time—high reflectance (or albedo) in the solar range of wavelengths and high emissivity in the atmospheric window; two simultaneous figures are then required to achieve acceptable performance in cool roofs [39], which can be considered highly selective materials.

As reported by Zhao et al. [23], cool roofs, in the form of coatings, tiles and other components, affect building performance by potentially reducing cooling loads, lowering electricity consumption, extending roof lifetime, and eventually reducing building impact on the heat island effect. According to ASHRAE [40], a cool roof must demonstrate a *solar reflectance index (SRI)* of 78 or higher. This figure of merit was defined by ASTM E1980 [41], as follows:

$$SRI = 123.97 - 141.35{\cdot}\chi + 9.655{\cdot}\chi^2$$

where the parameter χ depends on solar absorptance (α), on thermal emissivity (ε) and on the convective heat transfer coefficient (h_c) as well: the latest is evaluated for three wind conditions.

Energy saving associated with building integrated cool roofs have been studied in detail in several experimental and numerical works [42–44]. Akbari et al., in their study [42], reported that raising the solar reflectance of conventional dark roofs by 25% can reduce building cooling energy consumption by more than 10% and results in a lower ambient temperature. On the other hand, they found that raising roof reflectivity to 0.60 may reduce cooling energy use for buildings by about 20%.

A new prototyped cool clay tile on a traditional residential building was tested by Pisello et al. [45] to quantify summer benefits and winter penalties associated to the use of cool roofs in a temperate climate. The yearly analysis showed that the proposed technology achieved a decrease in indoor overheating by 4.7 °C, whereas the winter overcooling reduction is 1.7 °C. The substantial cooling benefits in summer are predominant with respect to the penalties in the winter season. Maximum year-round benefits were found for Palermo, in Southern Italy, with numerical simulations, corresponding to 14 kWh/m^2 per year. In fact, this technology may find application at low latitudes, in cooling dominated countries, where heating may be considered a lower concern for energy consumption in buildings.

A 67 μm thick roof surface under the mid-summer sun was fabricated by Gentle and Smith [46], suitable for large-scale production. Spectral selectivity of such materials allows for very high solar reflectance and emissivity, in the atmospheric window, to obtain net radiative power towards the clear sky. They met both of those criteria using specially chosen polyesters, deposited on a silver layer (200 nm). Such surface remained sub-ambient (2 °C) throughout a hot summer day (solar irradiance of 1060 W/m^2), with ambient air at 27 °C and high infrared intensity from the atmosphere of 400 W/m^2. Further numerical simulations showed that the reduction of cooling energy achievable using this new material would be 91 kWh/m^2 in Orlando (FL, USA).

Cool roofs with high reflectance may also produce visual discomfort, like glare. For this reason, cool-coloured roof tile coatings have been developed, with visible properties in the visible range (0.4–0.7 μm), with almost typical colour, like those of standard roofs but with much higher reflectance in the infrared wavelengths (0.7–2.5 μm).

Passive radiative cooling is well known to provide optimal figures in nocturnal hours, though the peak cooling load in buildings occurs in the daytime. Better performance than traditional cool roof materials can be achieved through recent results from photonics and optics, which enable the chance to use nanomaterials to improve both solar reflectance and infrared emissivity, with a useful diurnal net cooling power. With this regard, Baniassadi et al. [47] studied the effectiveness of engineered spectral properties of "super-cool" roofs showing reflectivity and emissivity values greater than 0.96 and 0.97, respectively, applied to a building rooftop in a numerical simulation activity in eight US cities. They claimed that the temperature rooftop remained lower than the ambient air temperature throughout the year, producing an average daily sensible flux of 30–40 W/m^2. Moreover, they found that this new technology could double the attainable cooling energy saving, compared to typical white roofs. At the basis of their simulation activities, they used the spectral properties of super-cool surfaces demonstrated by Jyotirmoy Mandal et al. [48]. They proposed hierarchically porous poly(vinylidene fluoride-co-hexafluoropropene) coatings with a high passive daytime radiative cooling capability: hemispherical solar reflectance R_{sol} = 0.96 ± 0.03 and long-wavelength infrared emissivity $\varepsilon_{LWIR} = 0.97 \pm 0.02$, with maximum sub-ambient temperature drops of ~6 °C and cooling powers of ~96 with 750 W/m^2 solar irradiance.

The approach used in daytime radiative coolers so far could be reconsidered according to Wang et al. [49], in terms of real cooling potentials, depending on location and real atmospheric conditions. If in a dry, sunny winter day in California, the multi-layered cooling system designed by Raman et al. [20] reported radiative cooling to nearly 5 °C below the ambient air temperature; numerical simulations by Lu et al. showed that when h_c equals 6.9 W/(m^{2}K) in Bangkok, the temperature reduction in summer was as low as

0.1 °C, with very low passive radiative cooling performance in the mid-latitudes. The same surface may perform even worse in damp tropic zones due to the reduced transparency of the atmospheric window, associated with water vapor. The above-reported considerations suggest that new radiative coolers should be customized according to location and climatic conditions. Moreover, photonic radiators generally require nanoscale-precision fabrication, based on physical vapor deposition techniques which may act as a limiting factor in terms of cost and mass production.

Photonic structures requires multilayer optical calculations [50] as well as complex and expensive fabrication processes. Raman's cooler consisted of seven alternate layers of HfO_2 and SiO_2, whose thicknesses were obtained after extensive numerical optimization. Similar structures adopted other dielectric materials, like Ta_2O_5 or TiO_2 [29,30]. The availability of simple, bottom-up processes, based on chemical approaches, would be more compatible with large scale production with lower production costs.

Wang et al. [33] reported a high-performance flexible hybrid membrane radiator based on polyvinylidene fluoride/tetraethyl orthosilicate, with 61 W/m^2 and a temperature decrease up to 6 °C, under a surface solar irradiance of 1000 W/m^2. Such a flexible hybrid membrane reflected ≈97% of solar radiation and exhibited an infrared emissivity higher than 0.96.

Zhai et al. [32] embedded dielectric microspheres randomly in a polymeric matrix, resulting in a metamaterial that is fully transparent to the solar spectrum while having an infrared emissivity greater than 0.93 across the atmospheric window due to phonon-enhanced Fröhlich resonances of the microspheres. They deposited silver on one side of the membrane, obtaining a noon-time radiative cooling power of 93 W/m^2 under direct sunshine. The described fabrication process is highly compatible with roll-to-roll manufacturing of this novel technology. Chen et al. [51] reported a membrane containing delignified biomass cellulose fibres and inorganic SiO_2 microspheres.

4.2. Integration of Passive Radiative Surfaces in Buildings and HVAC Systems

Daytime radiative cooling may represent a reliable mitigation strategy to counterbalance the impact of urban heat islands, using materials with high reflectivity [8] and chromogenics [52]. Daytime coolers can be fabricated in the form of multilayered photonic stacks, 2D–3D photonic structures—both require a top-down fabrication approach—paints or polymer structures embedding nanostructures and fabricated using a bottom-up approach. In theoretical terms, a surface area of 1 m^2 emits a maximum thermal power of 459.3 W, with a maximum radiation peaked at 9.67 μm, according to Wien's displacement law [53]. The amount of thermal power emitted by a blackbody at 300 K in the spectral range of the atmospheric window is roughly 32% of the total emitted power, i.e., 148 W. The latter represents the ideal reference value of the thermal power that can be emitted by a body at ambient temperature with a $\varepsilon = 1$. After Raman's seminal work [18,20], sub-ambient daytime radiative coolers have been demonstrated that they are capable of reaching 4 °C below ambient temperature when exposed to direct daylight and even 60 °C theoretical maximum temperature reduction. Nevertheless, external conditions dramatically affect the performance of a radiative cooler, apart from the amount of solar radiation impinging on the dissipating surface. The P_{atm} term of the energy balance may be dramatically influenced by the relative humidity of ambient air. In fact, experiments carried out in low humidity conditions or desert conditions in Israel have reported optimal results [54]. Atmospheric transmittance spectra were accurately studied by Berk et al. since the first decade of this century [55]. Thermal power emitted by the atmosphere may in fact be transmitted to the cooler surface according to its emittance properties, according to Kirchhoff's law.

Furthermore, the operation conditions of a radiative cooler should consider the great variability of the convective term, i.e., the $P_{non\ rad}$ term: the convective heat power could cancel out the radiative benefit if the air temperature and the convective heat transfer coefficient are high, depending on wind speed. To reduce convective heat gains, wind covers and windshields have been proposed in several works worldwide [49].

Carlosena et al. [56] fabricated and tested daytime radiative materials consisting of a Vikuiti substrate (0.97 reflectivity and 0.89 emissivity) covered with a 2 μm thick film of polymethylsesquioxane with embedded silica nanoparticles, which have a competitive cost of 0.3 €/m^2. The surfaces reached 7.32 °C temperature reduction in daytime conditions in moderate weather conditions, which is quite different from those used for experiments in ideal settings (no convection and low relative humidity). They demonstrated that only scalable and effective daytime radiative materials may be relevant in the context of future applications in constructions.

Radiative Sky Cooling (RSC) is used to reduce indoor thermal loads in building envelopes, both passively and actively. In active systems, unlike passive applications, a certain amount of energy is required to utilize free cooling. This is achieved using fans, pumps, or a combination of both in hybrid systems combined with radiative sky surfaces. Active application may involve integrating RSC directly into HVAC systems to improve overall cooling capacity and system efficiency. This emerging strategy demonstrates the significant potential of RSC technology in various building applications [34]. Typically, panels or surfaces that emit thermal radiation in the long-wavelength infrared spectrum are used in Active Radiative Sky Cooling systems (ARSC). The process of radiative cooling occurs during both day and night, but it is most effective at night: in both cases, the sky acts as a natural heat sink. By utilizing more effective technologies, ARSC systems may offer a sustainable and energy-efficient cooling method reducing active energy building loads.

A growing array of active cooling strategies has been proposed to enhance the efficiency of RSC applications within roofing systems. Active cooling roofs generate cold water or cold air through RSC mechanisms and subsequently convey it into indoor spaces for cooling in alignment with the specific requirements of the building. Bergman [57] demonstrated that spectrally selective surfaces for passive cooling could effectively match cooling loads, surface area for heat rejection and increase the amount of thermal power transferred from Earth to space. He adopted energy balances and heat transfer rate equations to predict of the performance of a novel active daytime radiative cooling concept, proposing the use of spectrally selective surfaces in conjunction with air conditioning and refrigeration systems.

Kousis et al. [58] discussed potential strategies and current progress in materials development of daytime radiative coolers, highlighting challenges and roadmaps to achieve effectiveness of their real-life applications. This gives rise to the need to design materials capable of achieving high performance at reasonable costs, thanks to advanced nanomanufacturing and synthesis processes.

Yuan et al. [59] demonstrated a daytime radiative cooling application using a selective polymer-based metamaterial to passively cool a full-scale model house, achieving a roof surface temperature of 2–9 °C below the ambient, continuously, for 72 h. Moreover, they claimed a further result: the indoor air temperature was also consistently below the ambient during the daytime.

Liu et al. [60] recently proposed a dual-layer film consisting of ethylene-tetra-fluoroethylene (25 μm) and silver (200 nm thick Ag layer), showing solar reflectivity of 0.94 and average total emissivity of 0.84, capable of reaching 1.6 °C below ambient air at daytime in Hefei (China).

Lim et al. [61] proposed high-energy band gap calcium carbonate ($CaCO_3$) for all-day radiative cooling, without any metal reflector. They adopted $CaCO_3$ microparticles with a diameter of 20–30 μm. The authors claimed cooling power of 93.1 W/m^2 and a sub-ambient temperature of 3.38 °C in daytime.

Huang et al. [62] reported a newly designed scalable-manufactured magenta-coloured daytime radiative cooler (60 cm wide and 500 cm a long) fabricated by a roll-to-roll deposition technique. Average temperature drops of 2.6−8.8 °C during the daytime were reported.

Yu et al. [63] fabricated a self-sustained insulated cooler with two features: daytime sub-ambient passive cooling (−3.3 °C in the field tests) and, on the other hand, the surface can harvest atmospheric water at night.

4.2.1. Air-Based Cooling Systems

A simple method of integrating RSC embodying a building cooling system is to use an air-cooling system, where air acts as the heat transfer medium. Such integration may be achieved through a simple air-based cooling system. This system includes an RSC surface, a propelling fan, and a connecting loop. The cooling process is facilitated through natural ventilation or fan-induced airflow, which directs the air through an RSC surface, positioned atop the building. The effectiveness of the air-based cooling systems depends on a restricted air passage beneath the radiator and a significant surface area, with the aim of maximizing the cooling of the exchanged air [49]. Moreover, the efficiency of these systems is usually determined by radiative cooling surface properties and climate conditions (e.g., ambient air temperature, humidity, wind speed).

According to Zhao et al. [23] three different air-based cooling system classes are currently reported in the literature (Figure 3). The first class (Figure 3a) represents a relatively simple air-based system, using the roof volume as a cooling sink, either directly into the room or via a duct system. The net cooling power of air-based systems is typically low, ranging from 20 to 40 W/m^2 [49]. The second class (Figure 3b) represents an active strategy combining HVAC hardware with an air-based system of ducts. The figure reports only daytime systems, though the first investigation of effective radiative heat coolers dates to the 1970's and is performed during night time. Some milestones in such activities have been reported hereafter.

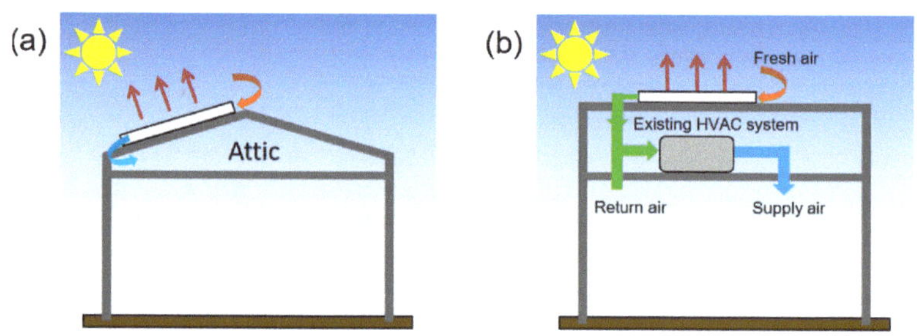

Figure 3. Active and passive strategies to exploit radiative coolers to cool fluids generally used in HVAC systems. (**a**) Air-based radiative sky cooling system using the roof as a radiator, (**b**) system integrated with HVAC system.

In 1977, Givoni introduced one of the first air-based applications named Roof Radiation Trap system. This system uses solar energy to heat buildings in winter and nocturnal radiation to cool them in summer [64]. This innovative approach involved the placement of a white-painted corrugated metal sheet on the roof to function as a nocturnal radiative cooler. The intervening space between the cooler and the roof was utilized for air storage. These corrugated metal sheets served the dual purpose of shielding against solar radiative heat during the day and cooling the air during the night. The interspace functioned as a cold reservoir, contributing to the cooling of the roof in subsequent days. Some vents connected the roof volume to the tested room. The results indicated that the roof radiation trap had the potential to yield a cooling capacity ranging from 29.7 to 55.8 W/m^2 under climatic conditions characteristic of Italy. The night sky radiant cooling performance of duct-type heat exchangers under different surface material and weather conditions was investigated by Liu et al. [65]. The differences between the inlet and outlet temperature of the night sky radiation heat exchanger resulted in a drop of 6.8 °C.

Along this line, Paker et al. [66] introduced the innovative "NightCool" roof concept designed to cool residential buildings. This approach involves establishing an attic area beneath a metal roof, with the internal building being interconnected or isolated based on airflow requirements within this space. The fan facilitated the airflow mechanism. The findings indicate that the "NightCool" roof resulted in an average cooling power of 5 to 10 W/m^2 overnight, particularly in the Florida climate.

Kimball et al. [67] have developed an air box convector using RSC technology. This system consists of a fan combined with two heat exchangers and is designed for cooling and dehumidification in tropical climates. Empirical results show that the innovative configuration using RSC has a potential energy saving of between 14% and 18% in the management of latent heat loads, contributing to the achievement of an optimal indoor environment.

Furthermore, Hollick [68] performed a series of performance tests on a nocturnal RSC roof. In this configuration, a sky radiative cooler was placed on the roof, and an air cavity was added underneath to aid air cooling. Cold air was then circulated into the room at night using a fan. The results showed that the RSC roof could lower the air temperature by approximately 4.7 °C compared to the ambient temperature.

Khedari et al. [69] designed and implemented four variants of roof radiators within an air-based system designed for a solar-powered school dormitory. Their experiments showed that during the night, the temperature of the roof radiators showed a reduction of 1 to 6 °C compared to the ambient air temperature, considering the prevailing weather conditions in Thailand.

Hu et. al. [70] introduced an innovative passive air conditioning module that integrates RSC and indirect evaporative cooling to boost efficiency. Results show that the system reduces the supply air temperature by 0.88 °C compared to the indirect evaporative cooling-only system. This corresponds to an 11.91% and 9.94% improvement in dew point efficiency and cooling energy gain, respectively.

Proper thermal insulation is also crucial to enhance the efficiency of ARSC system. Insulating materials prevent the reabsorption of heat from the surroundings, ensuring that the radiative cooling process remains effective [65].

Galvez et al. [71] tried to optimize the efficiency of an air-based RSC by analysing the performance of an air-based low-energy night-time radiative cooling system in Chile. The system is composed of a series of specialized radiative air collectors, a mechanical ventilation system, and a thermal storage wall, which is part of the indoor environment. The radiative air collector is cooled by radiative cooling due to radiative interaction with the sky. At night time, the fan is activated, and the cooled air is sent indoor to the gravel mass. At daytime, the fan is switched off and the gravel mass absorbs thermal energy that is accumulated indoor due to the internal loads and irradiation. Though this cannot be considered a daytime cooler, this technology is designed to operate during the daytime. This system aims at meeting the cooling demands of a building solely through natural ventilation or fans. Moreover, given the need for a narrow air duct and a large radiative cooling surface area, this air-based system is more suitable for use in detached houses or on the top floor of multi-storey buildings.

On the other hand, the cooling effect in daytime conditions is strongly affected by season (results presented in the first paper dealt with sunny days in winter conditions and in low humidity conditions [20]), radiation, location, humidity and air conditions. The non-radiative term of $P_{non\text{-}rad}$ is strongly affected by air speed. If the net nocturnal cooling of air systems reach values higher than 100 W/m^2 [9,72], the high convective heat transfer coefficients could hinder the performance in daytime conditions. Pollution also has a negative effect on coolers' performance since particulate emission absorbs (or emit) radiation, decreasing their cooling potential. According to Cui et al. [73], it is important to assess the influence of weather and geographical location on coolers' performance as well as to design radiative coolers capable of reducing the detrimental effects of wind, vapor, dirt, rain, and other external factors.

Zhao et al. [74] demonstrated the effectiveness of daytime cooling applied to a 1.08 m^2 system built using a wood framework and an aluminium plate (Figure 4). The cooling surface was protected with a polyethylene film acting as a convective shield. This prototype coupled radiative sky cooling with attic ventilation for temperature reduction: sub-ambient temperature reductions were between 5 and 8 °C, at nighttime, and between

3 and 5 °C during daytime hours. Daytime effective cooling technologies, compatible with passive air systems were demonstrated by Fu et al. [75] and by Torgerson et al. [76]. Wang et al. [77] reviewed typical materials and designs with adequate daytime radiative cooling capability: they can be grouped in several approaches, including nanophotonic materials obtained depositing alternating layers with suitable refractive properties, nano- and microparticle-based radiative materials, polymer-based radiative materials, and other radiative cooling designs.

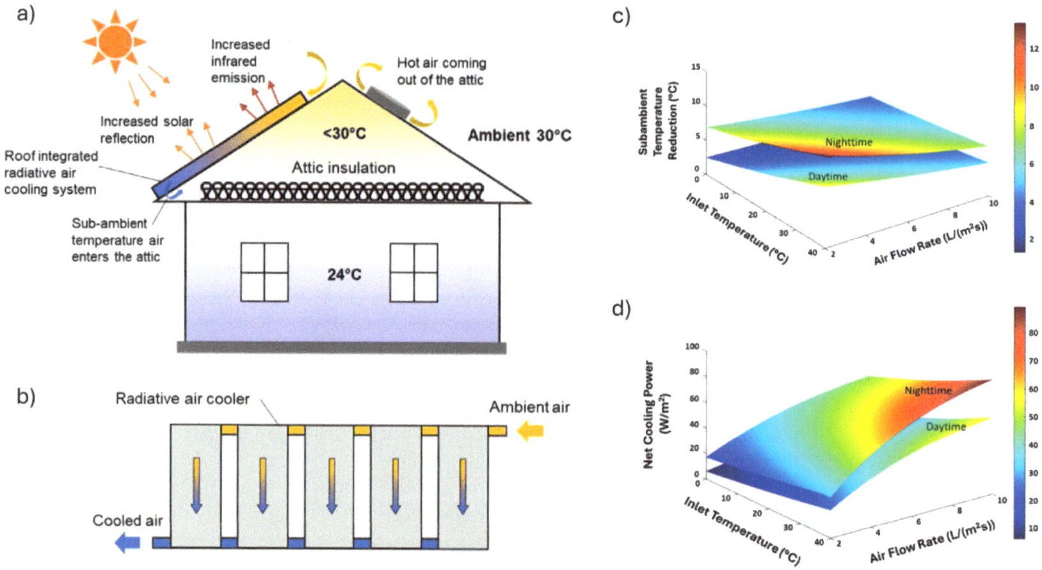

Figure 4. Radiative air coolers integrated atop a building (**a**,**b**); sub-ambient temperature reduction (**c**) and net cooling power attainable (**d**). [74].

Regarding the ideal component devoted to building integration of passive radiative coolers, roofs have generally been considered the most eligible surface to host such components. Nevertheless, Wu et al. [78] observed that roofs are generally responsible for only 10% to 20% energy consumption and such impact tends to be lower in high-rise buildings.

Passive radiative systems show limited effectiveness, highly depending on weather conditions, which may significantly harness their performance. Switchable properties may enhance their cooling capacity. In fact, if hot arid areas may benefit from passive coolers with static spectral properties almost throughout the year, due to high temperatures, their seasonal applicability is often hindered by lower temperatures and higher expenditures for heating in winter. Moreover, if the term regarding atmospheric power (P_{atm}) is strongly dependent on climate and weather conditions, the non-radiative term may be effectively attenuated by means of convective shields ($P_{non\ rad}$).

4.2.2. Water-Based Cooling Systems

Other systems use water as a heat transfer fluid, with higher specific heat capacity compared to air, using a cold storage unit, with net cooling power typically double, in the order of 40–80 W/m^2 [23].

Water-based cooling systems can be divided into two main types: open-loop and closed-loop. The open-loop water system typically involves a shallow roof pond as a key component, facilitating heat transfer to the environment through mechanisms such as convection, radiation and evaporation from the roof immersed in the pond water [79–82]. They have been largely investigated and used in arid areas of the planet. Conversely, the closed-loop water system differs from its open-loop counterpart by using water as the heat transfer fluid, with heat dissipation achieved through a flat plate radiator [69,83–85]. The heat carrier—water, in this case—flows in pipes embedded in a flat-plate and the system also includes an insulated water tank, a heat pump and the radiative plate.

The thermal energy derived from night-time radiative cooling, particularly in the form of chilled water, has the potential to be used as a cooling resource during the day. However, the typical low supply temperature of chilled water from conventional building air conditioning systems, such as 7 °C, may render the water from the radiative sky cooling system insufficiently chilled for direct use. Possible remedies include integrating the water-based cooling system with a radiant floor/ceiling system, which does not require an extremely low supply water temperature, typically in the range of 15–18 °C. Alternatively, the use of radiantly cooled water to pre-cool the heat transfer fluid before it enters the chiller is a viable solution.

Spanaki et al. [81] presented twelve different configurations of rooftop pools, providing a comprehensive analysis of their advantages and disadvantages, and discussed the criteria relevant to the design of rooftop pools.

Raeissi and Taheri [86] have shown that the implementation of a water-based open-loop system can lead to a significant reduction in the cooling load of the air conditioning system, through night-time radiative cooling, amounting to a 52% reduction during the summer season for a single-story house of 140.6 m^2 in Iran.

Meir et al. [87] proposed a radiative cooling system using water as a heat carrier, circulating in flat plate radiators. The system was equipped with a 280 L water reservoir. The cooling radiator consisted of sheets of polyphenylene oxide resin, arranged in twin-wall sheets. This system could cover a significant fraction of the cooling demand of a single-family house in Oslo (Norway).

Compared to the open-loop system, the closed loop results in higher efficiency. Zhou et al. [88] used an RSC system in conjunction with a heat exchanger to manage fresh air within the building environment. Cold water was generated using the radiative cooling panel and then fed to the heat exchanger to cool the fresh air. Cold storage significantly increases the drop in indoor air temperature, up to 12.7 °C. Any excess cooling capacity was stored in the thermal storage tank, not only meeting immediate cooling requirements but also providing additional thermal storage capacity.

Wang et al. [89] carried out numerical simulations for the HVAC system of a typical medium-sized office building integrating the use of the photonic radiative cooler. With reference to the cooling season, the free energy from the radiator addressed about 10% of the cooling load in Miami. A photonic radiative cooler encased in a polyethylene film in the air conditioning (AC) systems was designed to cool office buildings. Inside the cooler, water pipes were strategically placed to cool the water. The sky radiative cooler was positioned at the top of the building and connected to a water storage tank to form the RSC loop. The distance cooling loop, including the chiller, pump and cooling connections, were integrated with the RSC loop by heat exchange within the storage tank. This arrangement facilitated the adjustment of operating modes by means of valve switching (Figure 5).

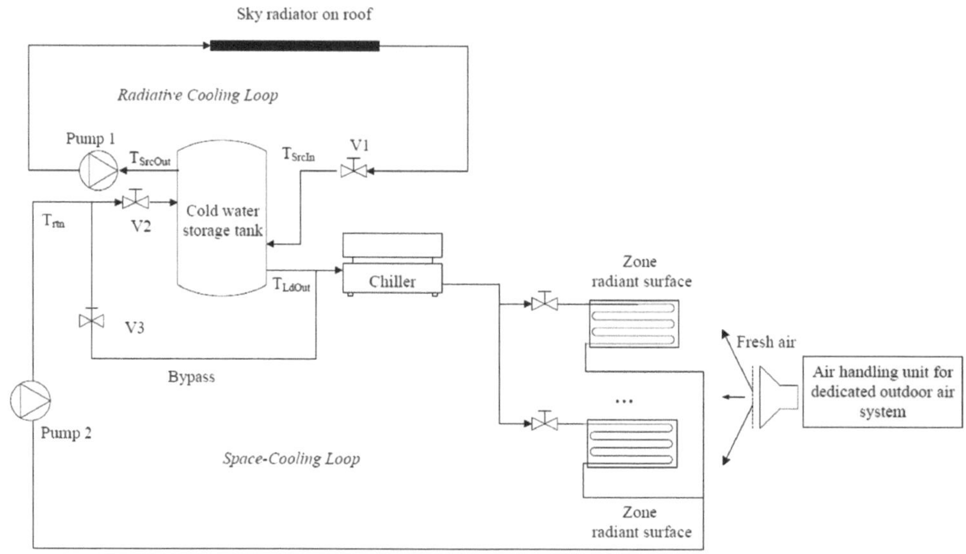

Figure 5. Schematic design of the integrated radiative cooling system [89].

A particularly promising roadmap for the application of RC modules is their integration with air conditioning systems to increase operational efficiency. Various studies have been carried out to explore the integration of either night or daytime RC structures with HVAC or other cooling systems. Within these hybrid systems, the use of chilled water from a radiative cooler helps to improve the efficiency of the air conditioning system by acting as a coolant for the cooling coils. Jeong et al. [10] described the design of an integrated radiative cooling, HVAC system with two different cooling coils—one supplied by radiative cooling and the other conventional. In this proposed cooling configuration, the conventional chillers used in the refrigeration cycle are replaced by an RC structure. The proposed cooling system has the potential to achieve a remarkable 35% reduction in energy consumption by complementing conventional space cooling equipment. Another study by Zhang et al. [90] investigated the integration of cold-water storage with radiative cooling to accumulate cooling energy. In addition, nighttime radiative coolers can be synergistically combined with evaporative cooling systems to facilitate pre-cooling, thereby increasing the overall systematic cooling efficiency. In this design, the radiative cooler generates cold water during the night, and during the day this cold water passes through the cooling coils to pre-cool the hot outdoor air before being subjected to a direct evaporative cooling process [91].

Recently, Yoon et al. [92] proposed a hybrid HVAC system that integrates a solar thermal collector and a radiative cooling panel as the heat source and heat sink, respectively. Annual power consumption is reduced by 3~29% as compared to a solar-assisted heat pump or radiative-cooling-assisted heat pump system.

With the spread of daytime passive radiative cooling, several studies [74,93] have shown that integrating such panels into an air conditioning system, known as a Radiative Cooling Assisted Heat Pump system, can reduce the electricity consumption in summer by 9–45%.

Along this line, Jia et al. [94] proposed a hybrid system that combines a ground source heat pump and all-day RSC radiators to improve the energy efficiency across China. It was found that the average annual cooling potential varies from 149.7 W/m^2 to 484.6 W/m^2 in all regions of China, with the highest cooling potential of 336.5 W/m^2 in cold northern China and the lowest cooling potential of 178.3 W/m^2 in tropical southern China.

In contrast to the air-based system, water-based implementations improve operability and controllability, thereby offering increased feasibility and efficiency. These systems can provide a cooling capacity tailored to the specific requirements of the building. However, their structural complexity is slightly increased, and their deployment is constrained by the available roof space for placement of the coolers.

Zhao et al. [93] reported a kilowatt-scale innovative module for radiative cooling named RadiCold, designed to cool water to 10.6 °C below ambient at daytime, and also investigated the effect of different weather conditions on its performance. The system surface area was 13.5 m^2 and a maximum cooling power of 1296 W was reported.

Goldstein et al. [95] reported their panels that harness radiative sky cooling to cool fluids 5 °C below the air temperature. Over three days of testing, we show that the panels cool water up to 5 °C, with a heat rejection power of 70 W/m^2. The authors also showed results of modelling activities, reporting that radiative surfaces, integrated on the condenser side of the cooling system of an office building, in a hot dry climate, reduced electricity consumption by 21% for cooling during the summer.

Table 1 shows that there are several technologies and materials that could lead to relevant results in terms of heat rejection. The determining factor, in the context of applications for the construction sector, will undoubtedly be that of cost (materials, installation, maintenance, disposal). According to Yu et al. [96], it is reasonable to expect that polymer-based porous structures and systems embodying randomly distributed particles (possibly without reflective silver metal layers) may perform well, being a promising approach for low-cost radiative cooling applications also in the construction sector, rather than nanophotonic-fabricated surfaces with a nanophotonic approach. For example, the cost of the RadiCold modules is estimated at about $25/$m^2$ [97]. Several technologies are still at a laboratory scale, and costs are only partially predictable. In general terms, polymers with a chemical-based bottom-up approach represent a viable alternative to expensive fabrication processes and will have more chances for building integration purposes. Apart from cost of materials, a consistent benefit related to radiative cooling technology is to save electricity used for cooling. A wide discussion of potential barriers to market adoption of radiative cooling materials has been reported in [98]. Their analysis revealed that the incremental cost of all components of the passive radiator should not exceed 8.25 $\$/m^2$ to 11.50 $\$/m^2$ of the total building floor area. It is quite easy to infer that materials with simple design criteria may cut fabrication costs. For example, Felicelli et al. [99] reported a low-cost, dual-layer system, comprising of a cellulose-based substrate as the bottom layer and a thin $BaSO_4$-based layer paint as the top layer. In the work by Zhang et al. [100], they estimated 26–46% saving of electricity consumption per year compared to split air conditioners in San Francisco, with a simple payback period of 4.8–8.0 years, and maximum acceptable incremental costs of 50.0–78.9 $\$/m^2$. Papers reviewed during the preparation of this work reported limited information about costs of the radiators due to the low level of technology readiness generally at a laboratory scale. Nevertheless, some works show that potential benefits may be compatible with building integration. Furthermore, it should be observed that polymer-based structures may be more prone to scalability, inherently lightweight, and easy to install.

Radiators with a non-dynamic behaviour allow passive radiative heat dissipation to be easily removed in the winter season since they could have undesirable effects in the winter season, leading to increased heating consumption. Energy savings attained in the cooling season may be counterproductive if the radiator is also used during winter. This aspect may be relevant according to the location where radiators are installed. Wu et al. [78] clearly reported that the location has a great impact in this balance between energy saving and increase in winter losses due to the radiators. They observed that in relatively high-temperature areas, the loss in winter is significantly lower than the energy benefits observed in summer; conversely, in most regions, the loss in winter may be even greater than the energy saved in summer. The location and season may dramatically affect the performance of radiators. Lu et al. [100] demonstrated that the maximum temperature depression of the radiator designed by Raman et al. may underperform in the mid-latitude region of Shanghai and work even worse in the hot and damp tropic areas.

Table 1. Main state of the art technologies and figures of merit of their performance.

Reference	Technology Adopted	Adaptiveness	Main Features
Raman et al. [20]	Integrated photonic solar reflector/thermal emitter (seven layers of HfO_2 and SiO_2).	No	R = 97% of incident sunlight; high selective emittance in the atmospheric transparency window. Exposed to direct sunlight >850 W/m^2 cools to 4.9 °C below ambient T_{air}; cooling power of 40.1 W/m^2 at T_{air}.
Pisello et al. [45]	Prototyped cool clay tile.	No	Tiles decrease summer peak indoor temperature of the attic by up to 4.7 °C. Winter maximum temperature reduction was 1.2 °C
Gentle et al. [46]	Birefringent polymer pairs, with high index and one with low index, for example poly-ethylene terephthalate (PET)/naphthalene dicarboxylate and polyethylene naphthalate.	No	For a 1000 W·m^{-2} incident, around 30 W/m^2 of solar energy is absorbed. Average thermal emittance of 0.96 from 7.9 to 14 μm. In the absence of a convective barrier, $T_{amb} - T_{roof}$ value is equal to 3 °C for peak midday conditions on a clear summer day, and to 7 °C at night.
Wang et al. [33]	300 μm thick flexible hybrid photonic membrane radiator. Polyvinylidene fluoride/ tetraethyl orthosilicate fibres with randomly distributed SiO_2 microspheres.	No	Average infrared emissivity > 0.96 and $R_{sol} \approx 97\%$. Average radiative cooling power of 61 W·m^{-2} and a temperature decrease up to 6 °C under solar intensity of 1000 W/m^2.
Zhai et al. [32]	Resonant polar dielectric microspheres randomly embedded in a polymeric matrix, backed with silver coating.	No	Daytime radiative cooling power of 93 W/m^2 under direct sunshine and economical roll-to-roll manufacturing of the metamaterial.
Zhang et al. [90]	Switchable radiative cooler composed of a radiative cooling coating and a temperature-responsive part.	Yes	The prepared radiative cooler shows $R_{sol} \approx 96\%$ and 95% infrared emission in the atmospheric transmittance window (8−13 μm). The temperature responsive part consists of nickel−titanium alloy springs working as a thermal switch, showing low thermal resistance (2.7 K·W^{-1}) in the ON state. The heat is blocked to escape the indoor space since a high thermal resistance (20 K·W^{-1}) is created when the thermal switch is in the OFF state.
Deng et al. [101]	Electro-controlled polymer-dispersed liquid crystal.	Yes	The device reaches near/sub-ambient temperature when the solar irradiance is below 400 W/m^2 and can dynamically manage daytime cooling efficiency by applying an external bias.
Carlosena et al. [56]	Vikuiti substrate covered with 2 um thick film of polymethylsesquioxane + SiO_2 microparticles.	No	Daytime radiative cooler with a cost of 0.3 €/m^2. >7 °C sub-ambient temperature in moderate weather conditions.
Zhao et al. [74]	"RadiCold" metafilm on top of an aluminum sheet. Polyethylene-based convective shield.	No	Sub-ambient temperatures of 3–5 °C during daytime.

Table 1. *Cont.*

Reference	Technology Adopted	Adaptiveness	Main Features
Yuan et al. [59]	Polymer-based spectrally selective metamaterial; R_{sol} about 96%.	No	When solar irradiation is 720 W/m^2, a surface temperature of the model house of 2–9 °C below the ambient during a 72 h experiment period is achieved.
Liu et al. [60]	Dual-layer film consisting of ethylene-tetra-fluoro-ethylene (25 μm) and silver.	No	R_{sol} = 0.94 and average total emissivity of 0.84, capable of reaching 1.6 °C below ambient air.
Lim et al. [61]	Single layer of a $CaCO_3$ composite without any metal reflector	No	Daytime sub-ambient temperatures of 3.38 °C and cooling power of 93.1 W/m^2.
Huang et al. [62]	Polymer–Tamm photonic structure.	No	Found theoretical thresholds for sub-ambient cooling through coloured coolers. Temperature drop of 2.6–8.8 °C during the daytime and 4.0–4.4 °C during the nighttime is achieved.
Yu et al. [63]	Porous polyethylene film at the top, an air layer in the middle, and poly(vinyl alcohol) hydrogel with lithium bromide at the bottom.	No	R_{sol} = 0.91; ε_{LWIR} = 0.96; daytime sub-ambient cooling ~3.3 °C in the field tests.
Wang et al. [89]	The photonic radiative cooler consists of seven alternating layers of two thin oxides, a couple of metal layers, and a silicon wafer substrate.	No	HVAC system integrating the use of the photonic radiative cooler was proposed and modelled using the whole energy simulation program, EnergyPlus. During the cooling season, the passive cooler addressed about 10% of the cooling load in Miami, 17–36% in Las Vegas, and 61–84% in Los Angeles.
Jia et al. [94]	Hybrid system that combines ground source heat pump and all-day radiative sky cooling radiators.	No	Annual average cooling power provided by radiative sky cooling radiators ranges between 149.7 and 484.6 W/m^2 across all regions of China.
Zhao et al. [93]	Polymer-based hybrid film (50 μm); silver (200 nm); adhesive (90 μm); and polycarbonate (0.5 mm).	No	A 10.6 °C sub-ambient cooling of a large mass of water in real experiments.
Goldstein et al. [95]	Visibly reflective extruded copolymer mirror (3M Vikiuiti ESR film) to achieve sub-ambient temperatures under sunlight on top of a silver reflective surface.	No	Panels cooling water down to a 5 °C sub-ambient temperature with water flow rates of 0.2 L/(min·m^2), resulting in an effective heat rejection flux of 70 W/m^2.

5. Switchable Materials and Systems Applied to Radiative Sky Cooling in Buildings

The persistent thermal radiation of cooling materials during winter seasons may increase the heating cost and consequently affect indoor thermal comfort. For this reason, great attention is being paid to the design of switchable radiative cooling technologies, to dynamically cool or heat objects, according to changing external conditions, to address internal requirements throughout the year time. They have been classified according to the external *stimuli* that activate their operation [102]. Switchable cooling technologies should possess a switching ability between cooling and heating (or even intermediate behaviours) modes: high solar reflectance and high infrared emissivity during summer and low solar reflectance and emissivity during cold seasons, as shown in Figure 6. An exhaustive review about switchable radiative cooling surfaces was also proposed by Myung Jin Yoo et al. [103] in 2023. Different mechanisms for switchable radiative cooling were investigated thoroughly: wetting/drying switching mechanism, mechanical switching mechanism, thermochromic switching mechanism and electrochromic switching mechanisms. Novel technologies make it possible to dynamically adjust the optical characteristics of materials surfaces, enabling smart thermal management.

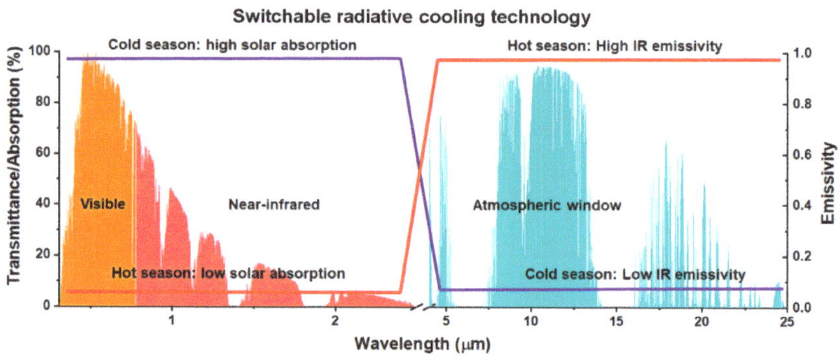

Figure 6. Ideal adaptiveness of spectral properties in selective radiative coolers.

The emittance modulation ($\Delta\varepsilon_{LWIR}$) is due to the change of the intrinsic molecular structure of materials: Yoo et al. [104] showed that some molecular bonds, namely C–O–C (peaked in the wavelengths 7936–9009 nm), C–OH (8071–9709 nm), and C–F_3 (8711 nm) affect vibrational energy resonance within the long-wavelength infrared range and cause an increase in infrared absorbing (emittance) features of the material. For example, the high value of ε_{LWIR} typically observed in polydimethylsiloxane at 9,813 nm and 11,455 nm is due to the resonating molecular bonds Si–O and Si–CH_3, respectively.

At high latitudes, a switching operation of roof properties would be welcome, with the ability to change reflectance properties when buildings switch from cooling to heating mode. Self-adaptive or smart radiative cooling materials should be designed for this purpose. Cool roofs, in fact, may unwantedly increase heating energy in winter. This point was accurately discussed by Testa et al. [104]. According to their studies, coatings with effectively switchable reflectance can improve annual energy savings by up to 6% compared to a static cool roof. Though static materials may produce overall energy savings in several climate conditions, switchable reflectance coatings may offer larger benefits, when applied to both roofs and opaque exterior walls.

Innovative dual-mode devices with both heating and cooling capacities can save energy of 236 GJ/year, 1.7 times more than cooling-only and 2.2 more than heating-only materials [102]. In this roadmap, passive adaptive cooling would be ideal to reduce maintenance and production costs. For example, the temperature-dependant behaviour of refractive indexes of VO_2 [105] would be compatible with the required operation of switchable passive coolers. Materials with stimulus-dependant characteristics represent a suitable field for further investigation. Such highly customized materials would enable adaptive regulation of IR emissivity and solar reflectance, creating an all-season energy-saving technology. In other words, switchable cool roofs should have the ability to change reflectance throughout the year, according to changes in external conditions.

Zhang et al. [106] proposed a switchable radiative cooler composed of a radiative cooling coating and a temperature-responsive part, based on nickel—titanium springs as temperature-responsive actuators, capable of changing thermal resistance from 2.7 K/W in the "on" state to 20 K/W in the "off" state, with possible temperature drops of 11 K in daytime.

Deng et al. [101] proposed a switchable polymer-dispersed liquid crystal to design a smart window showing on-demand passive radiative cooling efficiency that was designed and prepared by incorporating mid-infrared emitting monomers into the material matrix. They demonstrated the effectiveness of their device during a daytime experiment performed in Beijing, in spring.

In conclusion, the goal of an efficient radiative thermal management is related to the capability of a simultaneous, effective control of R_{sol} and ε_{LWIR}, and to avoid any unwanted performance in precise seasonal conditions. If reflectance control may be associated with a decrease in transmittance, one further benefit attainable may be the improvement of indoor visual comfort. The scattering observed in hydroxypropylcellulose and similar hydrogels, for example, may reduce high glare effects, in addition to increasing reflectance and emittance in hot and sunny summer days [107].

6. Conclusions

This work aims to quantify the effectiveness of radiative systems applied to the construction sector, with reference to their operation during daytime. This review provides the physical laws that evaluate the energy balance of passive radiative surfaces and provides the tools to quantify all the terms of these energy balances. Above all, the materials and technologies that allow—according to the current state of the art—for the production of passive radiative surfaces operating in daytime conditions and therefore under incident solar radiation, are reviewed, considering the effectiveness of the various technologies available today, both on the market and still at a laboratory scale. Different approaches in the fabrication of radiative systems have been reviewed, including photonic and polymeric membranes containing nanoparticles resonant with infrared radiation between 8 and 13 μm. Furthermore, different ways of integrating passive radiative systems inside buildings are shown to achieve air or water cooling in different system configurations. The most recent and high impact works appeared in this field of investigation were considered, with special attention to applications of switchable materials. Some references to the costs of available products are also provided, when available. The great scientific interest aroused by the possibility of producing surfaces intended for heat rejection, even under daylight, bodes well for the availability, in a few years, of several materials for applications in various fields. A separate discussion has been dedicated to the development of adaptive materials, which will give further degrees of freedom to a technology already endowed with disruptive potential. A driving element, in this intriguing roadmap, will be the cost of these materials, particularly in the construction sector, as well as the benefits that can be achieved. Indeed, in this review paper, we have reported the performance and cost of several materials, some of which are already patentable and commercially available, while others are still at a laboratory development level, with advantages and disadvantages. Some considerations have also been made in this aspect, when available in the scientific literature. Not every research group has reported cost considerations, and this represents a limit of this review but also of the field investigated, as well as the limited availability of LCA analysis on these types of materials and systems. These limitations, duly pointed out here, could be a warning to increase the level of information in the articles produced in this area of research.

Author Contributions: All authors have read and agreed to the published version of the manuscript.

Funding: This research received no external funding.

Conflicts of Interest: The authors declare no conflicts of interest.

References

1. EU Parliament. *Directive (EU) 2018/844 of the European Parliament and of the Council of 30 May 2018 Amending Directive 2010/31/EU on the Energy Performance of Buildings and Directive 2012/27/EU on Energy Efficiency*; EU Parliament: Strasbourg, France, 2018.
2. Li, T.; Zhai, Y.; He, S.; Gan, W.; Wei, Z.; Heidarinejad, M.; Dalgo, D.; Mi, R.; Zhao, X.; Song, J.; et al. A Radiative Cooling Structural Material. *Science* **2019**, *364*, 760–763. [CrossRef]
3. Granqvist, C.G.; Hjortsberg, A. Surfaces for Radiative Cooling: Silicon Monoxide Films on Aluminum. *Appl. Phys. Lett.* **1980**, *36*, 139–141. [CrossRef]

4. Granqvist, C.G.; Hjortsberg, A. Radiative Cooling to Low Temperatures: General Considerations and Application to Selectively Emitting SiO Films. *J. Appl. Phys.* **1981**, *52*, 4205–4220. [CrossRef]
5. Givoni, B. Options and Applications of Passive Cooling. *Energy Build.* **1984**, *7*, 297–300. [CrossRef]
6. Tsai, C.C.; Shi, N.; Pelaez, J.; Pierce, N.; Yu, N. Butterflies Regulate Wing Temperatures Using Radiative Cooling. In Proceedings of the 2017 Conference on Lasers and Electro-Optics, CLEO 2017—Proceedings 2017, San Jose, CA, USA, 14–19 May 2017; Volume 2017-Janua, pp. 1–2. [CrossRef]
7. Shi, N.N.; Tsai, C.C.; Craig, C.; Yu, N. Nano-Structured Wild Moth Cocoon Fibers as Radiative Cooling and Waveguiding Optical Materials. In Proceedings of the 2017 Conference on Lasers and Electro-Optics, CLEO 2017—Proceedings 2017, San Jose, CA, USA, 14–19 May 2017; Volume 2017-Janua, pp. 1–2. [CrossRef]
8. Shi, N.N.; Tsai, C.C.; Camino, F.; Bernard, G.D.; Yu, N.; Wehner, R. Keeping Cool: Enhanced Optical Reflection and Radiative Heat Dissipation in Saharan Silver Ants. *Science* **2015**, *349*, 298–301. [CrossRef]
9. Hossain, M.M.; Gu, M. Radiative Cooling: Principles, Progress, and Potentials. *Adv. Sci.* **2016**, *3*, 1500360. [CrossRef] [PubMed]
10. Jeong, S.Y.; Tso, C.Y.; Zouagui, M.; Wong, Y.M.; Chao, C.Y.H. A Numerical Study of Daytime Passive Radiative Coolers for Space Cooling in Buildings. *Build. Simul.* **2018**, *11*, 1011–1028. [CrossRef]
11. Ulpiani, G.; Ranzi, G.; Feng, J.; Santamouris, M. Expanding the Applicability of Daytime Radiative Cooling: Technological Developments and Limitations. *Energy Build.* **2021**, *243*, 110990. [CrossRef]
12. John, A.; Duffie, W.A.B. *Solar Engineering of Thermal Processes*; John Wiley & Sons, Inc.: Hoboken, NJ, USA, 2013.
13. Swinbank, W.C. Long-Wave Radiations from Clear Skies. *Q. J. R. Meteorol. Soc.* **1963**, *89*, 339–348. [CrossRef]
14. Bliss, R.W. Atmospheric Radiation near the Surface of the Ground: A Summary for Engineers. *Sol. Energy* **1961**, *5*, 103–120. [CrossRef]
15. Martin, M.; Berdahl, P. Summary of Results from the Spectral and Angular Sky Radiation Measurement Program. *Sol. Energy* **1984**, *33*, 241–252. [CrossRef]
16. Berdahl, P.; Martin, M. Emissivity of Clear Skies Permalink. *Energy Environ. Div.* **1984**. [CrossRef]
17. Ahmad, M.I.; Jarimi, H.; Riffat, S. *Nocturnal Cooling Technology for Building Applications*; Springer: Singapore, 2019; ISBN 9789811358340. [CrossRef]
18. Rephaeli, E.; Raman, A.; Fan, S. Ultrabroadband Photonic Structures to Achieve High-Performance Daytime Radiative Cooling. *Nano Lett.* **2013**, *13*, 1457–1461. [CrossRef] [PubMed]
19. Santamouris, M.; Yun, G.Y. Recent Development and Research Priorities on Cool and Super Cool Materials to Mitigate Urban Heat Island. *Renew. Energy* **2020**, *161*, 792–807. [CrossRef]
20. Raman, A.P.; Anoma, M.A.; Zhu, L.; Rephaeli, E.; Fan, S. Passive Radiative Cooling below Ambient Air Temperature under Direct Sunlight. *Nature* **2014**, *515*, 540–544. [CrossRef] [PubMed]
21. Fan, S.; Raman, A. Metamaterials for Radiative Sky Cooling. *Natl. Sci. Rev.* **2018**, *5*, 132–133. [CrossRef]
22. Heo, S.Y.; Lee, G.J.; Kim, D.H.; Kim, Y.J.; Ishii, S.; Kim, M.S.; Seok, T.J.; Lee, B.J.; Lee, H.; Song, Y.M. A Janus Emitter for Passive Heat Release from Enclosures. *Sci. Adv.* **2020**, *6*, eabb1906. [CrossRef] [PubMed]
23. Zhao, D.; Aili, A.; Zhai, Y.; Xu, S.; Tan, G.; Yin, X.; Yang, R. Radiative Sky Cooling: Fundamental Principles, Materials, and Applications. *Appl. Phys. Rev.* **2019**, *6*, 021306. [CrossRef]
24. Greffet, J.-J.; Bouchon, P.; Brucoli, G.; Sakat, E.; Marquier, F. Generalized Kirchhoff Law. *arXiv* **2016**, arXiv:1601.00312. [CrossRef]
25. Ji, M.; Xu, Y.; Mo, Y.; Zhang, Y.; Zhou, R.; Zhu, S. Land Surface Temperature Retrieval Method Based on UAV Thermal Infrared Remote Sensing and Synchronized Atmospheric Profiles. *J. Geo-Information Sci.* **2023**, *25*, 2456–2467. [CrossRef]
26. Vall, S.; Castell, A. Radiative Cooling as Low-Grade Energy Source: A Literature Review. *Renew. Sustain. Energy Rev.* **2017**, *77*, 803–820. [CrossRef]
27. Zhao, B.; Hu, M.; Ao, X.; Huang, X.; Ren, X.; Pei, G. Conventional Photovoltaic Panel for Nocturnal Radiative Cooling and Preliminary Performance Analysis. *Energy* **2019**, *175*, 677–686. [CrossRef]
28. Lim, X.Z. The Super-Cool Materials That Send Heat to Space. *Nature* **2020**, *577*, 18–20. [CrossRef] [PubMed]
29. Ko, B.; Lee, D.; Badloe, T.; Rho, J. Metamaterial-Based Radiative Cooling: Towards Energy-Free All-Day Cooling. *Energies* **2019**, *12*, 89. [CrossRef]
30. Kecebas, M.A.; Menguc, M.P.; Kosar, A.; Sendur, K. Passive Radiative Cooling Design with Broadband Optical Thin-Film Filters. *J. Quant. Spectrosc. Radiat. Transf.* **2017**, *198*, 1339–1351. [CrossRef]
31. Chen, Z.; Zhu, L.; Raman, A.; Fan, S. Radiative Cooling to Deep Sub-Freezing Temperatures through a 24-h Day-Night Cycle. *Nat. Commun.* **2016**, *7*, 13729. [CrossRef] [PubMed]
32. Zhai, Y.; Ma, Y.; David, S.N.; Zhao, D.; Lou, R.; Tan, G.; Yang, R.; Yin, X. Scalable-Manufactured Randomized Glass-Polymer Hybrid Metamaterial for Daytime Radiative Cooling. *Science* **2017**, *355*, 1062–1066. [CrossRef] [PubMed]
33. Wang, X.; Liu, X.; Li, Z.; Zhang, H.; Yang, Z.; Zhou, H.; Fan, T. Scalable Flexible Hybrid Membranes with Photonic Structures for Daytime Radiative Cooling. *Adv. Funct. Mater.* **2020**, *30*, 1907562. [CrossRef]
34. Yan, T.; Xu, D.; Meng, J.; Xu, X.; Yu, Z.; Wu, H. A Review of Radiative Sky Cooling Technology and Its Application in Building Systems. *Renew. Energy* **2024**, *220*, 119599. [CrossRef]
35. Hu, M.; Zhao, B.; Suhendri; Ao, X.; Cao, J.; Wang, Q.; Riffat, S.; Su, Y.; Pei, G. Applications of Radiative Sky Cooling in Solar Energy Systems: Progress, Challenges, and Prospects. *Renew. Sustain. Energy Rev.* **2022**, *160*, 112304. [CrossRef]

36. Pacheco, R.; Ordóñez, J.; Martínez, G. Energy Efficient Design of Building: A Review. *Renew. Sustain. Energy Rev.* **2012**, *16*, 3559–3573. [CrossRef]
37. Jia, S.; Weng, Q.; Yoo, C.; Xiao, H.; Zhong, Q. Building Energy Savings by Green Roofs and Cool Roofs in Current and Future Climates. *Npj Urban Sustain.* **2024**, *4*, 23. [CrossRef]
38. Feng, J.; Saliari, M.; Gao, K.; Santamouris, M. On the Cooling Energy Conservation Potential of Super Cool Roofs. *Energy Build.* **2022**, *264*, 112076. [CrossRef]
39. Chan, Y.H.; Zhang, Y.; Tennakoon, T.; Fu, S.C.; Chan, K.C.; Tso, C.Y.; Yu, K.M.; Wan, M.P.; Huang, B.L.; Yao, S.; et al. Potential Passive Cooling Methods Based on Radiation Controls in Buildings. *Energy Convers. Manag.* **2022**, *272*, 116342. [CrossRef]
40. ASHRAE. *Advanced Energy Design Guide for Small to Medium Office Buildings 50%*; ASHRAE: Peachtree Corners, GA, USA, 2014; ISBN 9781936504053.
41. *ASTM E1980-11(2019)*; Standard Practice for Calculating Solar Reflectance Index of Horizontal and Low-Sloped Opaque Surface. ASTM: West Conshohocken, PA, USA, 2019; Volume 04.04. [CrossRef]
42. Akbari, H.; Berdahl, P.; Levinson, R.; Wiel, S.; Miller, B.; Desjarlais, A. *Cool-Color Roofing Material. California Energy Commission, PIER Building End-Use Energy Efficiency Program*; CEC-500-2006-067; California Energy Commission: Sacramento, CA, USA, 2006.
43. Bamdad, K. Cool Roofs: A Climate Change Mitigation and Adaptation Strategy for Residential Buildings. *Build. Environ.* **2023**, *236*, 110271. [CrossRef]
44. Hernández-Pérez, I.; Xamán, J.; Macías-Melo, E.V.; Aguilar-Castro, K.M. *Reflective Materials for Cost-Effective Energy-Efficient Retrofitting of Roofs*; Elsevier Ltd.: Amsterdam, The Netherlands, 2017; ISBN 9780081011287.
45. Pisello, A.L.; Cotana, F. The Thermal Effect of an Innovative Cool Roof on Residential Buildings in Italy: Results from Two Years of Continuous Monitoring. *Energy Build.* **2014**, *69*, 154–164. [CrossRef]
46. Gentle, A.R.; Smith, G.B. A Subambient Open Roof Surface under the Mid-Summer Sun. *Adv. Sci.* **2015**, *2*, 2–5. [CrossRef] [PubMed]
47. Baniassadi, A.; Sailor, D.J.; Ban-Weiss, G.A. Potential Energy and Climate Benefits of Super-Cool Materials as a Rooftop Strategy. *Urban Clim.* **2019**, *29*, 100495. [CrossRef]
48. Mandal, J.; Fu, Y.; Overvig, A.C.; Jia, M.; Sun, K.; Shi, N.N.; Zhou, H.; Xiao, X.; Yu, N.; Yang, Y. Hierarchically Porous Polymer Coatings for Highly Efficient Passive Daytime Radiative Cooling. *Science* **2018**, *362*, 315–319. [CrossRef]
49. Wang, Z.; Liang, J.; Lei, D.; Jiang, C.; Yang, Z.; Yang, G.; Zhang, D.; Zhang, L.; Zhang, C.; Bai, Y. Temperature-Adaptive Smart Windows with Passive Transmittance and Radiative Cooling Regulation. *Appl. Energy* **2024**, *369*, 123619. [CrossRef]
50. Byrnes, S.J. Multilayer Optical Calculations. *arXiv* **2016**, arXiv:1603.02720.
51. Chen, Y.; Dang, B.; Fu, J.; Wang, C.; Li, C.; Sun, Q.; Li, H. Cellulose-Based Hybrid Structural Material for Radiative Cooling. *Nano Lett.* **2021**, *21*, 397–404. [CrossRef] [PubMed]
52. Cannavale, A.; Pugliese, M.; Giannuzzi, R.; Scarfiello, R.; Prontera, C.T.; Primiceri, V.; Mazzeo, M.; Martellotta, F.; Ayr, U.; Fiorito, F.; et al. Towards the Scale-up of Solid-State, Low-Emissive Electrochromic Films, Fabricated on a Single Substrate with Novel Electrolyte Formulations. *Sol. Energy Mater. Sol. Cells* **2022**, *241*, 111760. [CrossRef]
53. Çengel, Y.A. *Introduction to Thermodynamics and Heat Transfer*; McGraw-Hill: New York, NY, USA, 2009; Volume 960.
54. Erell, E.; Etzion, Y. Analysis and Experimental Verification of an Improved Cooling Radiator. *Renew. Energy* **1999**, *16*, 700–703. [CrossRef]
55. Berk, A.; Anderson, G.P.; Acharya, P.K.; Bernstein, L.S.; Muratov, L.; Lee, J.; Fox, M.; Adler-Golden, S.M.; Chetwynd, J.H., Jr.; Hoke, M.L.; et al. MODTRAN5: 2006 Update. *Algorithms Technol. Multispectral Hyperspectral Ultraspectral Imag. XII* **2006**, *6233*, 62331F. [CrossRef]
56. Carlosena, L.; Andueza, Á.; Torres, L.; Irulegi, O.; Hernández-Minguillón, R.J.; Sevilla, J.; Santamouris, M. Experimental Development and Testing of Low-Cost Scalable Radiative Cooling Materials for Building Applications. *Sol. Energy Mater. Sol. Cells* **2021**, *230*, 111209. [CrossRef]
57. Bergman, T.L. Active Daytime Radiative Cooling Using Spectrally Selective Surfaces for Air Conditioning and Refrigeration Systems. *Sol. Energy* **2018**, *174*, 16–23. [CrossRef]
58. Kousis, I.; D'Amato, R.; Pisello, A.L.; Latterini, L. Daytime Radiative Cooling: A Perspective toward Urban Heat Island Mitigation. *ACS Energy Lett.* **2023**, *8*, 3239–3250. [CrossRef]
59. Yuan, J.; Yin, H.; Yuan, D.; Yang, Y.; Xu, S. On Daytime Radiative Cooling Using Spectrally Selective Metamaterial Based Building Envelopes. *Energy* **2022**, *242*, 122779. [CrossRef]
60. Liu, J.; Xu, C.; Ao, X.; Lu, K.; Zhao, B.; Pei, G. A Dual-Layer Polymer-Based Film for All-Day Sub-Ambient Radiative Sky Cooling. *Energy* **2022**, *254*, 124350. [CrossRef]
61. Lim, H.; Chae, D.; Son, S.; Ha, J.; Lee, H. $CaCO_3$ Micro Particle-Based Radiative Cooling Device without Metal Reflector for Entire Day. *Mater. Today Commun.* **2022**, *32*, 103990. [CrossRef]
62. Huang, T.; Chen, Q.; Huang, J.; Lu, Y.; Xu, H.; Zhao, M.; Xu, Y.; Song, W. Scalable Colored Subambient Radiative Coolers Based on a Polymer-Tamm Photonic Structure. *ACS Appl. Mater. Interfaces* **2023**, *15*, 16277–16287. [CrossRef] [PubMed]
63. Yu, L.; Huang, Y.; Zhao, Y.; Rao, Z.; Li, W.; Chen, Z.; Chen, M. Self-Sustained and Insulated Radiative/Evaporative Cooler for Daytime Subambient Passive Cooling. *ACS Appl. Mater. Interfaces* **2023**, *16*, 6513–6522. [CrossRef]
64. Givoni, B. Solar Heating and Night Radiation Cooling by a Roof Radiation Trap. *Energy Build.* **1977**, *1*, 141–145. [CrossRef]

65. Liu, Z.; Tan, H.; Ma, G. Experimental Investigation on Night Sky Radiant Cooling Performance of Duct-Type Heat Exchanger. *Int. J. Vent.* **2017**, *16*, 255–267. [CrossRef]
66. Parker, D.S.; Sherwin, J.R. *Evaluation of the NightCool Nocturnal Radiation Cooling Concept: Annual Performance Assessment in Scale Test Buildings Stage Gate 1B*; Florida Solar Energy Center: Cocoa, FL, USA, 2008. [CrossRef]
67. Kimball, B.A.; Idso, S.B.; Aase, J.K. A Model of Thermal Radiation from Partly Cloudy and Overcast Skies. *Water Resour. Res.* **1982**, *18*, 931–936. [CrossRef]
68. Hollick, J. Nocturnal Radiation Cooling Tests. *Energy Procedia* **2012**, *30*, 930–936. [CrossRef]
69. Khedari, J.; Waewsak, J.; Thepa, S.; Hirunlabh, J. Field Investigation of Night Radiation Cooling under Tropical Climate. *Renew. Energy* **2000**, *20*, 183–193. [CrossRef]
70. Hu, M.; Seng Lee, P. Performance Evaluation of a Passive Air Conditioning Module Integrating Radiative Sky Cooling and Indirect Evaporative Cooling. *Appl. Therm. Eng.* **2024**, *244*, 122608. [CrossRef]
71. Gálvez, M.; García, J.; Barraza, R.; Contreras, J. Parametric Analysis of an Air-Based Radiative Cooling System Coupled to a Thermal Storage Wall for a Low-Income Household. *Energy Build.* **2021**, *252*, 111364. [CrossRef]
72. Al-Zubaydi, A.Y.T.; Dartnall, W.J. Design and Modelling of Water Chilling Production System by the Combined Effects of Evaporation and Night Sky Radiation. *J. Renew. Energy* **2014**, *2014*, 1–8. [CrossRef]
73. Cui, Y.; Luo, X.; Zhang, F.; Sun, L.; Jin, N.; Yang, W. Progress of Passive Daytime Radiative Cooling Technologies towards Commercial Applications. *Particuology* **2022**, *67*, 57–67. [CrossRef]
74. Zhao, D.; Aili, A.; Yin, X.; Tan, G.; Yang, R. Roof-Integrated Radiative Air-Cooling System to Achieve Cooler Attic for Building Energy Saving. *Energy Build.* **2019**, *203*, 109453. [CrossRef]
75. Fu, Y.; Yang, J.; Su, Y.S.; Du, W.; Ma, Y.G. Daytime Passive Radiative Cooler Using Porous Alumina. *Sol. Energy Mater. Sol. Cells* **2019**, *191*, 50–54. [CrossRef]
76. Torgerson, E.; Hellhake, J. Polymer Solar Filter for Enabling Direct Daytime Radiative Cooling. *Sol. Energy Mater. Sol. Cells* **2020**, *206*, 110319. [CrossRef]
77. Wang, C.; Chen, H.; Wang, F. Passive Daytime Radiative Cooling Materials toward Real-World Applications. *Prog. Mater. Sci.* **2024**, *144*, 101276. [CrossRef]
78. Wu, Y.; Zhao, H.; Sun, H.; Duan, M.; Lin, B.; Wu, S. A Review of the Application of Radiative Sky Cooling in Buildings: Challenges and Optimization. *Energy Convers. Manag.* **2022**, *265*, 115768. [CrossRef]
79. Hamza, A.; Ali, H. Passive Cooling of Water at Night in Uninsulated Open Tank in Hot Arid Areas. *Energy Convers. Manag.* **2006**, *48*, 93–100. [CrossRef]
80. Tang, R.; Etzion, Y. On Thermal Performance of an Improved Roof Pond for Cooling Buildings. *Build. Environ.* **2004**, *39*, 201–209. [CrossRef]
81. Spanaki, A.; Tsoutsos, T.; Kolokotsa, D. On the Selection and Design of the Proper Roof Pond Variant for Passive Cooling Purposes. *Renew. Sustain. Energy Rev.* **2011**, *15*, 3523–3533. [CrossRef]
82. Spanaki, A.; Kolokotsa, D.; Tsoutsos, T.; Zacharopoulos, I. Theoretical and Experimental Analysis of a Novel Low Emissivity Water Pond in Summer. *Sol. Energy* **2012**, *86*, 3331–3344. [CrossRef]
83. Hosseinzadeh, E.; Taherian, H. An Experimental and Analytical Study of a Radiative Cooling System with Unglazed Flat Plate Collectors. *Int. J. Green Energy* **2012**, *9*, 766–779. [CrossRef]
84. Okoronkwo, C.A.; Nwigwe, K.N.; Ogueke, N.V.; Anyanwu, E.E.; Onyejekwe, D.C.; Ugwuoke, P.E. An Experimental Investigation of the Passive Cooling of a Building Using Nighttime Radiant Cooling. *Int. J. Green Energy* **2014**, *11*, 1072–1083. [CrossRef]
85. Ferrer Tevar, J.A.; Castã No, S.; Marijuán, A.G.; Heras, M.R.; Pistono, J. Modelling and Experimental Analysis of Three Radioconvective Panels for Night Cooling. *Energy Build.* **2015**, *107*, 37–48. [CrossRef]
86. Raeissi, S.; Taheri, M. Skytherm: An Approach to Year-Round Thermal Energy Sufficient Houses. *Renew. Energy* **2000**, *19*, 527–543. [CrossRef]
87. Meir, M.G.; Rekstad, J.B.; LØvvik, O.M. A Study of a Polymer-Based Radiative Cooling System. *Sol. Energy* **2002**, *73*, 403–417. [CrossRef]
88. Zhou, K.; Miljkovic, N.; Cai, L. Performance Analysis on System-Level Integration and Operation of Daytime Radiative Cooling Technology for Air-Conditioning in Buildings. *Energy Build.* **2021**, *235*, 110749. [CrossRef]
89. Wang, W.; Fernandez, N.; Katipamula, S.; Alvine, K. Performance Assessment of a Photonic Radiative Cooling System for Office Buildings. *Renew. Energy* **2018**, *118*, 265–277. [CrossRef]
90. Zhang, K.; Zhao, D.; Yin, X.; Yang, R.; Tan, G. Energy Saving and Economic Analysis of a New Hybrid Radiative Cooling System for Single-Family Houses in the USA. *Appl. Energy* **2018**, *224*, 371–381. [CrossRef]
91. Pirvaram, A.; Talebzadeh, N.; Leung, S.N.; O'Brien, P.G. Radiative Cooling for Buildings: A Review of Techno-Enviro-Economics and Life-Cycle Assessment Methods. *Renew. Sustain. Energy Rev.* **2022**, *162*, 112415. [CrossRef]
92. Yoon, S.; Kim, M.; Seo, J.; Kim, S.; Lee, H.; Lee, J.; Lee, B.J. Performance Analysis of a Hybrid HVAC System Consisting of a Solar Thermal Collector and a Radiative Cooling Panel. *Energy Build.* **2021**, *241*, 110921. [CrossRef]
93. Zhao, D.; Aili, A.; Zhai, Y.; Lu, J.; Kidd, D.; Tan, G.; Yin, X.; Yang, R. Subambient Cooling of Water: Toward Real-World Applications of Daytime Radiative Cooling. *Joule* **2019**, *3*, 111–123. [CrossRef]
94. Jia, L.; Lu, L.; Chen, J. Exploring the Cooling Potential Maps of a Radiative Sky Cooling Radiator-Assisted Ground Source Heat Pump System in China. *Appl. Energy* **2023**, *349*, 121678. [CrossRef]

95. Goldstein, E.A.; Raman, A.P.; Fan, S. Sub-Ambient Non-Evaporative Fluid Cooling with the Sky. *Nat. Energy* **2017**, *2*, 17143. [CrossRef]
96. Yu, X.; Chan, J.; Chen, C. Review of Radiative Cooling Materials: Performance Evaluation and Design Approaches. *Nano Energy* **2021**, *88*, 106259. [CrossRef]
97. Fang, H.; Zhao, D.; Yuan, J.; Aili, A.; Yin, X.; Yang, R.; Tan, G. Performance Evaluation of a Metamaterial-Based New Cool Roof Using Improved Roof Thermal Transfer Value Model. *Appl. Energy* **2019**, *248*, 589–599. [CrossRef]
98. Fernandez, N.; Wang, W.; Alvine, K.J.; Katipamula, S. *Energy Savings Potential of Radiative Cooling Technologies*; Pacific Northwest National Lab. (PNNL): Richland, WA, USA, 2015; Volume 72.
99. Felicelli, A.; Wang, J.; Feng, D.; Forti, E.; Azrak, S.E.A.; Peoples, J.; Youngblood, J.; Chiu, G.; Ruan, X. Efficient Radiative Cooling of Low-Cost $BaSO_4$ Paint-Paper Dual-Layer Thin Films. *Nanophotonics* **2024**, *13*, 639–648. [CrossRef]
100. Lu, X.; Xu, P.; Wang, H.; Yang, T.; Hou, J. Cooling Potential and Applications Prospects of Passive Radiative Cooling in Buildings: The Current State-of-the-Art. *Renew. Sustain. Energy Rev.* **2016**, *65*, 1079–1097. [CrossRef]
101. Deng, Y.; Yang, Y.; Xiao, Y.; Xie, H.L.; Lan, R.; Zhang, L.; Yang, H. Ultrafast Switchable Passive Radiative Cooling Smart Windows with Synergistic Optical Modulation. *Adv. Funct. Mater.* **2023**, *33*, 2301319. [CrossRef]
102. An, Y.; Fu, Y.; Jian-Guo, D.; Yin, X.; Lei, D. Switchable Radiative Cooling Technologies for Smart Thermal Management. *Cell Rep. Phys. Sci.* **2022**, *3*, 101098. [CrossRef]
103. Yoo, M.J.; Pyun, K.R.; Jung, Y.; Lee, M.; Lee, J.; Ko, S.H. Switchable Radiative Cooling and Solar Heating for Sustainable Thermal Management. *Nanophotonics* **2024**, *13*, 543–561. [CrossRef]
104. Testa, J.; Krarti, M. A Review of Benefits and Limitations of Static and Switchable Cool Roof Systems. *Renew. Sustain. Energy Rev.* **2017**, *77*, 451–460. [CrossRef]
105. Cui, Y.; Ke, Y.; Liu, C.; Chen, Z.; Wang, N.; Zhang, L.; Zhou, Y.; Wang, S.; Gao, Y.; Long, Y. Thermochromic VO_2 for Energy-Efficient Smart Windows. *Joule* **2018**, *2*, 1707–1746. [CrossRef]
106. Zhang, H.; Huang, J.; Fan, D. Switchable Radiative Cooling from Temperature-Responsive Thermal Resistance Modulation. *ACS Appl. Energy Mater.* **2022**, *5*, 6003–6010. [CrossRef]
107. Cannavale, A.; Pugliese, M.; De Matteis, V.; Giannuzzi, R.; Mancarella, F.; Valenzano, V.; Ayr, U.; Liuzzi, S.; Maiorano, V.; Berardi, U.; et al. Multifunctional Hydrogel-Based Electrolytes for Thermoelectrochromic Devices. *J. Appl. Polym. Sci.* **2023**, *140*, e54330. [CrossRef]

MDPI AG
Grosspeteranlage 5
4052 Basel
Switzerland
Tel.: +41 61 683 77 34

Energies Editorial Office
E-mail: energies@mdpi.com
www.mdpi.com/journal/energies

www.ingramcontent.com/pod-product-compliance
Lightning Source LLC
Chambersburg PA
CBHW041352210926
51850CB00013B/3057

* 9 7 8 3 7 2 5 8 2 4 7 3 1 *